Lithic Materials and
Paleolithic
Societies

To
Kenneth and Irene Tuttle Adams and Ilona Matkovszki and her family:
Sandor Sr., Sandor Jr., and Ilona Sr.
And
Ralph and Madalyn Angel, for whom the World has been a source of endless fascination

Lithic Materials and
Paleolithic
Societies

Edited by

Brian Adams
Brooke S. Blades

A John Wiley & Sons, Ltd., Publication

Library of Congress Cataloguing-in-Publication Data

Lithic materials and Paleolithic societies / edited by Brooke Blades and Brian Adams.
    p. cm.
  Includes bibliographical references and index.
  ISBN 978-1-4051-6837-3 (hardcover : alk. paper) 1. Stone implements–Analysis. 2. Tools, Prehistoric–Analysis.
  3. Paleolithic period. 4. Land settlement patterns. 5. Social archaeology.
  I. Blades, Brooke S. II. Adams, Brian.
CC79.5.S76L577 2009
930.1′2–dc22

                                                 2008042549

A catalogue record for this book is available from the British Library.

Set in 9/11 pt Photina by Newgen Imaging Systems (P) Ltd, Chennai
Printed and bound in Singapore by Fabulous Printers Pte Ltd

1   2009

CONTENTS

LIST OF CONTRIBUTORS

Brian Adams
Public Service Archeology and Architecture
Program
University of Illinois at Urbana-Champaign
Anthropology Department
Urbana, IL, USA

William Andrefsky, Jr.
Chair, Department of Anthropology
Washington State University
Pullman, WA, USA

Ran Barkai
Institute of Archeology
Tel Aviv University
Ramat Aviv, Tel Aviv
Israel

Charlotte Beck
Department of Anthropology
Hamilton College
Clinton, NY, USA

Katalin T. Biró
Hungarian National Museum
Múzeum, krt. 14-16, H-1088 Budapest
Hungary

Brooke S. Blades
Archeologist/Principal Investigator
A.D. Marble and Company
Conshohocken, PA, USA

Laurence Bourguignon
INRAP, PACEA Direction interrégionale
Grand-Sud-Ouest
Centre d'activitiés ‹‹Les Échoppes››
Bâtiment F, Pessac
Talence, France

Parth R. Chauhan
Stone Age Institute and CRAFT Research Center
Indiana University
Bloomington, IN, USA

Stephen C. Cole
MACTEC Engineering and Consulting, Inc.
Knoxville, TN, USA

Anne Delagnes
CNRS, PACEA, Institut de Préhistoire et Geologie
de Quaternaire
Université de Bordeaux I
Avenue des Facultes
Talence Cedex, France

Viola Dobosi
Hungarian National Museum
Múzeum, krt. 14-16, H-1088 Budapest
Hungary

Jean-Philippe Faivre
IPGQ-PACEA UMR 5199 du CNRS
Université de Bordeaux I
Avenue des Facultes
Talence Cedex, France

Jehanne Féblot-Augustins
Préhistoire et Technologie
Maison de l'Archéologie et de l'Ethnologie
Université de Paris X
Nanterre Cedex, France

Ted Goebel
Center for the Study of the First Americans
Department of Anthropology
Texas A&M University
College Station, TX, USA

Avi Gopher
Institute of Archeology
Tel Aviv University
Ramat Aviv, Tel Aviv
Israel

Kelly E. Graf
Center for the Study of the First Americans
Texas A&M University
College Station, TX, USA

Linda T. Grimm
Department of Anthropology
Oberlin College
Oberlin, OH, USA

Sonia Harmand
UMR 7055 (CNRS), Laboratoire de Préhistoire et
Technologie
Maison de l'Archéologie et de l'Ethnologie
Nanterre Cedex, France

Peter Hiscock
School of Archeology and Anthropology
A.D. Hope Building
The Australian National University
Canberra ACT
Australia

George T. Jones
Department of Anthropology
Hamilton College
Clinton, NY, USA

Rebecca A. Kessler
Department of Anthropology
University of Washington
Seattle, WA, USA

Todd Koetje
Department of Anthropology
Western Washington University
Bellingham, WA, USA

Brad Koldehoff
American Bottom Survey Division
Illinois Transportation Archeological Research Program
University of Illinois
Urbana-Champaign
Belleville, IL, USA

Mikhail Konstantinov
Chita State Pedagogical University
Chita, Russia

Thomas J. Loebel
CAGIS Archeological Consulting Services
Department of Anthropology (m/c 027)
University of Illinois at Chicago
Chicago, IL, USA

Grant S. McCall
Department of Anthropology
Tulane University
New Orleans, LA, USA

Liliane Meignen
UNSA-CNRS, CEPAM
250 rue Albert Ainstein
Sophia Antipolis
Valbonne, France

Katragadda Paddayya
Department of Archaelogy
Deccan College
Pune
India

Michael D. Petraglia
Leverhulme Centre for Human Evolutionary Studies
Department of Biological Anthropology
University of Cambridge
Cambridge
UK

Ceri Shipton
Leverhulme Centre for Human Evolutionary Studies
Department of Biological Anthropology
University of Cambridge
Cambridge
UK

Karisa Terry
Department of Anthropology
Washington State University
Pullman, WA, USA

Alain Turq
Musée National de Prèhistoire des Eyzies
Les Eyzies de Tayac
France

Pawel Valde-Nowak
Institute of Archeology
Jagiellonian University
Kraków
Poland

INTRODUCTION: LITHICS, LANDSCAPES, AND SOCIETIES

Brooke S. Blades[1] and Brian Adams[2]

[1]Archeologis/Principal Investigator, A.D. Marble and Company
[2]Public Service Archeology and Architecture Program, University of Illinois at Urbana-Champaign

The evaluation of the relationship between landscape and social organization from a lithic perspective requires consideration of issues that lie at the very foundation of Paleolithic research. This volume examines the relationship with regards to archeological assemblages from five continents that date from the late Pliocene to the end of Pleistocene and earliest Holocene. The chapters have been arranged in broad themes reflecting varying scales of geographical analysis. However, several analytical approaches and interpretive issues are common to all of the themes.

Regardless of geographical location or time period, it may be argued that lithic materials reflect the knowledge and exploitation of the landscape and the social organization required to promote and facilitate that exploitation. A fundamental issue relates to the identification of lithic raw material sources, concerns that are common to research in both the Old and New Worlds (Blades 2006). Jehanne Féblot-Augustins (Chapter 3, this volume) emphasizes that expanded recognition of raw material sources in Western Europe has resulted in a greater appreciation for the distance of raw material movement during the early Upper Paleolithic, and to a lesser extent during the later Middle Paleolithic. These new data conform more closely to those suggested by Taborin (1993a, b) for the movement of shells from the Atlantic and Mediterranean coasts across and through southwestern France. Attempts to more accurately identify raw material sources have ranged

from physical (Biró and Dobosi 1991 and this volume, Chapters 4 and 8) to virtual lithic source guides (Elburg and van der Kroft 2004; Féblot-Augustins *et al.* 2005). Microscopic and geochemical analyses have permitted attributions to be made with greater confidence and are clearly the direction of future analyses. However, the subjective role of the researcher remains a consideration of considerable importance, as emphasized by Alain Turq (1992).

What are the temporal dimensions of the problem of understanding landscapes and societies from the perspective of lithic assemblages? Some of the earliest studies in East Africa have emphasized the importance of raw material selectivity. The geological survey of Olduvai Gorge conducted by Richard Hay (1976) unraveled not only the geological development of the landscape but the raw material sources for many of the stone tools at the various archeological loci in the basin. Sources between 2 and 4 km away were exploited during Beds I and II occupations as early as nearly 2 million years ago. Bed III materials were being transported 8 to 10 km, and distances had expanded to between 13 and 20 km by Bed IV occupations (Hay 1976; Reader 1988: 164). The hominin occupants were selecting finer-grain volcanic rocks and evidently transporting them over distances that increased through time, although not necessarily in a linear fashion. Evidence of raw material selectivity and transport across the African landscape and through time is explored in this volume by Sonia Harmand for the Turkana Basin and by Grant McCall for Namibia. The lithic distributions are seen as reflections of broader patterns of landscape utilization.

Lithic Materials and Paleolithic Societies, 1st edition. Edited by B. Adams and B.S. Blades, ©2009 Blackwell Publishing. ISBN 978-1-4051-6837-3.

The increasing developments of hominin physical and intellectual capabilities and related evidence of social interaction emerge as important temporal concerns in the studies. Ceri Shipton, Mike Petraglia, and Katragadda Paddayya present a strong argument for collective activity and intra-site organization during Early Paleolithic occupations in India. In so doing, they offer an alternative to prevailing opinions concerning the extent of planning and cognitive organization reflected in biface assemblages. The extent to which reduction technology and artifact forms result from predetermined concepts or continual reduction and resharpening is a major topic in Middle Paleolithic studies, as will be discussed subsequently. Kati Biró illustrates a general trend from primary utilization of local lithic raw materials to increased use of non-local materials between the Lower and Upper Paleolithic in Hungary.

The exploitation of the lithic landscape on varying scales, as explored by Parth Chauan for the Indian Subcontinent, Ran Barkai and Avi Gopher for a large quarry locus in southwestern Asia, Brian Adams for the Szeletian in central Europe, and Peter Hiscock in northern Australia, reveals that material variability may exert a profound effect on both the scale of raw material transport and morphological nature of lithic tool assemblages. Chauhan explores the Acheulean/ Soanian technological interface, while Barkai and Gopher argue for an active manipulation of the lithic extraction landscape throughout the Pleistocene. Adams exposes the extent to which raw material quality and post-depositional processes have influenced and even generated perceptions of acculturation during the Middle Paleolithic to Late Paleolithic transition. Hiscock describes the influence of reduction and reutilization on perceptions of Holocene assemblage variability.

The source locations and implied movement of raw materials from sources to sites are important considerations, but only provide a portion of the analytical framework. The technological and typological forms in which different raw materials appear on archeological sites provide the other elements that contribute to a broader understanding of hominin movement, group mobility, and underlying social actions. Two broad structures have emerged in Western scholarly literature for conceptualizing stone tool assemblages: the North American approach to technological organization and the French conception of chaîne opératoire. Despite some fundamental differences, both structures argue for a thorough examination of all lithic elements, not simply retouched tools.

All of the chapters in this volume owe a debt to these conceptual structures, but some do so in more explicit ways. Kelly Graf and Ted Goebel discuss spatial and temporal variability by comparing latest Pleistocene sites in northeastern Siberia and Alaska. Waste flake analyses are critical to the interpretations of Charlotte Beck, George Jones, and Rebecca Kessler for late Pleistocene and early Holocene occupations in the Great Basin of the western United States. The temporal analysis of Solvieux reveals an interesting application of technological organization to southwestern France by Linda Grimm and Todd Koetje.

The concept of chaîne opératoire had a long gestation in French anthropology. The application of this conceptual framework to the interpretation of Middle Paleolithic raw material assemblages by Jean-Michel Geneste (1985) has influenced scholars on both sides of the Atlantic. Geneste conducted his research during the late 1970s and early 1980s when a major emphasis of Paleolithic research in southwestern France focused on the relationship between raw material composition and the technological spectrum of the chaîne opératoire.

The influences of raw material variability, technological spectrum, and reduction on lithic typology – to say nothing of conflicts over the ultimate relationship between Neanderthals and anatomically modern humans – have combined to make the Middle Paleolithic in Europe one of the most controversial periods of Paleolithic research. (It is to be regretted that the Middle Stone Age and the Late Stone Age in Africa are not considered in this volume; several attempts were made to solicit chapters bearing on these periods, but to no avail.) The contribution by Liliane Meignen, Anne Delagnes, and Laurence Bourguignon summarizes Middle Paleolithic material utilization and lithic technological behavior by engaging data from several decades of research, including perspectives from Eric Boëda and Jacques Pelegrin, among others. The chapter by Peter Hiscock, Alan Turq, Jean-Philippe Faivre, and Laurence Bourguignon focuses on the Quina Middle Paleolithic industry to examine the reduction model promoted by Harold Dibble and the provisioning and reduction strategies of Steve Kuhn, in addition to other opinions on the analysis of lithic reduction.

Models to elucidate settlement patterns of "foragers and collectors" (Steward 1968; Binford 1980; Kelly 1995), "travelers and processors" (Bettinger 1991), or other frameworks often combine evolutionary ecology theory and lithic reduction analyses (Carr 1994). Thus the raw material composition and the variable technological spectra in which the different raw materials appear are critical elements in the overall interpretation of the nature and duration of settlement at a given site location. Grant McCall considers the landscape of mobility in the Early Paleolithic as reflected in statistical analyses. Steve Cole provides an examination of early Late Paleolithic temporal and spatial variability as reflected in "technological efficiency", which may be regarded as one approximation of broader energy efficiency, a common "currency" in optimal foraging theory. Brooke Blades examines the local lithic environment as a reflection of resources that the Aurignacian occupants of la Ferrassie would have encountered within the foraging radius around the shelter. He argues that core reduction may respond to the localized subsistence environment and direct one to more specialized landscape analyses. Brad Koldehoff and Tom Loebel contend that the late Pleistocene Clovis and Dalton traditions, at least in the middle of North America, may be interpreted as a contrast between "unbounded and bounded systems" as reflections of the density of the social landscapes.

Virtually all of these chapters address the question of social systems, although once again some do so more explicitly than others. Shipton, Petraglia, and Paddayya perceive evolutionary significance in the inference of Acheulean social organization. The contrast between "cooperation" as an ecological behavior reflected in many animal species and the "socially constituted purpose" of "sharing" as a human social interaction (Ingold 1988: 281, 284–285) is relevant to this analysis. Indeed, current research focusing on the Early Paleolithic, of which the topical papers in this volume are representative examples, underscores the point that there are many major questions to be addressed beyond those related to the earliest appearance of stone tools, the meaning of the Middle Paleolithic variability, and the nature of the transition between the Middle and Late Paleolithic.

Variability and evolutionary significance in the Late Paleolithic is implied in the research of Karisa Terry, Bill Andrefsky, and Mikhail Konstantinov in eastern Asia. Changes in patterns of raw material acquisition and utilization have been noted in other parts of the Old World. Cole and Adams both critically examine some of the assumptions about the Middle to Late Paleolithic transition. Féblot-Augustins provides exhaustive documentation of lithic raw material transfers and technological spectra for the late Middle Paleolithic and early

Late Paleolithic in central and western Europe. Viola Dobosi's chapter indicates temporal variability in the utilization of local and non-local lithic raw materials during the course of the Hungarian Gravettian, *ca.* 28 and 16 Ka.

One of the most difficult problems reflected in archeological lithic assemblages is determination of the means by which raw materials entered a given archeological site. Researchers have become much better at attributing materials to specific sources and in recognizing the relevance and components of technological spectra, but modes of acquisition remain problematic. David Meltzer (1989) emphasized some time ago that the problem of equifinality is partially to blame. Féblot-Augustins addresses the problem of distinguishing between direct and indirect material procurements with implications of seasonal/annual mobility associated with the former and socially-oriented exchange for the latter. The classic "down-the-line" or distance decay and "directional" or non-linear patterns of material distribution discussed by Colin Renfrew (1977) continue to provide an important avenue of research, but one that must be approached on a regional level. This approach was profitably exploited by Soffer (1985) for the Late Paleolithic on the Central Russian Plain. New World perspectives have been offered by Wilmsen (1970), Kelly and Todd (1988), and Meltzer (1989), among many others.

The problem of distinguishing patterns of movement and mobility from those of social exchange emerges principally from the reality that it often cannot and if possible should not be addressed solely as a lithic phenomenon (White 1989). The same archeological assemblage may preserve evidence of direct lithic procurement and acquisition of selected lithics and other truly exotic materials through indirect exchange (Soffer 1991). One of the fundamental distinctions between Middle Paleolithic and Late Paleolithic groups may lie in the realm of directional distributions and socially intensified networks that have generated such distributions. However, Late Paleolithic material distributions were by no means always directional in nature.

Certain papers offer provocative interpretations of human behavior. Barkai and Gopher contend that Early and Middle Paleolithic chert miners in effect engaged in landscape management practices to exploit the lithic resources at Sede Ilan. Paweł Valde-Nowak combines an important overview of newly-recognized lithic materials sources in southern Poland – one that expands on

the tradition established by Janusz Kozłowski (1973) –
with an analysis of the unique level VIII at Obłazowa
Cave as the material remnants of a Pavlovian ritual.
The latter interpretation is noteworthy as it derives
much of its argumentation from archeological context
that is sadly lacking for most Paleolithic manifestations
of symbolic behavior.

The editors wish to express their very sincere
appreciation to all of the contributing authors whose
extensive fieldwork, detailed research, and insightful
theoretical perspectives have made this volume a reality. The initial contributors were participants in a session at the 2004 Society for American Archaeology
meetings in Montréal and deserve special recognition
for their patience in waiting for the volume to appear.

The editorial staff at Blackwell Publishing demonstrated immediate and sustained interest in the concept
of the volume; our thanks are offered to Ian Francis,
Rosie Hayden, Delia Sandford, Kelvin Mathews, and
Rebecca Moore. In response to several very useful anonymous reviews, Ian suggested the volume be reorganized
and expanded to broaden the temporal and geographical
coverage. As a consequence, several authors agreed to
contribute specific chapters, many as a result of fateful
encounters with one of the editors at the 2006 Society
for Archaeology meetings in San Juan. These authors
are to be commended for fulfilling their commitments to
produce manuscripts within 6 months.

We all have contributed to and assembled this volume in the hope that it may provide a substantive contribution to the contemporary state of knowledge on
hominin lithic raw material utilization and technology and inferences on landscape and social organization prior to the Holocene interglacial. It seemed to the
authors and Blackwell Publishing that such a volume
was absent from the current literature and would provide important insights on the prehistoric past, but also
on the research traditions and perspectives that constitute modern Paleolithic research. We now commend
this work to you and ask that you recognize that biological and cultural evolution have collectively shaped
the curiosity, research perspectives, and traditional
prejudices that we have brought to its construction. We
encourage you to explore your own perspectives and
prejudices when reading and evaluating our efforts,
and thank you for doing so.

As a final note, we wish to express our deep affec-
tion for and appreciation to our families for their support: Meg Bleecker Blades and Emma Blades, and Ilona
Matovszki.

REFERENCES

Bettinger, R. (1991) *Hunter-Gatherers: Archaeological and Evolutionary Theory.* Plenum Press, New York.
Binford, L. (1980) Willow smoke and dogs' tails: Hunter-gatherer settlement systems and archaeological site formation. *American Antiquity* 45, 255–73.
Blades, B. (2006) Common concerns in the analysis of lithic raw material exploitation in the Old and New Worlds. In: Bressy, C., Burke, A., Chalard, P., and Martin, H. (eds) *Notions de territoire et de mobilité. Examples de l'Europe et des premières nations en Amérique du Nord avant le contact européen,* ERAUL 116, Liège, pp. 155–62.
Biró, K. and Dobosi, V. (1991) *Lithoteca Comparative Raw Material Collection of the Hungarian National Museum.* Hungarian National Museum, Budapest.
Carr, P. (1994) Technological organization and prehistoric hunter-gatherer mobility: Examination of the Hayes site. In: Carr, P. (ed.) *The Organization of North American Prehistoric Chipped Stone Tool Technologies,* International Monographs in Prehistory, Ann Arbor, pp. 35–44.
Elburg, R. and van der Kroft, P. (2004) www.flintsource.net.
Féblot-Augustins, J., Park, S.-J., and Delagnes, A. (2005) *Lithothèque du basin de la Charente* (www.alienor.org/Articles/lithoteque/index.htm).
Geneste, J.-M. (1985) *Analyse lithique d'industries moustériennes du Périgord: une approach technologique du comportement des groupes humains au Paléolithique moyen.* Thèse Sc., Université de Bordeaux I.
Hay, R. (1976) *Geology of the Olduvai Gorge.* University of California Press, Berkeley.
Ingold, T. (1988) Notes on the foraging mode of production. In: Ingold, T., Ricks, D., and Woodburn, J. (eds) *Hunters and Gatherers, Volume I: History, Evolution, and Social Change,* Berg Publishers, Oxford, pp. 269–85.
Kelly, R. (1995) *The Foraging Spectrum: Diversity in Hunter-Gatherer Lifeways.* Smithsonian Institution Press, Washington.
Kelly, R. and Todd, L. (1988) Coming into the country: Early Paleoindian hunting and mobility. *American Antiquity* 58, 231–44.
Kozłowski, J. (1973) The origin of lithic raw material used in the Carpathian countries. *Acta Archaeologica Carpathica* 13, 5–19.
Meltzer, D. (1989) Was stone exchanged among eastern North America Paleoindians? In: Ellis, C. and Lothrop, J. (eds) *Eastern Paleoindian Lithic Resource Use,* Westview Press, Boulder, pp. 11–39.
Reader, J. (1988) *Missing Links: The Hunt for Earliest Man,* 2nd Edition, Penguin Books, Harmondsworth.
Renfrew, C. (1977) Alternative models for exchange and spatial distribution. In: Earle, T. and Ericson, J. (eds) *Exchange Systems in Prehistory,* Academic Press, New York, pp. 71–90.

Soffer, O. (1985) Patterns of intensification as seen from the Upper Paleolithic of the Central Russian Plain. In: Price, T. and Brown, J. (eds) *Prehistoric Hunter-Gatherers*, Academic Press, New York, pp. 235–70.

Soffer, O. (1991) Lithics and lifeways – the diversity in raw material procurement and settlement systems on the Upper Paleolithic East European Plain. In: Montet-White, A. and Holen, S. (eds) *Raw Material Economies among Prehistoric Hunter-Gatherers*, University of Kansas Publications in Anthropology 19, Lawrence, pp. 221–34.

Steward, J. (1968) Causal factors and processes in the evolution of pre-farming societies. In: Lee, R. and Devore, I. (eds) *Man the Hunter*, Wenner-Gren Foundation and Aldine de Gruyter, New York, pp. 321–34.

Taborin, Y. (1993a) *La parue en coquillage au Palélithique.* 29th supplément à *Gallia Préhistoire*. CNRS, Paris.

Taborin, Y. (1993b) Shells of the French Aurignacian and Périgordian. In: Knecht, H., Pike-Tay, A., and White, R. (eds) *Before Lascaux: The Complex Record of the Early Upper Paleolithic*, CRC Press, Boca Raton, pp. 211–27.

Turq, A. (1992) *Le Paléolithique inférieur et moyen entre les vallées de la Dordogne et du Lot.* Thèse à l'Univérsite de Bordeaux I.

White, R. (1989) Toward a contextual understanding of the earliest body ornaments. In: Trinkhaus, E. (ed.) *The Emergence of Modern Humans: Biocultural Adaptation in the Later Pleistocene*, Cambridge University Press, Cambridge, pp. 211–31.

Wilmsen, E. (1970) *Litic Analysis and Cultural Inference: A Paleo-Indian Case.* Anthropological Papers of the University of Arizona 16, Tucson.

Part I

REGIONAL LANDSCAPE PERSPECTIVES

Chapter 1

RAW MATERIALS AND TECHNO-ECONOMIC BEHAVIORS AT OLDOWAN AND ACHEULEAN SITES IN THE WEST TURKANA REGION, KENYA

Sonia Harmand

Laboratoire de Préhistoire et Technologie, UMR 7055 (CNRS)
Maison de l'Archéologie et de l'Ethnologie, Nanterre

ABSTRACT

In East Africa, the significant increase in lithic studies on newly discovered Plio-Pleistocene sites provides renewed data on the emergence of early technological developments. The current techno-economic study of several rich and well-preserved Oldowan and Acheulean lithic assemblages from the West Turkana region in North Kenya opens new perspectives on how raw material procurement activities are economically structured in the Early Stone Age and documents their transformations during the Plio-Pleistocene time period. This study, carried out in combination with geological surveys, petrographic analyses, and lithic assemblage analyses, reveals distinctive characteristics in raw material procurement and exploitation patterns between 2.34 and 0.70 myrs in the West Turkana region. It demonstrates the antiquity of decision-making in ancient contexts and highlights substantial diachronic changes in the management of raw materials during the Plio-Pleistocene, which are related to the qualities and morphologies of the raw materials selected as well as to the way they were processed, rather than to variations in resource availability.

Lithic Materials and Paleolithic Societies, 1st edition. Edited by B. Adams and B.S. Blades, ©2009 Blackwell Publishing. ISBN 978-1-4051-6837-3.

INTRODUCTION

Raw materials, considered from the viewpoint of their provenance and use, have been the object of a long-standing interest in the African Lower Paleolithic. Over the past three decades, research addressing hominin behavior in relation to lithic procurement has been particularly active at Olduvai Gorge, Tanzania, and in the region of Koobi Fora, east of Lake Turkana, Kenya (e.g., Leakey 1971, 1975, 1994; Hay 1976; Isaac 1977; Isaac and Harris 1978; Jones 1979; Clark 1980; Toth 1982, 1987). Carried out on a regional scale and focusing mainly on hominin ranging and foraging behavior, raw material studies indicating that from 1.9 myr onwards site provisioning involved distances of several kilometers (overview in Féblot-Augustins 1997a) have furthered the construction of Early Pleistocene hominin activity models (e.g., "home base hypothesis", Leakey 1971; "routed foraging model", Binford 1980; "favored places model", Schick 1987; Schick and Toth 1993). The preferential use of a raw material during the Pliocene and the Early Pleistocene is generally interpreted as a result of local abundance (e.g., Merrick and Merrick 1976; Toth 1985; Schick 1987; Isaac et al. 1997). However, provenance studies based on bibliographical data and conducted from a techno-economic perspective have shown that distinct rock types from variously distant sources were introduced into sites under different forms (rough blocks/finished tools) and used for specific productions in the Acheulean, a pattern heralded in the Oldowan (Féblot-Augustins 1990, 1997b). More recently, lithic studies based on raw material characterization from new sites dated to the Late Pliocene have provided unexpected evidence for selection patterns as early as 2.6 myrs (Braun and Harris 2003; Plummer 2004; Stout et al. 2005). These involve a certain level of anticipation in the effects of raw material properties and further contradict the assumption generally made for the Early Stone Age of an opportunistic gathering of rocks.

In this chapter, we report the results of a techno-economic study that addresses the lithic procurement and exploitation patterns brought into play by the Plio-Pleistocene hominins of a recently investigated region of East Africa, the Nachukui Formation west of Lake Turkana, Kenya, and spans a wide chronological period ranging from 2.34 to 0.70 myrs (Harmand 2004, 2005).

One of the most noteworthy results ensuing from the study of several Nachukui Formation sites is new evidence of planning and foresight in raw material procurement and management from 2.34 myrs onwards, testified to by the selection of specific raw materials and cobble morphologies, and their stockpiling for future use. The in-depth techno-economic study has also highlighted significant diachronic changes and successive stages in the exploitation of raw materials between 2.34 and 0.70 myrs, related to an improvement in technical skills throughout the Plio-Pleistocene rather than to variations in resource availability (Harmand 2004, 2005).

PHYSICAL SETTING: THE NACHUKUI FORMATION, WEST TURKANA, KENYA

The Nachukui Formation lies on the western shore of Lake Turkana in the north of Kenya (Figure 1.1). The basin margin is bounded at the northwest by the Labur and Murua Rith ranges. The Plio-Pleistocene sedimentary deposits of West Turkana are aerially exposed in an area of about 700 km² and reach a thickness of about 730 m. The Nachukui Formation is divided into eight members (from 4 to 0.7 myrs) using widespread volcanic tuffs as bed markers (Harris et al. 1988; Feibel et al. 1989, 1991) (Figure 1.2). It is known as a source of significant paleontological finds, which have extended much of what is known about early hominin phylogeny and morphology in the Plio-Pleistocene (Brown et al. 1985; Walker et al. 1986; Feibel et al. 1989). Furthermore, this sequence is one of the longer and more complete in East Africa, and happens to be very rich in archeological sites of great antiquity. These sites were revealed by the West Turkana Archaeological Project, a project co-leaded by H. Roche and M. Kibunjia, and associating the National Museums of Kenya (Nairobi) and the *Mission Préhistorique au Kenya* (France). Since 1996, the Nachukui Formation has yielded a large number of rich and well-preserved archeological sites, geographically close and occupied by hominins between 2.34 and 0.70 myrs (Roche and Kibunjia 1994, 1996; Roche et al. 1999, 2003; Prat et al. 2003). Compared to the two other Plio-Pleistocene sedimentary formations from the Omo Group (Shungura Formation in the north of the Turkana Basin and Koobi Fora Formation in the west, dated from 4.5 to 0.7 myrs: Brown and Feibel 1988; Harris et al. 1988; Feibel et al. 1989, 1991), the Nachukui Formation is the only sedimentary formation that offers, on a regional scale, the opportunity

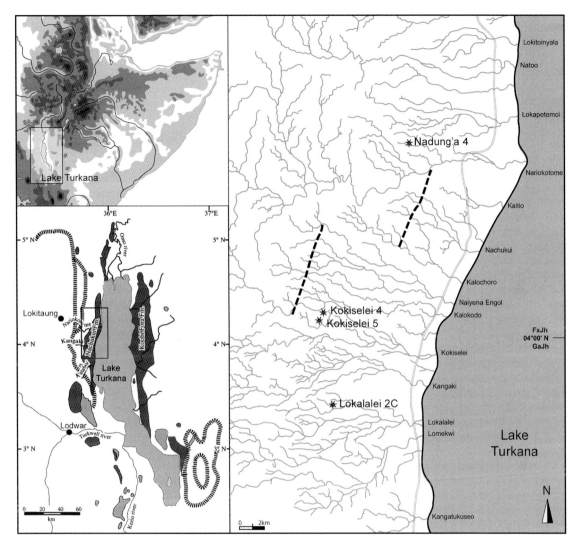

Figure 1.1   Counter-clockwise from upper left: East Africa; lake Turkana; archaeological sites discussed in the text. Modified after Roche *et al.* 2003.

to compare Early Stone Age lithic-oriented behaviors across a wide chronological time span, ranging from 2.34 to 0.70 myrs.

The assemblages under study in this chapter issue from four well-preserved archeological sites recently excavated in the Nachukui Formation, and dated through tuff-to-tuff correlations (Brown and Feibel 1988; Harris *et al.* 1988; Feibel *et al.* 1989). These sites, attributed to three Early Stone Age chronological groups and related to four cultural periods, are the Late

Pliocene site of Lokalalei 2C (Early Oldowan), the Early Pleistocene sites of Kokiselei 5 (Oldowan) and Kokiselei 4 (Early Acheulean), and the Early/Middle Pleistocene site of Nadung'a 4 (Acheulean) (Roche *et al.* 2003).

The Early Oldowan site of Lokalalei 2C (Figure 1.1) is one of the very few African Pliocene sites (Kibunjia *et al.* 1992; Delagnes and Roche 2005). It is located in the south of the Nachukui Formation (Figure 1.1), at the base of the Kalochoro Member, and dated at 2.34 myrs (Figure 1.2) (Roche *et al.* 1999; Brown and

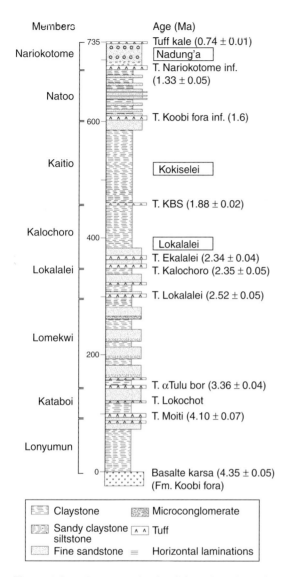

Members

Age (Ma)

Nariokotome

— 735

Tuff kale (0.74 ± 0.01)

Nadung'a

T. Nariokotome inf.
(1.33 ± 0.05)

Natoo

— 600

T. Koobi fora inf. (1.6)

Kaitio

Kokiselei

T. KBS (1.88 ± 0.02)

Kalochoro

400

Lokalalei
T. Ekalalei (2.34 ± 0.04)

Lokalalei

T. Kalochoro (2.35 ± 0.05)

T. Lokalalei (2.52 ± 0.05)

Lomekwi

200

Kataboi

T. αTulu bor (3.36 ± 0.04)
T. Lokochot
T. Moiti (4.10 ± 0.07)

Lonyumun

— 0

Basalte karsa (4.35 ± 0.05)
(Fm. Koobi fora)

Claystone   Microconglomerate

Sandy claystone   ^ ^ Tuff
siltstone

Fine sandstone   =   Horizontal laminations

Figure 1.2 Lithostratigraphical and chronological correlations of the Lokalalei, Kokiselei and Nadung'a archaeological sites within the Nachukui Formation. Modified from Harris *et al.* 1988.

Gathogo 2002; Delagnes and Roche 2005; Tiercelin *et al.* in prep.). A total of 17 m² was exhaustively excavated yielding 2629 lithic artifacts, cores, whole or broken flakes, a very few possibly retouched pieces, unmodified split cobbles and hammerstones. Despite the fact that the archaeological deposit is partly truncated by erosion, the preserved part of the site shows

clear evidence of good preservation, as indicated by a high ratio of very small elements and the freshness of the artifacts (Delagnes and Roche 2005). The Early Oldowan lithic material at Lokalalei 2C reflects organized and highly productive debitage[1] sequences, which result in a large production of relatively well standardized flakes (Delagnes and Roche 2005).

The Oldowan site of Kokiselei 5 (Figure 1.1) belongs to the upper part of the Kaitio Member (Harris *et al.* 1988) *ca.* 1.70 myr (Figure 1.2). It is situated in a low-energy setting, and its comprehensive excavation has revealed well-preserved features of the spatial organization (Roche *et al.* 2003). An area of 65 m² was excavated, yielding over 1600 Oldowan lithic artifacts including cores, flakes and fragments, unworked or roughly flaked pebbles and cobbles, and one trihedral tool (Texier *et al.* 2004; Texier *et al.* 2006). Owing to its chronological position, halfway between the Early Oldowan site of Lokalalei 2C and the Early Acheulean site of Kokiselei 4, it is a key site for understanding the transition between the Early Oldowan and the Early Acheulean.

Kokiselei 4 (Figure 1.1) is one of the oldest Early Acheulean sites in Africa, *ca.* 1.65 myr. It is a relatively eroded site where several trenches (19 m²) have been dug and material collected over a surface of 100 m². The lithic assemblage is quite small in number (*n* = 191) but yields large proto-handaxes, handaxes, picks, flakes (some of them very large), and cores.

Nadung'a 4 is located in the northern part of the Nachukui Formation (Figure 1.1) and is related to the end of the Early Pleistocene or to the very beginning of the Middle Pleistocene, *ca.* 0.70 myr (Figure 1.2) (Delagnes *et al.* 2006). The excavation covered an area of 53 m² from which an abundant lithic assemblage (*n* = 6797) *in situ* has been recovered in close association with the partial carcass of an elephant.

METHOD OF ANALYSIS: VOLCANIC PETROGRAPHY AND *CHAÎNE OPÉRATOIRE*

The lithic procurement and exploitation patterns brought into play by Plio-Pleistocene hominins are inferred from the techno-economic analysis of the aforementioned assemblages. The author drew upon the major publications concerning these assemblages (Roche *et al.* 1999; Delagnes and Roche 2005; Delagnes *et al.* 2006; Texier *et al.* 2006), and her own in-depth technological observations, which aimed more particularly to highlight the relationship between raw materials and tool production processes (Harmand 2005). This approach is based on the notion of *chaîne opératoire*, whereby each and every technical sequence and their mutual organization are identified (e.g., Leroi-Gourhan 1964, 1971; Geneste 1989, 1991). Artifacts are analyzed as the outcome of a process, that is to say the strategies of reduction and the technical skills involved in tool production.

The techno-economic analysis implemented here included as a prerequisite: (i) the systematic sourcing and sampling of raw materials by field surveys in order to assess the opportunities for procurement and the possibilities of natural transport, (ii) petrographic analyses, and (iii) rock mechanics tests through knapping experiments in order to evaluate the range of raw materials available in the conglomerates, and also to assess the relative abundance in these secondary deposits of each type of rock in relation to their characteristics (rock qualities and properties). This involved locating and mapping secondary sources, computing the distance from each archaeological site, and characterizing the types of outcrops and the nature of secondary deposits. As a result, the rocks were grouped into several categories, differing in terms of petrographic, structural, and granular patterns. In addition to the petrographic determination, the initial morphologies and sizes of all the collected cobbles were recorded and taken into consideration. Ultimately, building on the results of the technological study of the reduction sequences, the relationship between raw materials and desired end products has been determined and analyzed for each lithic assemblage.

AN ENDURINGLY LOCAL SITE PROVISIONING

The Nachukui Formation is located along the western shore of the Turkana Basin between the left bank of Lake Turkana and the Murua Rith and Labur Ranges, which border the basin to the west (Figure 1.1). The systematic sourcing of the poorly sorted debris-flow outcrops available in the vicinity of the archeological sites indicates the predominance of rounded boulders or cobbles of igneous and extrusive rocks all originating from flows of lava from these Miocene volcanic ranges. Paleogeographic reconstructions of the Turkana Basin indicate that during the last 4 myrs it was dominated by two hydrographic systems: a series of lakes acknowledged for different time periods, and a river system interpreted as the Paleo-Omo.

Prior to 2 myrs, the Turkana Basin was dominated by the Paleo-Omo, which flowed through the basin from the north and exited east into the Indian Ocean (Brown and Feibel 1988; Feibel *et al.* 1991; Rogers *et al.* 1994). After 2 myrs the basin was dominated by a lake system, which waxed and waned at different time periods represented in the basin. Even then the river continued to flow into the lake system and exited the basin to the east until the damming of the exit by the uplifting of Mt Kulal to the southeast. In the Nachukui Formation, the archeological sites are all located in close proximity to small and ephemeral east-flowing streams, joining the main axial river system that flows from north to south (Lokalalei 2C site), or the paleo-lake Turkana (Kokiselei 5, Kokiselei 4 and Nadung'a 4 sites) (Brown and Feibel 1988). From the development of these east-flowing streams, debris-flow outcrops were accumulated near the archeological sites, between ten to a few hundred meters, and stratigraphically a few meters below the archeological layers or within the same layers, providing the only ready sources of boulders, cobbles, and pebbles for the knappers[2].

These secondary deposits yielding the available rocks result from the filling of small channels flowing from west to east according to the paleogeographic reconstructions of the hydrographic systems that dominated the Turkana basin between 4.5 and 0.7 myrs (Brown and Feibel 1988, 1991; Feibel *et al.* 1991). Throughout the Plio-Pleistocene in the studied area, raw material procurement consisted of collecting and carrying volcanic rocks from exclusively local debris-flow outcrops or dry riverbeds available in the immediate vicinity of the sites, at an average distance comprising between ten to a hundred meters. As a result, the cost of search, acquisition, and transport of raw materials was minimal, and this local provisioning seems to have undergone no changes between 2.34 and 0.70 myrs.

VARIABLE ROCK COMPOSITION OF LOCAL SOURCES

The volcanic rocks carried by the network of rivers, nowadays dry most of the time, display distinctive physical features in terms of color, grain, texture, homogeneity, and veining. These are indicative of micro- to cryptocristalline groundmasses, fine- to coarse-grained fabrics, and aphyritic to porphyritic textures, resulting in distinct knapping and functional properties and varying initial morphologies and sizes.

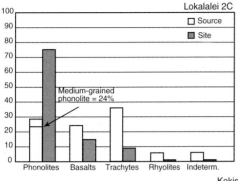

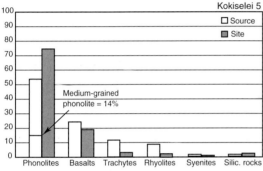

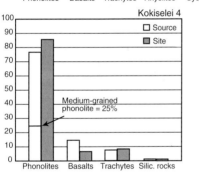

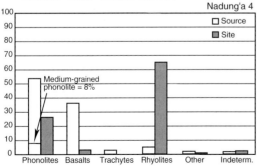

Figure 1.3   Raw material composition at geological sources and at archaeological sites in the Nachukui Formation.

For instance, medium-grained phonolite is an excellent raw material in terms of its flaking qualities.

In addition, the detailed petrographic and morphometric characterization of the secondary deposits through sampling of the conglomerates revealed variations in the proportions of the several types of lavas available in the vicinity of the different archeological localities (Figure 1.3). The randomly sampled outcrops in the Kokiselei 5, Kokiselei 4, and Nadung'a 4 areas are dominated by a dark gray aphyritic phonolite (54%, 77%, and 54%, respectively) (Figure 1.3), which accounts for only 28% of the raw materials at the randomly sampled outcrops around Lokalalei 2C. Among the phonolites, the medium-grained type is predominant in the randomly sampled outcrops at Lokalalei 2C (24% of the raw materials), while it is a minority at Kokiselei 5 (14% of the raw materials), at Kokiselei 4 (25% of the raw materials), and at Nadung'a 4 (8% of the raw materials). Dark black porphyric or aphyric basalts account for 15–36% of the rocks sampled on the outcrops. Various other rocks occur in smaller proportions: light brown trachytes (1.6–37%), red and green rhyolites (between 0 and 8%), siliceous rocks (0–1.5%), and large-grained rocks such as syenites (0–1%) (Figure 1.3).

RAW MATERIAL SELECTION AND EXPLOITATION PATTERNS: DIACHRONIC CHANGES THROUGHOUT THE PLIO-PLEISTOCENE

Selective raw material provisioning as early as 2.34 myrs

At the Late Pliocene Early Oldowan site of Lokalalei 2C, the relative proportions of the different categories of artifacts suggest that the site was a knapping spot where large quantities of raw materials were brought from a nearby channel (no further than 50m away) for on-site subsistence-related activities (Delagnes and Roche 2005). 190 cobbles or fragments of cobbles were carried to the site along with a few unmodified split cobbles. The knappers favored medium-grained phonolite (52% of the on-site raw materials, Figure 1.4), a type of raw material that was most suitable in terms of flaking quality and morphology for the production of large amounts of flakes, considering the technical skills they had developed (60% of the flakes and flake fragments, $n = 745$). At the sources, this involved selecting

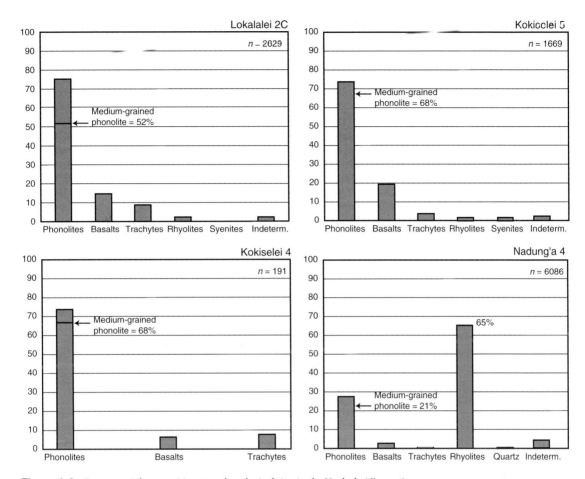

Figure 1.4   Raw material composition at archaeological sites in the Nachukui Formation.

phonolites in general over basalts and trachytes, but within the phonolite group itself there was little selection since the medium-grained type is predominant (see Figure 1.3).

Medium-grained phonolite is a rock with good flaking qualities, because it displays a parallel mineral orientation that gives the rock a natural foliation. It has the mechanical advantage of breaking easily along the foliation plane when using direct hard hammer percussion, and therefore offers a measure of predictability in terms of fracture orientation (Harmand 2004, 2005). At the sources, this type of phonolite occurs mainly as rounded to sub-rounded cobbles, and more rarely as angular cobbles.

Because lithic technical skills at 2.3 myrs remained strongly constrained[3] by the original shape of the

available raw materials (Delagnes and Roche 2005), the initial morphology of the blocks had significant repercussions on the condition in which raw materials were introduced onto the site (whole or split cobbles, Harmand 2005). Medium-sized *angular* cobbles or fragments of cobbles (12 cm maximum dimension[4]) of the medium-grained phonolite with *naturally* serviceable striking surfaces ($n = 95$) were preferentially exploited for their suitability to be knapped without any preparation (Delagnes and Roche 2005). Flakes, cores, and fragments are mainly made on these angular specimens of medium-grained phonolite. The low representation of such specimens at the sources is probably one of the reasons for the deliberate breakage of large-sized (>15 cm) to medium-sized (8–15 cm) rounded cobbles of medium-grained phonolite into several pieces to obtain suitable blanks, probably prior to transport and perhaps at the source, where the raw material was collected.

To a lesser extent, debitage was also conducted on poorer quality cobbles or fragments of cobbles of aphyritic or porphyritic basalts (14%, Figure 1.4). The presence of numerous phenocrysts of olivine and pyroxene within the porphyritic basalts significantly lessens the predictability of flake sizes and morphologies. A series of unmodified cobbles (n = 54), mostly rounded cobbles (tested or not) of medium-grained trachyte or fine-grained basalt, were also brought to the site and probably stockpiled as resistant, massive, and difficult to break "manuports". Eighteen heavy and medium-sized rounded cobbles of a resistant medium-grained trachyte bear signs of percussion damage and are interpreted as hammerstones used for knapping most of the cores flaked at the site (Delagnes and Roche 2005). These hammerstones were selected among the cobbles most appropriate for percussion in terms of mass, size, and shape, within the supply of raw materials brought to the site (Harmand 2004).

The differential selection and management of raw materials at the Late Pliocene site of Lokalalei 2C suggests sensitivity to the quality of raw materials and an empirical knowledge of the mechanics of rock fracture from 2.3 myrs onwards. This is in keeping with the new and unexpected results obtained at the Early Oldowan localities of Gona in Ethiopia and Kanjera in Kenya (Plummer *et al.* 1999; Semaw 2000; Semaw *et al.* 2003; Plummer 2004; Stout *et al.* 2005). On the other hand, the selection of particular morphologies (angular clasts) at Lokalalei 2C is tied to the level of skill mastered by the knappers.

MORE ABOUT SELECTION IN THE PLEISTOCENE

Oldowan

The selective attitude towards raw materials highlighted for the Early Oldowan of Lokalalei 2C is documented to an even higher degree for the Oldowan of Kokiselei 5, *ca.* 1.70 myr, where the same medium-grained phonolite accounts for a higher proportion of the on-site raw materials (68% against 52%, Figure 1.4). Moreover, selection at the sources within the phonolite group was more important, since the medium-grained type of phonolite is a minority

(see Figure 1.3). In addition, two types of productions, flakes and "heavy-duty tools" (*sensu* M. Leakey 1971), are identified, each of them involving the selection of specific sizes of clasts at the nearby raw material sources, from where an estimate of 170 cobbles or fragments of cobbles were brought to the site.

Small to medium-sized cobbles (angular and rounded) or fragments of cobbles[5] of the medium-grained phonolite were used for debitage reduction sequences to obtain flakes (74% of the flakes and flake fragments, n = 998). To a lesser extent, debitage was also conducted on lower quality cobbles or fragments of cobbles of aphyritic or porphyritic basalts (20% of the on-site raw materials, Figure 1.4), as well as on the poorly represented trachytes, rhyolites, syenites, and siliceous rocks (a total of 6.5%, Figure 1.4).

Larger blocks of medium-grained phonolite, fragmented prior to their transport to the site, were selected for the manufacture of heavy-duty tools (22 cm maximum dimension, 2.5 kg) (only 1% of the lithic assemblage). In addition, the lithic assemblage includes 138 cobbles consisting mostly of rounded cobbles (tested or not) of fine-grained aphyric or porphyric basalts, possibly stockpiled to serve as resistant and massive "manuports" and/or hammerstones (Harmand 2005).

Thus, around 1.7 myr in the Nachukui Formation, there is some evidence that distinct provisioning patterns came into play for distinct morpho-(functional?) ends. Furthermore, the debitage systems appear less constrained by the initial morphology of the raw materials available at the sources. Indeed, at Kokiselei 5, the knappers display for the first time the ability to modify the initial morphology of the raw material by creating new striking platforms, rather than always using **naturally** serviceable ones (Texier *et al.* 2006). This is probably why they did not find it necessary to select only angular cobbles at the sources, or to conduct debitage on the deliberately fragmented larger blocks of medium-grained phonolite, as at Lokalalei 2C: these fragments were transformed into heavy-duty tools. Finally, the Kokiselei 5 knappers had the ability to tackle small volumes, as testified to by the flaking of a chert block – a very unusual raw material in this assemblage.

EARLY ACHEULEAN

The Early Acheulean assemblage of Kokiselei 4, *ca.* 1.65 myr, is characterized by the appearance of large handaxes or proto-handaxes and picks, alongside a flake production (Texier *et al.* 2006). New volumes are exploited to create specific tools through bifacial shaping,[6] involving a measure of anticipation in the selection of raw materials suitable for such a purpose in terms of quality, size, and morphology.

Accordingly, the site was provisioned mainly with large flat slabs (length >20 cm) of a medium-grained phonolite (68% of the on-site raw materials, Figure 1.4). Selection at the sources was high since such rocks and such dimensions are poorly represented in the nearby conglomerates (see Figure 1.3). The initial selection of a hard-wearing high-grade raw material from which blanks with a specific morphology (large flat elongated flakes) could be easily obtained, enabled the knappers to maximize tool sizes and to produce large handaxes and picks (*n* = 35) through a shaping reduction system. As a result, the flakes produced have relatively high dimensions (20 cm maximum length) and display long sharp cutting edges, serviceable without any transformation by retouch.

Because of the disturbed archaeological context and the small amount of material recovered (191 pieces), it is difficult to assess the relative importance of the shaping and debitage systems in the assemblage. Firm evidence of the latter is however provided by 11 cores, made from small cobbles of medium-grained phonolite.

ACHEULEAN

In the assemblage of Nadung'a 4 the large cutting-tools characteristic of the Acheulean, handaxes and cleavers, are lacking, and the heavy-duty component is small (10 "worked pebbles"). However, the assemblage is remarkable in many respects, among which is the use of raw materials hitherto barely represented – rhyolites – for the exclusive manufacture of a new toolkit of "light-duty tools" (*sensu* M. Leakey 1971).

The high proportion of small elements (length <2 cm), the significant quantity of cores, and the presence of hammerstones are consistent with knapping activities carried out on the spot. An estimate of 170 blocks and cobbles were introduced into the site from the nearby sources.

Raw material type frequencies show a marked increase in the amount of rhyolites (65% of the on-site raw materials (Figure 1.4), which were rarely exploited during Late Pliocene and Early Pleistocene time period. This involves a high degree of selection at the sources,

since rhyolites are quite uncommon (see Figure 1.3). The medium-grained phonolite was also used, to a lesser degree (21%, Figure 1.4), and this also testifies to selection at the sources, where this raw material is a very small minority (see Figure 1.3).

At Nadung'a 4, the knappers judiciously exploited the rocks according to their petrographic properties, for the purpose of carrying out various processing tasks, probably partly connected with the presence of an elephant carcass (Delagnes *et al.* 2006). The study of the relationship between raw materials and artifacts highlights different behaviors depending on the type of raw material used.

On the one hand, rhyolite was exclusively used to produce flakes and obtain cutting edges. Mostly red, sometimes green, and more seldom gray, brown, or yellow, rhyolite is a fine-grained raw material. Despite the fact that this rock frequently occurs as small diaclasic angular blocks, less compact and less homogeneous than phonolites and basalts owing to frequent internal fissures (Harmand 2005), it is very suitable for obtaining hard-wearing and potentially functional active edges (sharp cutting edges resistant to abrasion), using direct hard hammer percussion. At Nadung'a 4, a better controlled production of flakes (Delagnes *et al.* 2006) accounts for the choice of this fine-grained but less homogenous raw material, from which numerous sharp flakes were produced (61% of the flakes and flake fragments >2 cm, *n* = 2196), serviceable without any transformation by retouch. In addition, notches and denticulates are numerous in the lithic assemblage (*n* = 410) (Delagnes *et al.* 2006), and these light-duty tools were probably also partly devoted to tasks related to meat processing and consumption activities such as cutting and scraping.

On the other hand, the medium-grained phonolite was equally used for flake production according to the same principles of debitage documented for rhyolite (Delagnes *et al.* 2006), and for manufacturing some heavy-duty tools by simple shaping.

The remaining rocks are those usually found in the region. Represented by large and compact cobbles of phonolite, basalt, or trachyte (maximum length: 24 cm; maximum width: 12 cm), they were used for heavy-duty tools, suitable for activities such as breaking an elephant carcass. The few hammerstones found in the site are medium-size cobbles of quartz, basalt, or phonolite, appropriate for hammering hard rocks.

Around 0.7 myr in the Nachukui Formation, the procurement and exploitation patterns point to more highly selective raw material provisioning. Furthermore, the selection of different types of rocks appears very closely linked to specific types of products. This is interpreted as the result of higher requirements in terms of technical efficiency, as indicated by the presence of notches and denticulates, requirements that could best be met by the use of rhyolite (Delagnes *et al.* 2006).

CONCLUSION

Usually focusing on more recent periods of Prehistory, the present techno-economic study applied to rich and well-preserved Early Oldowan, Oldowan, Early Acheulean, and Acheulean lithic assemblages from the West Turkana region has proved relevant for investigating how raw material procurement activities are economically structured in the Early Stone Age, and for monitoring behavioral differences between complexes from the Late Pliocene, the Early Pleistocene, and the very beginning of the Middle Pleistocene. Early hominin provisioning behaviors in the Nachukui Formation show the antiquity of decision-making in ancient contexts (selection of raw materials for their flaking quality, size, and/or morphology as early as 2.3 myrs), and the existence of a higher degree of technological planning than previously acknowledged for the Early Oldowan. Furthermore, the techno-economic analysis has highlighted significant changes in raw material procurement and exploitation during the Plio-Pleistocene. In the Pliocene, selection is largely determined by the knappers' technological limitations. During the Pleistocene, selection bears increasingly on rocks best suited in terms of quality and morphologies to more varied templates (light-duty/heavy-duty tools). These could take shape as a result of the development of higher technical skills, and possibly higher cognitive capacities. The results of this study are consistent with the idea of a gradual improvement in technical efficiency throughout the Plio-Pleistocene related to distinct hominin species or genera with different levels of technical capabilities (*Paranthropus*, *H. aff. habilis*, *H. erectus*). They lend strong support to recent investigations on early hominin lithic production, which reveal a much more complex picture of the first technological manifestations, suggestive of temporal variability in the technical systems documented for the African Oldowan (Roche 2000; Martinez-Moreno *et al.* 2003; Roche *et al.* 2003;

de la Torre *et al.* 2003; de la Torre 2004; Delagnes and Roche 2005).

At present, evidence of temporal changes in techno-economic patterns for African Oldowan and Acheulean productions remains limited. More studies based on detailed and comparable techno-economic analyses from a larger number of sites are needed to sketch an overview of potentially different responses to resource availability by Oldowan and Acheulean hominins in terms of raw material selection and management, to make regional assessments and ultimately to generate a clearer picture of the nature and significance of technical changes in the Early Stone Age.

ACKNOWLEDGMENTS

I am grateful to H. Roche for her helpful suggestions on an earlier draft of this chapter and am greatly indebted to J. Féblot-Augustins for significantly improving the style and content of the final version.

ENDNOTES

1 The word debitage is used throughout this chapter in its original French meaning as a reflection of a specific reduction process, not as the residue of production.
2 The size, morphology, and cortex of the lavas sampled in primary contexts are very different from those of the cobbles in derived sources. There is no indication in the archaeological assemblages that any raw material was collected in primary context.
3 As evidenced by knapping accidents, which reflect technical deadlocks due to the lack of extensive edges with appropriate striking angles (Delagnes and Roche 2005: 443).
4 Dimensions inferred from refittings (see Delagnes and Roche 2005).
5 Dimensions inferred from refittings (see Texier *et al.* 2006).
6 Shaping refers to a "knapping operation carried out for the purpose of manufacturing a single artifact by sculpting the raw material in accordance with the desired form" (Inizan *et al.* 1999: 138).

REFERENCES

Binford, L.R. (1980) Willow smoke and dog's tails: Hunther-gatherer settlement systems and archaeological site formation. *American Antiquity* 45(1), 4–20.

Braun, D.R. and Harris, J.W.K. (2003) Technological developments in the Oldowan of Koobi Fora: Innovative techniques of artifact analysis. In: Moreno, J.M., Torcal, R.M., and Sainz, I.T. (eds) *Oldowan: Rather More than Smashing Stones*, Centre d'Estudis del Patrimoni Arqueologic de la Prehistoria, Bellaterra, pp. 117–44.

Brown, F.H. and Feibel, C.S. (1988) "Robust" hominids and Plio-Pleistocene paleogeography of the Turkana Basin, Kenya and Ethiopia. In: Grine, F.E. (ed.) *Evolutionnary History of the "Robust" Australopithecines*, Aldine de Gruyter, New York, pp. 325–41.

Brown, F.H. and Feibel, C.S. (1991) Stratigraphy, depositional environments, and paleogeography of the Koobi Fora Formation. In: Harris, J.M. (ed.) *Koobi Fora Research Project, vol. 3, The Fossil Ungulates: Geology, Fossil Artiodactyls, and Palaeoenvironments*, Clarendon Press, Oxford, pp. 1–103.

Brown, F.H. and Gathogo, P.N. (2002) Stratigraphic relation between Lokalalei 1A and Lokalalei 2C, Pliocene archaeological sites in West Turkana, Kenya. *Journal of Archaeological Science* 29, 699–702.

Brown, F.H., Harris, J., and Leakey, R.E.F. (1985) Early *Homo erectus* skeleton from West Lake Turkana, Kenya. *Nature* 316, 788–92.

Clark, J.D. (1980) Raw material and African lithic technology. *Man and Environment* 4, 44–55.

de la Torre, I. (2004) Omo Revisited. Evaluating the technological skills of Pliocene Hominids. *Current Anthropology* 45(4), 439–65.

de la Torre, I., Mora, R., Dominguez-Rodrigo, M., Luque, L., and Alcala, L. (2003) The Oldowan industry of Peninj and its bearing on the reconstruction of the technological skills of Lower Pleistocene hominids. *Journal of Human Evolution* 44, 203–24.

Delagnes, A. and Roche, H. (2005) Late Pliocene hominid knapping skills: The case of Lokalalei 2C, West Turkana, Kenya. *Journal of Human Evolution* 48, 435–72.

Delagnes, A., Brugal, J.-Ph., Harmand, S., Lenoble, A., Prat, S. Tiercelin, J.-J., and Roche, H. (2006) Interpreting pachyderm single carcass sites in the African Middle Pleistocene record: A multidisciplinary approach in the site of Nadung'a 4 (Kenya). *Journal of Anthropological Archaeology* 25, 448–65.

Féblot-Augustins, J. (1990) Exploitation des matiè res premières dans l'Acheuléen d'Afrique: Perspectives comportementales. *Paléo* 2, 27–42.

Féblot-Augustins, J. (1997a) *La circulation des matières premières au Paléolithique. Synthèse des données. Perspectives comportementales*, vol. 75, ERAUL, Liège, 275 p.

Féblot-Augustins, J. (1997b) Anciens hominidés d'Afrique et premiers témoignages de circulation des matières premières. *Archéo Nil* t. 7, 17–45.

Feibel, C.S., Brown, F.H., and McDougall, I. (1989) Stratigraphic context of fossil hominids from the Omo group deposits: Northern Turkana Basin, Kenya and Ethiopia. *American Journal of Physical Anthropology* 78, 595–622.

Feibel, C.S., Harris, J.M., and Brown, F.H. (1991) Paleoenvironmental context for the Late Neogene of the Turkana Basin. In: Harris J.M. (ed.) *Koobi Fora Research Project, vol. 3, The Fossil Ungulates: Geology, Fossil Artiodactyls, and Palaeoenvironments*, Clarendon Press, Oxford, pp. 321–70.

Geneste, J.-M. (1989) Economie des ressources lithiques dans le Moustérien du sud-ouest de la France. In: Otte M. (ed.) *L'Homme de Néandertal: La subsistance* vol. 6, ERAUL, Liège, pp. 75–97.

Geneste, J.-M. (1991) L'approvisionnement en matières premières dans les systèmes de production lithique: La dimension spatiale de la technologie. In: Mora, R., Terradas, X., Parpal, A., and Plana, C. (eds) *Tecnologia y cadenas operativas liticas*, Barcelona: Universitat Autonoma de Barcelona, Treballs d'Arqueologia I, pp. 1–36.

Harmand, S. (2004) Raw materials and economic behaviour of Late Pliocene Hominids: The case of Lokalalei 2C and Lokalalei 1 sites, West Turkana (Kenya). Poster presented at the Paleoanthropology Society Annual Meeting, Montreal (Canada), March 30–31.

Harmand, S. (2005) *Matières premières lithiques et comportements techno-économiques des homininés plio-pléistocènes du Turkana occidental, Kenya*. Thèse, Université Paris X-Nanterre, 317 p.

Harris, J.M., Brown, F.H., and Leakey, M.G. (1988) Stratigraphy and paleontology of pliocene and pleistocene localities west of lake Turkana, *Kenya: Contribution Science* 399, 1–128.

Hay, R.L. (1976) *Geology of the Olduvai Gorge: A Study of Sedimentation in a Semi-arid Basin*. University of California Press, Berkeley.

Inizan, M.-L., Ballinger, M., Roche, H. and Tixier, J. (1995) *Technology and Terminology of Knapped Stone*, vol. 5, CREP, Nanterre.

Isaac, G.L. (1977) *Olorgesailie: Archaeological Studies of a Middle Pleistocene lake basin in Kenya*. University of Chicago Press, Chicago, 272 p.

Isaac, G.L. and Harris, J.W.K. (1978) Archaeology. In: Leakey, M.G. and Leakey, R.E. (eds) *Koobi Fora Research Project. The Fossil Hominids and an Introduction to their Context, 1968–1974*. Clarendon Press, Oxford, pp. 64–85.

Isaac, G.L., Harris, J.W.K., and Kroll, E. (1997) The stone artifact assemblages: A comparative study. In: Isaac, G.L. (ed.) *Koobi Fora Research Project, vol. 5: Plio-Pleistocene Archaeology*. Clarendon Press, Oxford, pp. 262–99.

Jones, P. (1979) Effects of raw material on biface manufacture. *Science* 204, 835–6.

Kibunjia, M., Roche, H., Brown, F.H., and Leakey, R.E. (1992) Pliocene and Pleistocene archaeological sites west of Lake Turkana, Kenya. *Journal of Human Evolution* 23, 431–8.

Leakey, M.D. (1971) *Olduvai Gorge: Excavations in Beds I and II, 1960–1963, vol. III.* Cambridge University Press, Cambridge, 172 p.

Leakey, M.D. (1975) Cultural patterns in the Olduvai sequence. In: Butzer, K.W., and Isaac, G.L. (eds) *After the Australopithecines*, Mouton, pp. 477–93.

Leakey, M.D. (1994) *Olduvai Gorge: Excavations in Beds III, IV and the Masek Beds, 1968–1971*, vol. V. Cambridge University Press, Cambridge, 327 p.

Leroi-Gourhan, A. (1964) *Le geste et la parole. Technique et langage*. Albin Michel, Paris.

Leroi-Gourhan, A. (1971) *Evolution et technique. L'homme et la matière*. Albin Michel, Paris.

Martinez-Moreno, J., Mora Torcal, R., and de la Torre, I. (2003) Oldowan, rather more than smashing stones: An introduction to The Technology of First Humans. In: Moreno, J.M., Torcal, R.M., and Sainz, I.T. (eds) *Oldowan: Rather More than Smashing Stones*, Centre d'Estudis del Patrimoni Arqueologic de la Prehistoria, Bellaterra, pp. 11–35.

Merrick, H.V. and Merrick, J.P.S. (1976) Recent archaeological occurrences of earlier Pleistocene age from the Shungura Formation. In: Coppens, Y., Howell, F.C., Isaac, G.L., and Leakey, R.E. (eds) *Earliest Man and Environments in the Lake Rudolf Basin*. Chicago University Press, Chicago, pp. 574–84.

Plummer, T. (2004) Flaked stones and old bones: Biological and cultural evolution at the dawn of technology. *Yearbook of physical anthropology, Wiley Interscience* 47, 118–64.

Plummer, T., Bishop, L.C., Ditchfield, P., and Hicks, J. (1999) Research on Late Pliocene Oldowan sites at Kanjera South, Kenya. *Journal of Human Evolution* 36, 151–70.

Prat, S., Brugal, J.-Ph., Roche, H., and Texier, P.-J. (2003) Nouvelles découvertes de dents d'hominidés dans le Membre Kaitio de la Formation de Nachukui (1.65–1.9 Ma), Ouest du lac Turkana (Kenya). *Comptes Rendus Palévol* 2, 685–93.

Roche, H. (2000) Variability of Pliocene lithic productions in East Africa. *ACTA Anthropologica Sinica* 19, 98–103.

Roche, H. and Kibunjia, M. (1994) Les sites archéologiques plio-pléistocènes de la Formation de Nachukui, West Turkana, Kenya. *Comptes rendus de l'Académie des sciences* 318, série II, 1145–51.

Roche H. and Kibunjia, M. (1996) Contribution of the West Turkana Plio-Pleistocene sites to the archaeology of the lower Omo/Turkana Basin. *Kaupia* 6, 27–30.

Roche, H., Delagnes, A., Brugal, J.-Ph., Feibel, C.S., Kibunjia, M., Mourre, V., and Texier, P.-J. (1999) Early hominid stone tool production and knapping skill 2.34 Myr ago in West Turkana. *Nature* 399, 57–60.

Roche, H., Brugal, J.-Ph., Delagnes, A., Feibel, C.S., Harmand, S., Kibunjia, M., Prat, S., and Texier, P.-J. (2003) Les sites archéologiques plio-pléistocènes de la Formation de Nachukui (Ouest Turkana, Kenya): Bilan préliminaire 1997–2000. *Comptes Rendus Palévol* 2(8), 663–3l.

Rogers, M.J., Harris, J.W.K., and Feibel, C.S. (1994) Changing patterns of land use by Plio-Pleistocene hominids in the lake Turkana basis. *Journal of Human Evolution* 27, 139–58.

Schick, K. (1987) Modelling the Formation of Early Sone Age artifact concentrations. *Journal of Human Evolution* 16, 789–808.

Schick, K. and Toth, N. (1993) *Making Silent Stones Speak. Human Evolution and the Dawn of Technology*. A Touchstone Book, Simon and Schuster, New York.

Semaw, S. (2000) The world's oldest stone artifacts from Gona, Ethiopia: Their implications for understanding stone technology and patterns of human evolution between 2.6–1.5 million years ago. *Journal of Archaeological Science* 27, 1197–214.

Semaw, S., Rogers, M.J., Quade, J., Renne, P.R., Butler, R.F., Dominguez-Rodrigo, M., Stout, D., Hart, W.S., Pickering T., and Simpson, S.W. (2003) 2.6-Million-year-old stone tools and associated bones from OGS-6 and OGS-7, Gona, Afar, Ethiopia. *Journal of Human Evolution* 45, 169–77.

Stout, D., Quade, J., Semaw, S., Rogers, M., and Levin, N. (2005) Raw material selectivity of the earliest stone toolmakers at Gona, Afar, Ethiopia. *Journal of Human Evolution* 48, 365–80.

Tiercelin, J.-J., Schuster, M., Roche, H., Thuo, P., Brugal, J.-Ph., Trentesaux, A., Barrat, J.-A., Bohn, M. Prat, S., Harmand, S., Davtian, G., and Texier, P.-J. (in prep.) Stratigraphy and paleoenvironments of the 2.3–2.4 Myr Lokalalei archaeological sites, Nachukui Formation, West Turkana, Northern Kenya Rift.

Texier, P.-J., Roche, H., and Harmand, S. (2006) Kokiselei 5, formation de Nachukui, West Turkana, Kenya: Un témoignage de la variabilité ou de l'évolution des comportements techniques au Pléistocène ancien ? Actes du XIVème Congrès de l'Union Internationale des Sciences Préhistoriques et Protohistoriques, Liège Septembre 2–8, 2001.

Texier, P.-J., Roche, H., Brugal, J.-Ph., and Harmand, S. (2004) KS5, a transition site between the oldowan (KS1) and the early acheulean (KS4) from Kokiselei site complex (West Turkana, Kenya)? Paper presented at the Paleoanthropology Society Annual Meeting, Montreal (Canada), March 30–31.

Toth, N. (1982) *The Stone Technologies of Early Hominids at Koobi Fora, Kenya: An Experimental Approach*. PhD dissertation, University of California, Berkeley.

Toth, N. (1985) The oldowan reassessed: A close look at early stone artifacts. *Journal of Archaeological Science* 12, 101–20.

Toth, N. (1987) Behavioral inferences from early stone artifact assemblages: An experimental model. *Journal of Human Evolution* 16, 763–87.

Walker, A., Leakey, R.E.F., and Harris, J. (1986) 2.5 Myr *Australopithecus boisei* from west of Lake Turkana, Kenya. *Nature* 322, 517–22.

Chapter 2

PATTERNS OF LITHIC MATERIAL PROCUREMENT AND TRANSFORMATION DURING THE MIDDLE PALEOLITHIC IN WESTERN EUROPE

Liliane Meignen[1], Anne Delagnes[2], and Laurence Bourguignon[3]

[1]UNSA-CNRS, CEPAM, Sophia Antipolis, France
[2]CNRS, PACEA, Talence, France
[3]INRAP, PACEA, Talence, France

ABSTRACT

Analyses of Middle Paleolithic technological behaviors – and by extension of Neandertal cognitive capacities and mobility organization – have been revolutionized by theoretical perspectives devised from lithic technological and raw material investigations. Binary oppositions such as "Levallois/non-Levallois" are increasingly regarded as oversimplifications since differing reduction strategies are apparent in the same archaeological assemblages. Various factors including raw material quality and cyclical reduction are reflected in multiple lithic matrices. Detailed analyses of *chaînes opératoires* provided insights on the structures of lithic technological behaviors. The authors refer to extensive data from recent excavations in France and Belgium to argue that Middle Paleolithic tool provisioning and management strategies show clear organization and planning depth, even if those strategies differ from those of the Upper Paleolithic.

Lithic Materials and Paleolithic Societies, 1st edition. Edited by B. Adams and B.S. Blades, ©2009 Blackwell Publishing. ISBN 978-1-4051-6837-3.

Defined in a general way as "a period of flake-based industry characterized by an important stability in tool types", the Middle Paleolithic has been considered for a long time as a remarkably unified whole over long periods. If it is true that the technological changes are slower than those of later periods, the Middle Paleolithic of Western Europe, and more particularly the French Middle Paleolithic on which the present discussion focuses, is actually highly diversified. This notion of diversity was already implicit in the typological approach advocated by Bordes (1950), which led to the recognition of five Mousterian facies based on the relative proportions of retouched tool types (Bordes 1953, 1981). Recently the perception of this diversity in Western Europe has been greatly refined by the development of technological approaches.

Over the last two decades, the technological approaches (the so-called *chaînes opératoires* approach in the French literature (Cresswell 1982; Karlin *et al.* 1991; Lemonnier 1986)) have allowed scholars not only to recognize various stages in lithic tool making (a topic not developed in this chapter) but also to investigate the basic conceptual processes that underlay the sequence of manufacturing steps in stone tool production.

Different core volume organizations (i.e., *conceptions volumétriques* in the French literature) have been identified as well as their respective end-products and by-products. The pioneering work carried out by Boëda (1986, 1994) was focused on the Levallois concept, and its variability as expressed in different reduction modalities. Blanks can be produced following different modalities: the recurrent methods (with unidirectional, bidirectional, or centripetal removals), through which several predetermined flakes are produced from the same flaking surface, and the preferential method in which a single Levallois blank is produced from each flaking surface. Less standardized and more diversified end-products result from the centripetal recurrent method, while the preferential methods lead to more predetermined shapes (large oval Levallois flakes or Levallois points, depending on how the core was initially prepared).

This work was followed later by the identification of other production methods: "discoidal method" (Boëda 1993), "alternating platform technique" (Ashton 1992) or "Clactonian method" (Forestier 1993), "Quina method" (Bourguignon, 1996, 1997), "laminar production system" (Boëda 1990; Revillion 1994), the "Kombewa-like Les Tares method" (Geneste and Plisson 1996), the "Pucheuil-type method" (Delagnes 1993), and the bifacial shaping method (*chaîne opératoire de façonnage bifacial*).

Since then, many lithic assemblages have been studied within this framework, expanding our knowledge of the internal variability within each concept (Geneste 1988; Turq 1989; Boëda *et al.* 1990; Boëda 1991; Delagnes 1992; Jaubert 1993; Meignen 1993a; Locht and Swinnen 1994; Jaubert and Farizy 1995; Texier and Francisco-Ortega 1995; Delagnes and Ropars 1996; Geneste *et al.* 1997, among others). In fact, these lithic production systems are far from rigid and due to their inherent flexibility, the various flaking modalities recognized do not always match up to the limits of our conventional technological categories.

However, it clearly means that the traditional binary opposition of "Levallois/non-Levallois" or "elaborated/ non-elaborated debitage" needs to be abandoned. It is worth noting that different reduction strategies are in some cases associated in the same assemblage (Table 2.1), each of them presumably associated with different functional purposes, as for instance at Riencourt-les-Bapaume in level CA (Beyries 1993). Moreover, in some assemblages, the coexisting flaking methods occur on different types of raw material. Examples include the following: Sclayn layer 5 in Belgium, where Discoidal, Quina, and Levallois products have been manufactured from three different types of raw materials (Moncel 1998b), at Coudoulous I, les Fieux, La Borde with Discoidal debitage mostly on quartz/quartzite and Levallois method on flint (Jaubert and Farizy 1995; Jaubert and Mourre 1996). These assemblages clearly illustrate the high variability of the lithic technical systems among the Western European Middle Paleolithic.

But in the context of the debate on the behaviours of Neandertals versus those of modern humans, the levels of complexity and elaboration are often inferred from lithic production systems of the Middle and Upper Paleolithic. In fact, Acheulean lithic production systems reflecting elaborated production of bifaces existed adjacent to other techniques that required much less elaboration. These data therefore suggest possible complementary approaches to simple and complex production systems (Roche and Texier 1991; Texier and Roche 1995). The coexistence in numerous periods of less complex systems with more sophisticated ones – a phenomenon emphasized by Jacques Pelegrin (2004) – contradicts the idea of diachronic evolution of technological production in a linear fashion

Table 2.1 Qualitative evaluation of the degree of sophistication for the four main Middle Palaeolithic debitage systems (Levallois, Discoidal, Quina, Laminar)

	Core shaping and/or maintenance	Predetermination of desired end-products	Normalization of desired end-products	Potential for resharpening	Ramification	Productivity
Levallois preferential	+++	+++	+++	++	---	---
Levallois recurrent						
uni/bidirect	++	++	++	+--	++	++
centripetal	+	+--	+--	+--	+--	+++
Discoidal	+--	+--	+	+--	++	+++
Quina	---	+	+--	+++	+++	+++
Laminar	+	+	++	+--	---	++

from simple to more complex. These different levels of elaboration should rather be understood within the context of group socio-economic organization and the functional complementarity of their activities.

For the Middle Paleolithic, this point is illustrated by the ramified nature of *chaînes opératoires* – pieces as unretouched debitage blanks or as those already retouched for tools are subsequently used as cores – a phenomenon recently identified in all production systems known for this period: *débitage des Tares*, Levallois, Discoidal, and Quina (Delagnes 1996; Geneste and Plisson 1996; Bourguignon *et al.* 2004a). In each of these systems, one observes in effect one or several secondary productions depending upon the products or sub-products arising from the primary production. This is indicative of planning in the organization of the *chaînes opératoires* implemented with the aim of obtaining differentiated blanks.

By presenting these systems of productions on the basis of a series of qualitative criteria, we have been able to reveal different degrees of elaboration in the organization of production. The technical "investment" seems related either to the debitage phase or to those of tool production and maintenance. Initial criteria characterize the degrees of predetermination and of normalization[1] of the products emerging from the debitage sequence. They result from the more or less constraining characteristics of core preparation and/or maintenance. A second range of criteria concerns the phase of preparation and management of the tool component estimating the potential for resharpening the produced supports, the importance

of the phenomenon of ramification, and finally the productivity of the system in quantitative terms.

The results obtained (Table 2.1) show that the studied systems of production present different degrees of sophistication/predetermination, but that none appears really more complex than the others. In fact, two broad groupings may be defined:

• The first comprises systems that present an important investment related to the phase of core shaping/maintenance, such as Levallois debitage for the preferential blank, for example. This initial investment involves a strong predetermination and standardization of the end-products but, on the other hand, these systems have a low productivity.

• The second unit includes debitage systems, such Quina debitage, which show a weak technical investment in the phase of core shaping, generating little marked standardization of the produced blanks. On the other hand, in these systems, the debitage leads to a strong productivity which is very often increased by ramification. Moreover, the lifespan of the tool can be prolonged considerably because its morphology permits repeated resharpenings.

Thus, standardization of Levallois products arises from the strong investment in the core shaping phase, while the standardization of the tools of the Quina system results from the adjustment of only slightly predetermined flakes. The installation of the morphology of the tools thus intervenes, according to systems, at various moments in the *chaîne opératoire*.

However, between these two extremes, a whole series of productions developed with the less accentuated tendencies. Within unidirectional recurrent Levallois debitage, for example, the technical investment may relate according to lithic assemblages rather than to the debitage phases for obtaining relatively normalized blanks, as seen, amongst others, in the assemblages of Biache-St-Vaast IIA (Boëda 1988a) and Vaufrey, levels VII and VIII (Geneste 1988). In these cases, the Levallois products are either slightly retouched (a standard of retouch that strengthens the original morpho-functional attributes) or unretouched (and probably used as such). Conversely, the technical investment may relate rather to the phases of final improvement of the Levallois products, arranged in a panoply of varied retouched tools, for instance in the assemblages of Abri Suard at La Chaise-de-Vouthon (Delagnes 1991, 1995). Centripetal recurrent Levallois debitage has been characterized as highly productive in spite of predetermination, normalization of the end-products are reduced, as are their potential for resharpening and ramification. In the same way, Discoidal debitage presents, when directed towards the production of pseudo-Levallois points in series, a marked normalization of the products in spite of a limited investment in the phase of core shaping/maintenance (Bourguignon and Turq 2003).

It is clear that the concept of a dichotomy between simple and complex systems is not relevant in Middle Paleolithic lithic production. In fact, one observes different degrees of sophistication according to systems, with a technical investment that is more pronounced at various points in the *chaîne opératoire*, either during the phase of production or during the phase of management of the tools.

These various options fall under structured techno-economic systems, revealing the methods of acquisition/resource management by Middle Paleolithic men, the organization of the territories, and the adopted patterns of mobility. In the field of lithic production, the pioneering work of Jean-Michel Geneste in the years 1985–90 and much research since then has made it possible to recognize real strategies of acquisition/management of lithic raw materials from the earliest Middle Paleolithic, not only in the nearby area most frequently traversed (0–5 km) but also in a wider area, with different behaviors recognized according to the zones, both distant (30–80, even 100 km) and intermediate (5–20 km) (Geneste 1985, 1988; Féblot-Augustins 1997, 1999; Turq 2000). The technical forms in which these raw materials circulate vary according to the zones from which the materials were obtained. High mobility (long distances) is recognized for the technically elaborate objects reflecting the final stages of the *chaîne opératoire*, such as predetermined products, retouched tools, and bifaces, while the nuclei and cortical products are, in this model, moved only infrequently, if at all, from the places of production. The fragmentation of the *chaîne opératoire* across the landscape seems to reflect a desire to reduce the energetic cost of transport. These behaviors thus reveal a strategy of landscape management based on a conception of future needs (Geneste 1991, 1992).

Data from more recent work on raw material mobility, while confirming this concept of "anticipated productions", also serve to moderate this model. The new research shows that the categories of products transferred over long distances – that is, beyond 20 km – are more diversified than once thought. A certain number of recently studied lithic assemblages show nuclei or cortical flakes imported from such distances, along with tools, bifaces, and predetermined products. The industry of the level 8GHF of Artenac in Charente contains 8% of pieces on Senonian flint from Dordogne sources at least 42 km away. This distant raw material was not only introduced in the form of finished products such as large Levallois flakes, but also in the form of preformed nuclei. The latter are not present in the assemblage (exported?), but their reduction on site is attested by numerous core management flakes (Delagnes *et al.* 2006). The Senonian flint coming from west of the Charente Basin – more than 70 km away – was also introduced in the form of cores in another layer (6c), but in this case as an isolated piece, a core on flake. The study of the Mousterian series in the cave of Néron (levels II and III), in the Rhone Valley, also illustrates core introductions from sources 20–40 km away, whereas at Champ-Grand the exotic materials, primarily in the form of finished products, come from very distant sources, about 230–250 km away (Slimak 2004).

All these products transferred over long distances seem to function as "matrices", that is, as a potentially exploitable volume of raw material for debitage production or for direct use either before or after retouch. The flexible nature of these forms, which may have been supports for blank production, shaping, and/or retouch, undoubtedly encouraged Paleolithic individuals to select these types of products for transport. They constituted, in effect, raw material supplies that were constantly available and easily transportable, making it possible to meet varied and immediate needs (anticipation for general rather than specific needs).

Certain large imported flakes fulfill this matrix role – flakes-matrices – since they constitute raw material reserves for one or more production cycles (ramified *chaînes opératoires*; see Bourguignon *et al.* 2004a, figures 1–4), and one finds them in lithic tool assemblages in the form of cores on flakes of distant raw materials. The Levallois flake on Grand-Pressigny flint discovered in the occupation at La Folie in Poitiers that had been exploited on the lower face to obtain two Kombewa-type flakes illustrates this point (Bourguignon *et al.* 2002). Certain flakes/matrices allow long cycles of tool utilization, with several sequences of rejuvenation and recycling. In the Quina Mousterian site of Marillac (Charente), scrapers with scaly retouch or notched scrapers of exotic flints reflect tool exhaustion, a testament to a long life-cycle (Meignen and Vandermeersch 1986; Meignen 1988; Bourguignon 1997, 2001; Costamagno *et al.* 2006). Other tools/matrices at this same site ensured two successive sequences (tools, then cores): certain notched scrapers provided resharpening and/or recycling flakes[2] that were ultimately used as blanks for small retouched scrapers. Among these rejuvenation flake scrapers, some seem to have been introduced in this form. This phenomenon, illustrated by a long cycle of ramification on exotic materials matrices coming this time from more than 100 km distant, has also been observed in the sites of Espagnac and Champ-Grand (Jaubert 2001; Slimak 2004).

This double sequence is also proven for certain bifaces/matrices, which are simultaneously general-purpose tools for long-term use and also sources of flakes. It is not uncommon, primarily within Mousterian of Acheulean Tradition (MTA) industries, to find biface reduction flakes of imported materials reshaped as tools without the corresponding bifaces being present at the site. This phenomenon is seen, for example, in the sites of Rochette, Les Fieux c.KS, and Abri Brouillaud (Geneste 1985; Soressi 2002; Faivre 2006). These flakes were either obtained initially from a biface that was later transported away – and thus are the only reflection of this biface/matrix at the site – or were introduced in their current form.

However, the versatility of matrices does not seem to be systematic throughout the Middle Paleolithic technocomplex. Whereas the matrices that constituted the raw material reserves – matrices of production – for one or more cycles of exploitation (ramified *chaînes opératoires*) were recognized in all Middle Paleolithic production systems (flake production and bifacial shaping), the matrices that permitted long cycles of tool use (resharpening and recycling of matrices/tools) seem limited to systems of Quina production and bifacial industries (e.g., MTA or Micoquian).

In addition, it seems that these techno-economic behaviors, derived from a rational organization of generalized matrices, were applied to both local and distant raw materials. Indeed, one observes in certain cases the same modes of circulation for the matrices, but this time within the local territory. Such is the case at the MTA site of Combe Brune 1, near Bergerac, where more than 100 different blocks were exploited to obtain the lithic assemblage. More than 90% were obtained in the micro-regional area, as they reflected facies variations of Bergerac flint. The pieces show evidence of diverse modes of introduction: in the form of matrices/tools (predetermined Levallois flakes, bifaces, retouched tools) or in the form of matrices of production. These matrices are identical to those coming from more distant locations, such as those in the Isle or Vézère valleys. This occupation thus illustrates a single techno-economic mode of management in the transport of materials whatever the distances of circulation (Bourguignon *et al.* 2004b; Bourguignon *et al.* in press).

One may also cite the example of the industry in level 8GHF at Artenac (Charente) that is dominated by Levallois debitage for preferential flakes. The presence of only partial *opératoires* sequences (management products and end-products) and the absence of cores that seem to have been carried away after brief exploitation, characterize reduction of local raw materials (middle Jurassic) as well as imported materials (upper Cretaceous) (Delagnes *et al.* 2006). This same spatial/temporal sequencing of the *chaînes opératoires* is also observed in the Mousterian occupation at l'Abri du Musée aux Eyzies de Tayac (Bourguignon, to appear). At Pié-Lombard (Alpes-Maritimes), the lithic assemblage consists almost entirely of finished products on flints from local (<5 km), neighboring (5–20 km), or distant (20–50 km) sources. Such evidence seems, in this case, related to short durations of occupation and undoubtedly to site function (Porraz 2005).

Finally, current data on the Quina Mousterian occupation in layer 22 at the site of Jonzac provide an interesting example of the generalized matrix comparable to that in the Quina Mousterian at

Marillac. This time, however, the raw materials were local ones: introduction to the site of large flakes shaped into wide Quina scrapers that served simultaneously as a matrix/tool and matrix/core for production of flakes through shaping of the edge and lower face (Soressi 2004; Lenoir 2005). The retouch/resharpening flakes were later eventually transformed into retouched tools.

These examples are, nevertheless, limited in number. Recent research tends to show that they are undoubtedly indicative of specific site contexts such as short occupations or specific purpose sites (Porraz 2005) and that future research should focus in this direction. Current data indicate that the Middle Paleolithic lithic assemblages reflect most often a local component that is always dominant, with a weaker component of imported materials from distances beyond 15 to 20 km in various forms or matrices, as has been discussed. In general, these imported products show greater intensities of retouch and reutilization than seen on local materials. This phenomenon would be a reflection through these matrices of the "curation strategy" described in Anglo-American literature (Binford 1973, 1977, 1979).

Ethnographic data show that the transport of personal equipment or "personal gear" (Binford 1979) during movements seems to be a more or less universal phenomenon among hunter-gatherer populations (Kuhn 1995 and references therein) that were, by definition, mobile societies. The composition of the equipment varies according to the geographical distribution of the raw material sources, the patterns of residential mobility of the groups, the destinations, the times of occupation, and the activities undertaken at those locations. The equipment must address all of these considerations within the constraint of portability, and corresponds in fact to the compromise between optimal potential of use and the energetic cost of transport as a function of weight and distance (Kuhn 1992 and cited references). Also, the transported equipment generally consists of a kit of generalized tools ("general-purpose tools" Kuhn 1995: 22 and cited references) that are very adaptable, unless the anticipated activities are well-defined and to be undertaken quickly, such as a hunting expedition. In this case, the equipment adapted to this specific activity is prepared in advance and then transported.

The matrices—sometimes generalized, sometimes specialized – that we have described previously may be an expression of this provisioning strategy ("provisioning of individuals" (Kuhn 1992, 1995) or supporting activities with technical equipment transported by the group). Middle Paleolithic hunter-gatherer groups would have carried a limited number of non-specialized artifacts to meet their requirements. This limited tool-kit, with a potentially long life-span as tools and/or cores, nevertheless would be supplemented by tools produced on the spot. On the other hand, diversified activities during long-term occupations would have derived the needed tool assemblages from different provisioning strategies, mainly from large quantities of local raw materials ("provisioning of place" (Kuhn 1992, 1995) or introduction of blocks, preformed cores, but also possibly large flakes), that were reduced on site. The foreign products that were introduced in limited numbers ("mobile toolkit") when a group arrived at a long-term residential location will thus be obscured in the much larger quantity of local waste materials resulting from the debitage on site (Kuhn 1995). This process explains the small or even negligible proportions of distant materials found at sites of long-term occupation.

CONCLUSION

These analyses of technical productions and levels of knowledge reveal that the Middle Paleolithic is characterized by diversified production systems that may not be categorized as a dichotomy between "simple" and "complex". Differences in the degrees of sophistication and predetermination reflect only specific moments of technical investment, relating to different stages of the *chaînes opératoires* according to the debitage systems.

It seems that the realm in which Middle Paleolithic behavioral complexity is best illustrated is resource planning and the degree of mobility of the populations of hunter-gatherers within the territory. Various responses are seen according to the available resources and the activities to be carried out. Some broad outline emerges, however, which makes it possible to characterize the Middle Paleolithic:

• The technical equipment transported by the Middle Paleolithic hunter-gatherers reveals quite varied forms, both when introduced to sites and when carried away from sites of occupation and/or production.

• Middle Paleolithic circulations seem to be for the most part generalized matrices of various types (flakes, predetermined products, retouched tools, bifaces, and – more recently recognized – preformed cores), that is, objects with a broad range of potential uses for production of blanks and/or tools, and not specialized products. They do not seem to be systematically the outcome of a particular investment in their production but, on the other hand, were maintained in a fashion that ensured a long use life (exploitation until exhaustion, a lengthy cycle of ramification, successive recycling/resharpening – that equate to a "curation strategy").

• These matrices were traditionally of exotic raw materials and circulated over long distances (often greater than 30–40 km, and in some cases as much as 150/250 km). However, several archaeological examples seem to indicate that identical behaviors were undertaken on strictly local materials. The linkages of such behaviors with site location and mobility structure merit further exploration.

During the Middle Paleolithic, acquisition of unaltered materials is reflected almost solely in local raw materials, even if they were of poor quality. However, the data discussed herein emphasize a capacity to anticipate needs, resulting in the transport of generalized toolkits in the form of tools to fulfill a broad and diverse range of potential needs For populations of hunter-gatherers that were mobile by definition, such behavior would have freed them during movement from constraints imposed by specialized toolkits. Many additional studies that have not been reviewed herein document, on the basis of interdisciplinary work, the complementary nature of site locations within the landscape, establishing a relationship with the various mobility patterns derived from recent hunter-gatherers (Kuhn 1995; Hovers 1997; Marks and Chabai 2001; Slimak 2004; Meignen *et al.* 2005; Porraz 2005; Costamagno *et al.* 2006). Even if the tool provisioning and management strategies differ from those of the Upper Paleolithic, we are far from the caricatured images of the period 1980–90 that argued for the incapacity of Neandertals to manage their subsistence (Trinkaus 1986; Binford 1989).

ACKNOWLEDGMENTS

We are very grateful to Brooke S. Blades for his masterful translation of this chapter and claim responsibility for any remaining echo of the initial French formulation.

ENDNOTES

1 Normalization implies one or several characteristics – for example, thickness, overall form, etc. – while standardization concerns production that is made strictly identical, such as bladelets produced by pressure.
2 Type IV and V flakes, supports with particular morpho-technical characteristics (Bourguignon 1997, 2003).

REFERENCES

Ashton, N.M. (1992) The High Lodge flint industries. In: Ashton, N.M., Cook, J., Lewis, S.G., and Rose, J. *High Lodge. Excavations by G. de G. Sieveking, 1962–8 and J. Cook, 1988,* British Museum Press, London, pp. 124–68.

Beyries, S. (1993) Analyse fonctionnelle de l'industrie lithique du niveau CA : rapport préliminaire et directions de recherché. In: Tuffreau, A. (ed.) *Riencourt-lès-Bapaume, Pas-de-Calais : un gisement du Paléolithique moyen, DAF,* Ed. de la MSH, Paris, pp. 53–61.

Binford, L.R. (1973) Interassemblage variability – the Mousterian and the "functional" argument. In: Renfrew, C. (ed.) *Explanation of Cultural Change – Models in Prehistory,* Duckworth, London, pp. 227–54.

Binford, L. (1977) Forty-seven trips: a case study in the character of archaeological formation processes. In: Wright, R. (ed.) *Stone Tools as Cultural Markers.* Australian Institute of Aboriginal Studies, Canberra, pp. 24–36.

Binford, L. (1979) Organization and formation processes: looking at curated technologies. *Journal of Anthropological Research* 35, 255–73.

Binford, L. (1989) Isolating the transition to cultural adaptations: an organizational approach. In: Trinkaus, E. (ed.) *The Emergence of Modern Humans. Biocultural Adaptations in the Late Pleistocene.* Cambridge University Press, Cambridge, pp. 18–41.

Boëda, E. (1986) *Approche technologique du concept Levallois et évaluation de son champ d'application.* Doctorat, Paris X- Nanterre.

Boëda, E. (1988) Biache-St-Vaast. Analyse technologique du débitage du niveau IIA. In: Tuffreau, A. and Sommé, J. (eds) *Le gisement paléolithique moyen de Biache-St-Vaast (Pas-de-Calais),* vol. 21, *Mémoires de la SPF.* SPF, Paris, pp. 185–214.

Boëda, E. (1990) De la surface au volume. Analyse des conceptions des débitages Levallois et laminaire. In: Farizy, C. (ed.) *Paléolithique moyen récent et Paléolithique supérieur ancien en Europe.,* vol. 3, Nemours: Mémoires du Musée de Préhistoire d'Ile-de-France, pp. 63–8.

Boëda, E. (1991) Approche de la variabilité des systèmes de production lithique des industries du Paléolithique inférieur et moyen : chronique d'une variabilité attendue. *Techniques et culture* 17–18, 37–79.

Boëda, E. (1993) Le débitage discoïde et le débitage Levallois récurrent centripète. *Bulletin de la Société Préhistorique Française* 90, 392–404.

Boëda, E. (1994) *Le concept Levallois : variabilité des méthodes*, Vol. 9. *Monographie du CRA*, CNRS Editions, Paris.

Boëda, E., Geneste, J.M., and Meignen, L. (1990) Identification de chaînes opératoires lithiques du Paléolithique ancien et moyen. *Paléo* 2, 43–80.

Bordes, F. (1950) Principes d'une méthode d'étude des techniques de débitage et de la typologie du Paléolithique ancien et moyen. *L'Anthropologie* 54, 19–34.

Bordes, F. (1953) Essai de classification des industries "moustériennes". *Bulletin SPF* 50, 457–67.

Bordes, F. (1981) Vingt-cinq ans après: le complexe moustérien revisité. *Bulletin de la Société Préhistorique Française* 78, 77–87.

Bourguignon, L. (1996) La conception de débitage Quina. *Quaternaria Nova* VI, 149–64.

Bourguignon, L. (1997) *Le Moustérien de type Quina: Nouvelle définition d'une entité technique*. Université Paris X-Nanterre.

Bourguignon, L. (2001) Apports de l'expérimentation et de l'analyse techno-morpho-fonctionnelle à la reconnaissance du processus d'aménagement de la retouche Quina. In: Bourguignon, L., Ortega, I., and Frère-Sautot, M.C., (eds) *Préhistoire et Approche expérimentale*, M. Mergoil, Montagnac, pp. 35–66.

Bourguignon, L., Bidart, P., and Turq, A. (sous presse) Etude technologique et techno-économique des industries lithiques de Combe-Brune 1. In: Bidart, P., Bourguignon, L., Ortega, I., Sellami, F., Rios, J., Guibert, P., and et Turq, A. (eds) *Le gisement Moustérien de Tradition Acheuléenne de Combe Brune 1, Déviation Nord de Bergerac*, Rapport final d'opération, INRAP.

Bourguignon, L., Ortega, I., Sellami, F., Brenet, M., Grigoletto, F., Vigier, S., Daussy, A., and Casgrande, F. (2004a) Les occupations paléolithiques découvertes sur la section Nord de la déviation de Bergerac: Résultats préliminaires obtenus à l'issue des diagnostics. *Bulletin Préhistoire du Sud-Ouest* 11, 155–72.

Bourguignon, L, Sellami, F., Deloze, V., Sellier-Segard, N., Beyries S., Emery-Barbier A. (2002) L'habitat Moustérien de "La Folie" (Poitiers, Vienne): Synthèse des premiers résultats, *Paleo* 14, 29–48.

Bourguignon, L. and Turq, A. (2003) Une chaîne opératoire de débitage discoïde sur éclat du Moustérien à denticulés aquitain: Les exemples de Champ Bossuet de Combe-Grenal c. 14. In: Peresani, M. (ed.) *Discoid Lithic Technology: Advances and Implications*, vol. 1120, BAR International Series, Archaeopress, Oxford, pp. 131–52.

Bourguignon, L., Turq, A., and Faivre, J.-P. (2004b) Ramifications des chaînes opératoires: Spécificité du Moustérien?, *Paléo* 15, 37–48.

Costamagno, S., Meignen, L., Maureille, B., and Vandermeersch, B. (2006) Les Pradelles (Marillac-le-Franc, France): A Mousterian reindeer hunting camp? *Journal of Anthropological Archaeology* 25, 466–84.

Cresswell, R.C. (1982) Transferts de techniques et chaînes opératoires. *Techniques et Culture* 2, 143–63.

Delagnes, A. (1991) Mise en évidence de deux conceptions différentes de la production lithique au Paléolithique moyen. In: *25 ans d'études technologiques en préhistoire: Bilan et perspectives*: Actes des XIe recontres internationales d'archéologie et d'histoire d'Antibes, Éditions APDCA, Juan-les-Pins, pp. 125–37.

Delagnes, A. (1992) *L'organisation de la production lithique au Paléolithique moyen : approche technologique à partir de l'étude des industries de La Chaise-de-Vouthon (Charente)*. Thèse de Doctorat, Université Paris X - Nanterre.

Delagnes, A. (1993) Un mode de production inédit au Paléolithique moyen dans l'industrie du niveau 6e du Pucheuil (Seine-Maritime). *Paléo* 5, 111–20.

Delagnes, A. (1995) Faible élaboration technique et complexité conceptuelle: Deux notions complémentaires, *Cahier noir* 7, 101–10.

Delagnes, A. (1996) Le site d'Etoutteville (Seine-Maritime): L'organisation technique et spatiale de la production laminaire à Etoutteville. In: Delagnes, A. and et Ropars, A. (eds) *Paléolithique moyen en pays de Caux (Haute-Normandie): Le Pucheuil, Etoutteville : Deux gisements de plein air loessique.*, vol. 56, D.A.F., Maison des Sciences de l'Homme, Paris, pp. 164–228.

Delagnes, A., Féblot-Augustins, J., Meignen, L., and Park, S.J. (2006) L'exploitation des silex au Paléolithique moyen dans le Bassin de la Charente: Qu'est-ce qui circule, comment ... et pourquoi? *Bulletin de l'Association des Archéologues de Poitou-Charentes* 35, 15–24.

Faivre, J.-P. (2006) L'industrie moustérienne du niveau Ks (locus 1) des Fieux (Miers, Lot): Mobilité humaine et diversité des compétences techniques. *Bulletin de la Société Préhistorique Française* 103/1, 16–32.

Féblot-Augustins, J. (1997) *La circulation des matières premières au Paléolithique*. ERAUL 75, Liège.

Féblot-Augustins, J. (1999) Raw material transport patterns and settlement systems in the European Lower and Middle Palaeolithic: Continuity, change and variability. In: Roebroeks, W. and Gamble, C. (eds) *The Middle Palaeolithic Occupation of Europe*. University of Leiden, Leiden, pp. 193–214.

Forestier, H. (1993) Le Clactonien : mise en application d'une nouvelle méthode de débitage s'inscrivant dans la variabilité des systèmes de production lithique du Paléolithique ancien. *Paléo* 5, 53–82.

Geneste, J.-M. (1991) Systèmes techniques de production lithique: Variations techno-économiques dans les processus de réalisation des outillages paléolithiques. *Techniques et culture* 17–18, 1–35.

Geneste, J.-M. (1985) *Analyse lithique d'industries moustériennes du Périgord: Une approche technologique du comportement des groupes humains au Paléolithique moyen.* Thèse Sc., Université de Bordeaux I.

Geneste, J.-M. (1988) Les industries de la grotte Vaufrey: Technologie du débitage, économie et circulation de la matière première. In: Rigaud, J.-P. (ed.) *La grotte Vaufrey-Paléoenvironnement, chronologie, activités humaines.* Mémoires de la SPF, Paris, pp. 441–517.

Geneste, J.-M. (1992) L'approvisionnement en matières premières dans les systèmes de production lithique: La dimension spatiale de la technologie. In: Mora, R., Terradas, X., Parpal, A., and Plana, C., (eds) *Technologia y Cadenas Operativas Liticas.* Treballs d'Arqueologia, Universitat Autonoma de Barcelona, Barcelona, pp. 1–36.

Geneste, J.-M. and Plisson, H. (1996) Production et utilisation de l'outillage lithique dans le Moustérien du sud-ouest de la France: Les Tares, à Sourzac, Vallée de l'Isle, Dordogne. *Quaternaria Nova VI,* 343–68.

Hovers, E. (1997) *Variability of Levantine Mousterian Assemblages and Settlement Patterns: Implications for Understanding the Development of Human Behavior.* PhD Dissertation, Hebrew University, Jerusalem.

Jaubert, J. (1993) Le gisement paléolithique moyen de Mauran (Haute-Garonne) : techno-économie des industries lithiques. *Bulletin SPF* 90, 328–35.

Jaubert, J. (2001) Un site moustérien de type Quina dans la vallée du Célé (Pailhès à Espagnac-Sainte-Eulalie, Lot). *Gallia-Préhistoire* 43, pp. 1–99.

Jaubert, J., and Farizy, C. (1995) Levallois debitage: exclusivity, absence or coexistence with other operative schemes in the Garonne Basin, Southwestern France. In: Dibble, H.L. and Bar-Yosef, O. (eds) *The Definition and Interpretation of Levallois Technology,* vol. 23, *Monographs in World Archaeology.* Prehistory Press, Madison, Wisconsin, pp. 227–48.

Jaubert, J. and Mourre, V. (1996) Coudoulous, Le Rescoundudou, Mauran : diversité des matières premières et variabilité des schémas de production d'éclats. *Quaternaria Nova VI,* 313–41.

Karlin, C., Bodu, P., and Pelegrin, J. (1991) Processus techniques et chaînes opératoires. Comment les préhistoriens s'approprient un concept élaboré par les ethnologues. In: Balfet, H. (ed.) *Observer l'action technique: des chaînes opératoires, pour quoi faire?* Editions du CNRS, Paris, pp. 101–17.

Kuhn, S. (1992) On planning and curated technologies in the Middle Paleolithic. *Journal of Anthropological Research* 48, 185–213.

Kuhn, S. (1995) *Mousterian Lithic Technology. An Ecological Perspective.* Princeton University Press, Princeton.

Lenoir, M. (2004) Les industries des niveaux du Paléolithique moyen. Les racloirs des niveaux moustériens-quelques observations, In: *Le site Paléolithique de Chez-Pinaud à Jonzac, Charente-Maritime.* Préhistoire du Sud-ouest, suppl. 8, pp. 61–78.

Locht, J.L. and Swinnen, C. (1994) Le débitage discoïde du gisement de Beauvais (Oise) : aspects de la chaîne opératoire au travers de quelques remontages. *Paléo* 6, 89–104.

Marks, A. and Chabai, V. (2001) Constructing Middle Paleolithic settlement systems in Crimea: Potentials and limitations. In: Conard, N. (ed.) *Settlement Dynamics of the Middle Paleolithic and Middle Stone Age.* Kerns Verlag, Tübingen, pp. 179–204.

Meignen, L. (1988) Un exemple de comportement technologique différentiel selon les matières premières: Marillac couches 9 et 10. In: Otte, M. (ed.) *L'Homme de Néandertal : La Technique.* Liège, ERAUL, pp. 71–79.

Meignen, L. (ed.) (1993) *L'abri des Canalettes. Un habitat moustérien sur les grands Causses (Nant, Aveyron),* Vol. 10. *Monographie du CRA n°10.* CNRS Editions, Paris.

Meignen, L., Bar-Yosef, O., Speth, J., and Stiner, M. (2005) Changes in settlement patterns during the Near Eastern Middle Paleolithic. In: Hovers, E. and Kuhn, S., (eds) *Transitions before the Transition: Evolution and Stability in the Middle Paleolithic and Middle Stone Age.* Springer, New York, pp. 149–70.

Meignen, L. and Vandermeersch, B. (1986) Le gisement moustérien de Marillac (Charente), couches 9 et 10. Caractéristiques des outillages, économie des matières premières. *111° Congrès des Sociétés Savantes, Pré et Protohistoire,* Editions du CTHS, pp. 135–44.

Moncel, M.H., Patou-Mathis, M., and Otte, M. (1998) Halte de chasse au chamois au Paléolithique moyen : la couche 5 de la grotte Scladina (Sclayn, Namur, Belgique). In: Brugal, J.P., Meignen, L., and Patou-Mathis, M. (eds) *Economie préhistorique : les comportements de subsistance au Paléolithique,* Editions APDCA, Sophia-Antipolis, pp. 291–308.

Pelegrin, J. (2004) Le milieu intérieur d'André Leroi-Gourhan et l'analyse de la taille de pierre au Paléolithique. In: Audouze, F. and Schlanger, N. (eds) *Autour de l'homme: Contexte et actualité d'André Leroi-Gourhan.* Editions APDCA, Antibes, pp. 149–62.

Peresani, M.E. (2003) *Advancements and implications in the study of the Discoid Technology,* vol. 1120. *BAR International Series,* Archaeopress, Oxford.

Porraz, G. (2005) En marge du milieu alpin- Dynamiques de formation des ensembles lithiques et modes d'occupation des territoires au Paléolithique moyen, Thèse de Doctorat, Université de Provence.

Revillion, S. and Tuffreau, A. (1994) *Les industries laminaires au Paléolithique moyen,* vol. 18. *Dossier de Documentation Archéologique.* CNRS Editions, Paris.

Roche, H. and Texier, P.J. (1991) La notion de complexité dans un ensemble lithique: Application aux séries acheuléennes d'Isenya, Kenya, In: *25 ans d'études technologiques en préhistoire : Bilan et perspectives.* Editions APDCA, Juan-les-Pins, pp. 99–108.

Slimak, L. (2004) *Les dernières expressions du Moustérien entre Loire et Rhône*. Thèse de Doctorat, Université de Provence.

Soressi, M. (2002) *Le Moustérien de tradition acheuléenne du sud-ouest de la France – Discussion sur la signification du faciès à partir de l'étude comparée de quatre sites: Pech de l'Azé I, Le Moustier, la Rochette et la Grotte XVI*, Université Bordeaux 1.

Soressi, M. (2004) Les industries des niveaux du Paléolithique moyen. L'industrie lithique des niveaux moustériens (fouilles 1998–99). Aspects taphonomiques, économiques et technologiques. In: *Le site Paléolithique de Chez-Pinaud à Jonzac, Charente-Maritime*. Préhistoire du Sud-ouest, suppl. 8, pp. 79–95.

Texier, P.J. and Roche, H. (1995) The impact of predetermination on the development of some acheulean chaînes opératoires. In: Bermudez De Castro, J.M. (ed.) *Human Evolution in Europe and the Atapuerca Evidence*, Junta de Castilla y Leon, Valladolid, pp. 403–20.

Trinkaus, E. (1986) Les Néandertaliens, *La Recherche*, pp. 1040–7.

Turq, A. (1989) Approche technologique et économique du faciès Moustérien de type Quina: étude préliminaire. *Bulletin de la SPF* 86, 244–56.

Turq, A. (2000) *Paléolithique inférieur et moyen entre Dordogne et Lot*. Paléo, suppl no. 2, SAMRA, Les Eyzies.

Chapter 3

REVISITING EUROPEAN UPPER PALEOLITHIC RAW MATERIAL TRANSFERS: THE DEMISE OF THE CULTURAL ECOLOGICAL PARADIGM?

Jehanne Féblot-Augustins

Préhistoire et Technologie, Maison de l'Archéologie et de l'Ethnologie, Université de Paris X

ABSTRACT

In this chapter, the author revisits, updates, and modifies an earlier study. Building on very current data pointing to the existence of raw material transfers over far longer distances than previously recorded for the Upper Paleolithic of Western Europe, in globally less exacting environments than those acknowledged for Eastern Central Europe, the chapter investigates the relevance of the cultural ecological paradigm for addressing issues relating to man's dealings with space on a macro-regional scale. Additionally, behavioral and cognitive implications for Paleolithic transitions are explored. While the recently collected data lead to toning down synchronic differences between western and eastern macro-regions, they reinforce previously stated diachronic differences between the Middle and Upper Paleolithic. Transfers greater than 100 km remain exceptional in the European Middle Paleolithic, suggesting differences of a socio-economic nature. Moreover, the recorded long-distance occurrences remain associated with a strategy of provisioning individuals, even when continuity in mobility patterns can be argued for the same region. Nevertheless, the fact that Middle Paleolithic people did not provision places with raw materials from afar need not necessarily be indicative of specific cognitive capacities, as opposed to those of Upper Paleolithic people.

Lithic Materials and Paleolithic Societies, 1st edition. Edited by B. Adams and B.S. Blades, ©2009 Blackwell Publishing. ISBN 978-1-4051-6837-3.

INTRODUCTION

In the context of late Pleistocene societies, raw material provenance studies have proved useful for investigating the way man interacts with his environment and other human communities. Whether plotted transfers are interpreted as correlates of group mobility (Rensink *et al.* 1991) or as proxies for the extent of interaction networks expected as part of the survival and social process (Gould 1980; Wiessner 1982; Gamble 1986; Whallon 1989), the issues are generally addressed in relation to the cultural ecological paradigm (but see Gamble 1999).

Behavioral models devised to assess the size of exploited territories and the extent of spatial networks of connection among groups are mainly based on ethnographic accounts of present day hunter-gatherers and rely heavily on ecological approaches (e.g., Binford 1980; Kelly 1983; Keeley 1988; Meltzer 1988). The nature, abundance, and accessibility of resources are seen as directly influencing hunter-gatherer subsistence strategies, and these in turn result in different patterns, rates, and scales of movement across a landscape. In particular, the magnitude of group mobility appears indicative of adaptation to different environmental contexts, with the longest moves being recorded in the most exacting environments. Risk-minimizing strategies in high risk environments where resources are liable to important fluctuations also promote the creation of spatially extensive interaction networks through which information about vital resources is dispensed. The exchange of goods forms an integral part of these strategies.

Building on very current data pointing to the existence of raw material transfers over far longer distances than previously recorded for the Upper Paleolithic of Western Europe, this chapter investigates the relevance of the cultural ecological paradigm for addressing issues relating to man's dealings with space on a macro-regional scale.

PROVENANCE STUDIES IN THE 1980S AND EARLY1990S: IN WESTERN EUROPE SHORTER IS BETTER

The fact that in Western Europe, and more particularly in France until quite recently, long-distance (>100 km) transfers of lithic raw materials have seldom been recognized – or acknowledged when they had been identified for the Paleolithic – can be ascribed to a number of converging reasons.

A major factor is that attention focused primarily on the site and on the raw materials introduced into it, rather than on the larger-scale distribution of particular types of raw materials. As a result, systematic geological surveys were conducted on a regional basis, in order to identify potential sources of raw materials; first close to the site, then further away (mainly up to 40–50 km and sometimes more), but always within the confines of a region, generally construed as the catchment area of a major watercourse, or as a major sedimentary basin. However, this was done with very little or no contact between researchers operating in different areas, such as, for instance, the Paris, Aquitaine, Charente, and Creuse basins, or the Pyrenees, and the various lithotheques remained strictly personal possessions. The amount and quality of this earlier work is fantastic, and has been the cornerstone of numerous provenance studies. It is unfortunate that its regional character has precluded the identification of materials introduced from other areas, although this was probably suspected when artifacts did not have their counterpart in the regional geological database.

By contrast, research on raw material procurement in Eastern Central Europe has not developed along exactly the same lines as in France (see Biró, this volume, Chapter 4, for a review of provenance studies in Hungary). Initially, attention centered primarily on the deposits themselves and on the distribution from the sources of distinctive types of raw materials (chocolate flint, Świeciechów flint, obsidian, Szeletian felsitic porphyry, among others). As a result, greater emphasis has been laid on recognizing long to very long-distance (300–400 km) transfers (e.g., Dobosi, Valde-Nowak, this volume, Chapters 8 and 14). This approach was complemented in the early 1980s by systematic sourcing of siliceous raw materials, and more prominence given to the characterization by physical and chemical methods of the raw materials recovered at archeological sites. Moreover, in Hungary, the collection of representative comparative samples and their constitution into a lithotheque has taken place at State level, and their publication on a web-page makes the knowledge widely available abroad.

In France in the very early1980s, another major factor pertained to the reluctance with which results were met that were based on other methods than those associating a thorough knowledge of the regional range of raw materials with their macroscopic (as opposed to microscopic) characterization by color, texture, and visible inclusions. In 1981, A. Masson, a researcher with training in geology, used a combination of methods derived from petrology, mineralogy, and micropaleontology to source artifacts from various sites of the eastern part of the Massif central, in the Auvergne region. The results were surprising: in all Upper Paleolithic sites, a varying and often very high proportion of the raw materials could be traced to the Cher valley and the Grand-Pressigny area, some 200–300 km north following the course of the Allier and Loire rivers. The scientific arguments were backed by ethnographic data and by comparisons with similar distances already known for Central Europe. One of the shortcomings of the study was the summary nature of the geological surveys conducted in the area, which is admittedly poor in high quality flint but does yield a variety of raw materials. Masson's results contradicted previous ones by a researcher from Bordeaux (Torti 1980), who had investigated regional resources but had, however, failed to source one of the "exotic" flints and had erroneously characterized – and therefore sourced – another. They also went against conclusions drawn from a leading area in provenance studies, the Aquitaine basin, where transfers were shown not to exceed 100 km (Demars 1980); incidentally, these conclusions were also backed by ethnographic data, but obviously from different sources. Partly because Bordeaux University occupied a prominent position in archeology, but mainly because her study was not grounded in fieldwork, severe criticism was leveled against Masson; her results were entirely disregarded, and, consequently, given a wide berth by anyone dealing with raw material transfers.

INITIAL LINES OF ENQUIRY INTO MIDDLE AND UPPER PALEOLITHIC RAW MATERIAL TRANSFERS

Such is the context in which this author (Féblot-Augustins 1997a,b) collected bibliographical data between 1987 and 1993 for a survey of raw material transfers in the Paleolithic of Western Europe and Western and Eastern Central Europe, analysed from the standpoint of scale of mobility and techno-economic patterns of procurement. The purpose was to monitor change versus continuity and macroregional variability in past hunter-gatherers' dealings with space, combined with an attempt to single out the extent of interaction networks (Féblot-Augustins 1999). In this connection, special attention was paid to the longest transfers, since a cross-cultural comparison of ethnographic records of exchange (Féblot-Augustins and Perlès 1992) has shown that highly valued items travel extreme distances (600–3000 km) by down-the-line trade, which mode of exchange can be archeologically recognized provided that "relay" sites containing samples of a specific raw material are identified between the source of that raw material and its furthest known occurrence.

Scale of mobility

Globally, in comparison with the Middle Paleolithic, a dramatic increase in transfers exceeding 100 km appeared for the entire European Upper Paleolithic. However, the results summarized in Table 3.1 also showed an enduring clinal relationship across the Middle-Upper Paleolithic divide between greater scales of group mobility and increasing continentality. Considered in a cultural ecological perspective, this relationship was explained in terms of environmentally driven responses: in more continental zones, such as Eastern Central Europe, where both seasonal and annual variations in productivity are important, it was assumed that hunter-gatherers had to move seasonally over larger territories to reduce risk, an added factor being the larger size and lesser compression of the different ecological zones the fauna had to travel through. A similar explanation in terms of risk-minimizing strategies accounted for more extensive spatial networks of connection among groups. Whereas in the Eastern Central European Upper Paleolithic the existence of such networks was commonly revealed through down-the-line trade of very small quantities of finished products in remarkable raw materials over distances vastly exceeding 300 km (e.g., obsidian, accounting for 48% of such transfers), as well as through the large scale movement of fossil shell ornaments, in Western Europe and Western Central Europe it was more consistently disclosed by down-the-line trade of ornaments. Considering down-the-line trade as the mechanism responsible for the transfer of finished products or fossils shells over the longest distances does not, however, imply that interaction networks could not operate within shorter distances.

Table 3.1 Initial propositions about the scale of group mobility and the extent of interaction networks in the European Middle and Upper Paleolithic, based on transfers of lithics (L) and fossil shell personal ornaments (O). (After Féblot-Augustins 1999: table 3.)

	Western Europe	Western Central Europe	Eastern Central Europe
Middle Palaeolithic			
Group mobility	100–120 km	110 km	100 km ⇒ 200–300 km (exceptional)
Interaction networks	?	?	200–300 km (exceptional)
Upper Palaeolithic			
Group mobility	100–125 km (L)	160 km (L)	200–250 km (L) (common)
Interaction networks	mode at 200 km (O)	⇒ 850 km (O)	• >300 km… 450 km… 700 km (L)
			• <300 km (L) (probable)
			• ⇒450 km… 1400 km (O)

Western Europe: assemblages mainly from the Aquitaine, Tarn, Creuse, and Paris basins (France), and from the Meuse basin (Belgium).
Western Central Europe: assemblages from the Rhineland, Southwestern Germany, Switzerland.
Eastern Central Europe: assemblages from northeastern Austria, Thuringia, Poland, the Czech and Slovak Republics, Hungary.

Techno-economic patterns of procurement

While former conclusions relating to the scale of mobility are liable to change as a result of the incorporation of new data, it should be emphasized that nothing has emerged so far that contradicts the following overview of techno-economic patterns of procurement.

Contrary to the scale of mobility, techno-economic patterns of procurement do not vary macro-regionally and exhibit similar characteristics in the whole of Europe for the Middle and Upper Paleolithic respectively. However, as for the overall scale of mobility, there are important differences in procurement patterns across the Middle–Upper Paleolithic divide. In the Upper Paleolithic, the transport over long distances (140–160 km to 250 km) of substantial quantities (>50% in the Gravettian) of unworked nodules or preformed cores (with or without a complement of blanks or tools) points explicitly to the building up of supplies. This pattern, which Kuhn (1992) terms **provisioning places**, can be identified in most cases where the poor to medium quality local raw material is not suitable for Upper Paleolithic blade production, although admittedly the selective attitude towards raw materials fluctuates through time and seems

dependant – among other factors – on the particular requirements of a given culture's lithic technology. Unsurprisingly, where high-grade local raw material is available only small quantities are introduced from afar, most often blanks or tools, sometimes complemented by small cores – Binford's (1979) portable **personal gear**. In the Middle Paleolithic, provisioning places occurs within a far smaller radius, which rarely exceeds a dozen kilometers; longer distances are associated with the transport of personal gear, the strategy for **provisioning individuals** as they go along (Kuhn 1992). Interestingly, the exceptions relate to areas yielding only poor to medium quality stone, thus arguing for the existence of a selective attitude towards raw materials in the Middle Paleolithic. However, even in such cases, the distances are much shorter and the quantities far smaller than in the Upper Paleolithic. This is confirmed by recently documented examples: 14% of Senonian flint transported as preformed cores over 30–45 km to sites located in the Jurassic area of the Charente (Delagnes et al. 2006), 38% of Oligocene flint and 20% of Bedoulian flint transported as preformed cores over 20 km and 35–40 km respectively to the Grotte de Néron cave site in the lower Rhone valley where local raw materials consist of siliceous limestone cobbles (Slimak 2004).

RECENT DEVELOPMENTS IN PROVENANCE STUDIES: THE LONGER THE BETTER

Over the last few years, a number of papers have been published stating far longer lithic transfer distances than previously recorded for all periods of the French Upper Paleolithic (Table 3.2). This outburst should not be construed as a sudden precipitation process, but as the outcome of the progressive development of provenance studies combined with a change in mentalities – as well as a reflection of the inevitable time-lag between intuition and validation, and between research and publication.

Ever since the 1980s, when scholars realized the huge potential of this theme of research, addressed in a techno-economic perspective (e.g., Geneste 1985), provenance studies by Paleolithicians, Mesolithicians, and Neolithicians have multiplied, and more and more areas are being covered. In addition, earlier investigations into regional stone resources (e.g., the Périgord, the Charente, the Pyrenees) have been updated and complemented through ever more systematic sourcing and sampling. Some work has also seen a shift from site-centered studies to ones focusing on the distribution of particular types of raw materials (e.g., Primault 2003a for Grand-Pressigny flints; Riche 1998 for Barremo-Bedoulian Vercors flint). Most studies remains regional, but there is now a greater awareness of the need to exchange information and flint samples, which has led to closer interaction between researchers working in different areas, and consequently to the acknowledgment of transfers exceeding the confines of a region. Thus, the most recurrently identified exotic materials are early Maastrichtian Bergeracois flints from the Périgord in northern Aquitaine, Turonian Grand-Pressigny and Cher flints from the Touraine, various Turonian and Santonian flints from the Charente basin ("blond" Turonian, "silex des Vachons," "grain de mil"), and "flysch" and late Maastrichtian Chalosse flints from the Pyrenees. Interaction occurs at the individual rather than the institution level, and no lithotheque has yet been constituted at State level. However, there is some incentive to pool regional data: a few lithotheques dealing with raw materials from the Charente basin (Féblot-Augustins *et al.* 2005), the Bugey (Féblot-Augustins 2005), and the Touraine (Primault 2006) have already been published on web-pages, and similar projects are in progress for Southern France.

The breakthrough in provenance studies is also closely linked to a development in characterization methods. Discussing these falls outside the scope of this chapter, but it should be stressed that identification of flints is increasingly based on the combined analysis of their macroscopic characters and their sedimentary microfacies, complemented by micropaleontological analyses on thin sections. Never dissociated from the systematic sampling and analysis of geological specimens collected in the field, this method, first advocated by M. and M.-R. Séronie-Vivien (1987), has been successfully used to validate earlier intuitions, to accurately characterize and source materials displaying convergent macroscopic features, and to confirm earlier attributions. It has in particular been used to vindicate Masson's attribution of the artifacts from the Auvergne sites to the Turonian of the Cher and Grand-Pressigny regions (Surmely *et al.* 1998).

The method, however, is not infallible, in particular when applied to patinated material. This latter difficulty arises for "grain de mil" flint, which derives its evocative name from the fact that it was first distinguished in archeological material by the distinctive flecked aspect of its patina. Late Santonian material with a similar patina is well known in the west of the Charente basin (Jonzac), quite close to the Atlantic coast, but looks quite different when it is fresh. Paradoxically, this is the material with which the longest transfers are associated, and analyses are under way to ascertain that no other material closer to the sites where "grain de mil" was identified can fit the bill.

The last few years have witnessed a shift from excessive caution to what may be over-enthusiasm in the identification of long to very long transfer distances. This is perhaps no surprise, since research has always been subject to conflicting influences, which successively gain in popularity. The current trend definitely calls into question the relevance of the cultural ecological paradigm. Whether this paradigm must be wholly abandoned or whether it can be salvaged to some degree is investigated in the following sections.

CURRENT UPPER PALEOLITHIC RAW MATERIAL TRANSFERS, WITH A FOCUS ON WESTERN EUROPE

The database

The initial database for the European Upper Paleolithic, beginning with the Aurignacian, comprised 279 lithic assemblages (134 for Western Europe, 43 for Western

Table 3.2 Western European sites with new evidence of raw material transport over distances exceeding 100 km. Shorter distances are not included here. Some sites (*) already figure in Féblot-Augustins 1997a.

Sites	Country	Region (department)	Industry	Distance from raw material sources (km)	References
Le Piage*, c.F	France	North Aquitaine (46)	Early Aurignacian	240–200–130–160	Séronie-Vivien 2003; Le Brun-Ricalens and Séronie-Vivien 2004; Bordes et al. 2005
Roc-de-Combe*, c.7	France	North Aquitaine (46)	Early Aurignacian	200–160–130	Bordes 2002; Bordes et al. 2005
Caminade	France	North Aquitaine (24)	Early Aurignacian	210	Séronie-Vivien 2003; Bordes et al. 2005
Castanet*	France	North Aquitaine (24)	Early Aurignacian	230–190–130	Morala, Turq, Bordes (n.d.); Pelegrin (n.d.)
Font-Yves*	France	North Aquitaine (19)	Early Aurignacian	200	Primault 2003b; Bordes et al. 2005
Font-Robert	France	North Aquitaine (19)	Early Aurignacian	200	Primault 2003b; Bordes et al. 2005
Hui*	France	North Aquitaine (47)	Early Aurignacian	160	Bordes et al. 2005
Pataud, c.7	France	North Aquitaine (24)	Later Aurignacian	120	Chiotti et al. 2003
Abri des Roches, c.5	France	Touraine (36)	Later Aurignacian	220	Primault 2003a
Les Vachons, abri 2, c.1	France	Charente basin (16)	Early Aurignacian	150	Primault 2003a,b
Fontaury	France	Charente basin (16)	Early Aurignacian	160	Primault 2003a,b
Le Châtelperron	France	Auvergne (03)	Early Aurignacian	220	Surmely and Pasty 2003
Gr. des Hyènes (Brassempouy)	France	Pyrenees (40)	Early Aurignacian	160–150	Bon 2002; Bordes et al. 2005
La Tuto de Camalhot, nv. inf.	France	Pyrenees (09)	Early Aurignacian	320–220–200–190	Bon 2002; Bon et al. 2005; Bordes et al. 2005
Régismont-le-Haut	France	Languedoc (34)	Early Aurignacian	380–270–220–220–130	Bon 2002; Bordes et al. 2005
Les Roches, c.3	France	Touraine (36)	Gravettian	220	Primault 2003a,b
Les Vachons, abri 2, c.5	France	Charente basin (16)	Gravettian	150	Primault 2003b
Les Vachons, abri 2, c.4	France	Charente basin (16)	Gravettian	150	Primault 2003a,b
Grotte du Renne, c.5 (Arcy)	France	Bourgogne (89)	Gravettian	250	Klaric 2003; Primault 2003a,b
La Vigne-Brun, unité KL19	France	Auvergne (42)	Gravettian	230–180	Digan 2003, 2006
Le Blot	France	Auvergne (43)	Gravettian	280	Bracco 1995; Surmely and Pasty 2003
Le Sire	France	Auvergne (63)	Gravettian	250	Surmely and Pasty 2003

Site	Country	Region	Period	Distance	Reference
Abri du Rond	France	Auvergne (43)	Protomagdalenian	300	Bracco 1995; Digan 2003
Le Blot, nv. 27	France	Auvergne (43)	Protomagdalenian	280	Masson 1981; Surmely and Pasty 2003
Le Blot, nv. 23	France	Auvergne (43)	Protomagdalenian	280	Masson 1981; Surmely and Pasty 2003
Le Cuzoul	France	North Aquitaine (46)	Solutrean	300–210–110	Renard (n.d.)
Laugerie-Haute*	France	North Aquitaine (24)	Solutrean	250	Piel-Desruisseaux 2004 (figured in, pers.obs. JFA)
Pégourié	France	North Aquitaine (46)	Badegoulian	200	Séronie-Vivien 1995, 2003
Le Silo	France	Touraine (36)	Badegoulian	230	Primault 2003b
Le Blot	France	Auvergne (43)	Badegoulian	280	Digan 2003; Surmely and Pasty 2003
La Roche à Tavernat	France	Auvergne (43)	Badegoulian	300	Bracco 1994; Digan 2003; Surmely and Pasty 2003
Rond-du- Barry, nv F2	France	Auvergne (43)	Badegoulian	300	Masson 1981; Bracco 1995
Enval-Durif	France	Auvergne (63)	Late Magdalenian	240	Masson 1981; Surmely 2000
Blanzat	France	Auvergne (63)	Late Magdalenian	200	Masson 1981; Surmely 2000; Surmely and Pasty 2003
La Goutte-Roffat	France	Auvergne (42)	Magdalenian	230	Digan 2003
Le Blot	France	Auvergne (43)	Late Magdalenian	280	Bracco 1995; Surmely 2000; Surmely and Pasty 2003
Le Sire	France	Auvergne (63)	Magdalenian	250	Surmely and Pasty 2003
Thônes	France	Auvergne (63)	Middle Magdalenian	200	Surmely 2000; Surmely and Pasty 2003
Pont-de-Longues	France	Auvergne (63)	Late Magdalenian	240	Surmely 2000; Surmely and Pasty 2003
Longetraye	France	Auvergne (43)	Magdalenian	300	Masson 1981; Bracco 1995; Surmely 2000
Rond-du- Barry, nv E1	France	Auvergne (43)	Late Magdalenian	300	Masson 1981; Bracco 1995; Surmely 2000
Enlêne, nv. inf.	France	Pyrenees (09)	Middle Magdalenian	300–230–200–180–170–170–160–160	Lacombe 1998
Troubat, nv. 8	France	Pyrenees (65)	Late Magdalenian	270–220–200–200–150–110	Lacombe 1998
Cardina I, c.4 base	Portugal	Coa Valley	Gravettian	210–190–170–130	Aubry and Mangado Llach 2003
Olga Grande 4	Portugal	Coa Valley	Gravettian	210–190–170–130	Aubry and Mangado Llach 2003

Central Europe, and 102 for Eastern Central Europe), displaying 158 occurrences of transfers over distances exceeding 100 km (8, 34, and 116 respectively for each macro-region). The recent data presented in Tables 3.2–3.4 include 46 new lithic assemblages, bringing the total to 325, and complement previous information about 17 lithic assemblages that were part of the initial database. In all, an additional 110 occurrences of transport over more than 100 km have been recorded (80, 19, and 11 respectively for each macro-region). Attention has focused on Western Europe, and almost exclusively on the French Upper Paleolithic where the research effort concerning provenance studies has been particularly important. New data for Western Central Europe relate to southwestern Germany, and are all the more welcome since, with the exception of Hahn's Geissenklösterle monograph (Hahn 1988), accurate information about the raw materials of this area was sorely lacking. Undoubtedly, there is a wealth of new data on long-distance transfers in Eastern Central Europe (see Dobosi, this volume, Chapter 8), which confirm previous trends. However, only the current data mentioned in the volume papers are incorporated in Table 3.4, since the aim here is to assess the repercussions of recent Western European data on interpretive frameworks rather than to present an exhaustive overview of all transfers in each of the three macro-regions. For this reason, and also because the threshold of transfers for Western Europe was previously of *ca.* 100 km, only distances greater than 100 km are mentioned in Tables 3.2–3.4 and will chiefly be discussed hereafter.

Magnitude of transfers

Two comparative series of histograms have been drawn up, based on two types of data, which each highlight a particular point. The first two histograms (Figures 3.1 and 3.2) show the frequency distribution of all recorded transfers over distances greater than 100 km for the lithic assemblages included in the initial and current databases, with respective totals of 158 and 268. The second pair of histograms (Figures 3.3 and 3.4) show the frequency distribution of maximum transport distances (MTD) only – including MTDs up to 100 km. The overall number of occurrences therefore equals the number of lithic assemblages under study (279 and 325 respectively); 17 MTDs from the initial database have been modified as a result of additional

information about raw material provenances. Emphasis is placed below on differences between Western Europe and Eastern Central Europe, the two extremes along the continentality gradient, while a separate section is devoted to Western central Europe.

Compared with Figure 3.1, Figure 3.2 shows a dramatic increase in the proportion of transfers over distances ranging between 100 and 300 km for Western Europe. In addition, distances exceeding 300 km are now also represented in this macro-region. They are, however, exceptionally rare ($n = 2$) in comparison with Eastern Central Europe ($n = 23$). Their range is also smaller since they do not exceed 380 km, whereas in Eastern Central Europe distances well over 400 km are recorded (4 occurrences at 420–450 km, one at 600 km and one at 700 km). This tends to show that although under 300 km the gap between Western Europe and Eastern Central Europe has closed, the overall magnitude of lithic transfers remains greater in the latter macro-region. A preliminary conclusion is that the relationship between increasing continentality and greater scale of mobility is considerably weakened, but not wholly invalidated.

The latter point is supported by the frequency distributions of MTDs for both databases. Although the proportion of Western European assemblages displaying MTDs greater than 100 km has increased (Figure 3.4 and see Figure 3.3), in particular for the]200–300] km procurement class ($n = 31$ against $n = 1$), this should not obscure the fact that assemblages with MTDs shorter than 100 km are still by far the most frequent. If MTDs are considered as proxies for the maximum size of territories or interaction networks, the persisting predominance of short MTDs suggests that in most cases either territories or networks were smaller in Western Europe than in Eastern Central Europe. By contrast, the Eastern Central European distribution is quite balanced, and, as expected, displays relatively important percentages associated with the highest values (>300 km).

INTERPRETING THE CURRENT WESTERN EUROPEAN TRANSFERS: FROM GROUP MOBILITY TO INTERACTION NETWORKS

Insofar as previous analyses (Table 3.1) have suggested that the magnitude of initially recorded lithic transfers for Western Europe represented the scale of group

Table 3.3 Western Central European sites with new evidence of raw material transport over distances exceeding 100 km. Shorter distances are not included here. Some sites (*) already figure in Féblot-Augustins 1997a.

Sites	Country	Region	Industry	Distance from raw material sources (km)	References
Geissenklösterle nvx inf.*	Germany	Southwest	Aurignacian	180–160	Burkert and Floss 2005
Geissenklösterle nvx sup.*	Germany	Southwest	Aurignacian	180–160	Burkert and Floss 2005
Vogelherd*	Germany	Southwest	Aurignacian	220–120	Burkert and Floss 2005
Hohlenstein-Stadel*	Germany	Southwest	Aurignacian	400–220	Burkert and Floss 2005
Bockstein-Törle	Germany	Southwest	Aurignacian	160	Burkert and Floss 2005
Geissenklösterle*	Germany	Southwest	Gravettian	180–160	Burkert and Floss 2005
Bockstein-Törle	Germany	Southwest	Gravettian	120	Burkert and Floss 2005
Hohle Fels*	Germany	Southwest	Magdalenian	180–160	Burkert and Floss 2005
Brillenhöhle*	Germany	Southwest	Magdalenian	180–160	Burkert and Floss 2005
Buttentalhöhle	Germany	Southwest	Magdalenian	240	Burkert and Floss 2005
Burkhardtshöhle*	Germany	Southwest	Magdalenian	175	Burkert and Floss 2005
Hohlenstein-Bärenhöhle	Germany	Southwest	Magdalenian	120	Burkert and Floss 2005

Table 3.4 Eastern Central European sites with new evidence of raw material transport over distances exceeding 100 km. Shorter distances are not included here. Some sites (*) already figure in Féblot-Augustins 1997a.

Sites	Country	Region	Industry	Distance from raw material sources (km)	References
Oblazowa	Poland	North-Carpathians	Gravettian	350-200–200	Valde-Nowak (this volume)
Püspökhatvan-Öregszőlő	Hungary	Cserhát Mts	Gravettian	230–180	Dobosi (this volume)
Mogyorósbánya*	Hungary	Danube bend	Ságvárian	250–150	Dobosi (this volume)
Pilismarót-Dios*	Hungary	Danube bend	Epigravettian	400	Dobosi (this volume)
Pilismarót-Pálrét*	Hungary	Danube bend	Epigravettian	400	Dobosi (this volume)
Hłomcza	Poland	North-Carpathians	Magdalenian	220–120	Valde-Nowak (this volume)

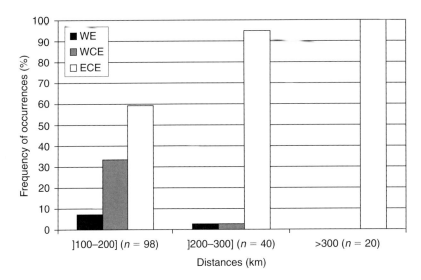

Figure 3.1 Initial database. Frequency distribution of all recorded transfers greater than 100 km for Western Europe (WE), Western Central Europe (WCE), and Eastern Central Europe (ECE).

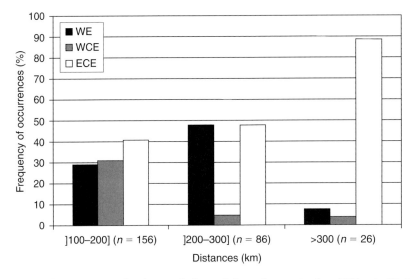

Figure 3.2 Current database. Frequency distribution of all recorded transfers greater than 100 km for Western Europe (WE), Western Central Europe (WCE), and Eastern Central Europe (ECE).

mobility within the economic territory, the current distributions raise questions about the significance of distances greater than 100 km. Do such distances correspond to the maximum size of territories or to the extent of interaction networks, and what arguments can be put forth to distinguish between the causes underlying the distributions? Several regional and chronological examples developed below will help to discuss these issues.

Inferences about the nature of procurement behaviors rely partly on the techno-economic patterns of procurement illustrated in Figure 3.5. Quantities

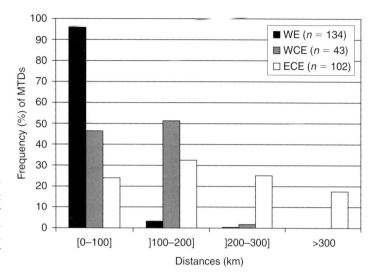

Figure 3.3 Initial database. Frequency distribution of the maximum transport distances (MTD) per lithic assemblage for Western Europe (WE), Western Central Europe (WCE), and Eastern Central Europe (ECE). There is a single MTD per assemblage.

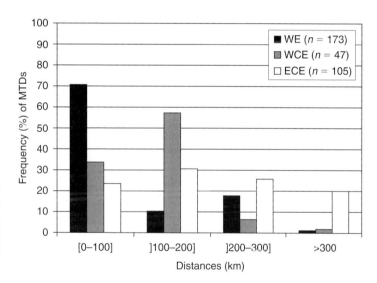

Figure 3.4 Current database. Frequency distribution of the maximum transport distances (MTD) per lithic assemblage for Western Europe (WE), Western Central Europe (WCE), and Eastern Central Europe (ECE). There is a single MTD per assemblage.

transported fall into two groups: high, that is ranging between 20% and 95% with one exception at 8%, and low, that is less than 1% with two exceptions at 1.5% and 1.8%. They are also associated with different modes of transport.

High quantities (*n* generally >1000) are always associated with the transport of unworked nodules and raw materials in more or less advanced stages of processing, such as preformed cores, cores made from flakes, complemented by a variable amount of blanks

or retouched tools in the same raw materials. Cores are generally present in the assemblages and there is evidence of abundant on-site reduction. As previously argued, and following Kuhn (1992), this points to the building up of supplies according to a strategy of provisioning places to fulfill anticipated needs at a particular location. However, alternative explanations can be suggested for the context within which this strategy operated, as illustrated by examples drawn from the Auvergne and the Pyrenees regions.

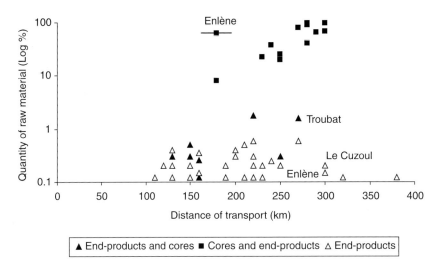

Figure 3.5 Techno-economic procurement patterns: known relative quantities and modes of transport (per type of raw material) in relation to distances >100 km for Western European assemblages. Full squares denote the transport of unworked nodules, preformed cores, or cores worked from flakes, with a variable complement of end-products (blanks or tools). With the exception of the full crossed square for Enlène, all full squares relate to assemblages from the Auvergne region. The full crossed square for Enlène (Pyrenees) refers to the global proportion of raw materials from various sources located 160 to 200 km north of the site, in the Aquitaine basin. Full triangles denote the transport of end-products (mainly blade blanks) complemented by small bladelet cores. Open triangles denote the transport of end-products alone (mainly blade tools).

Low quantities are systematically associated with the transport of either end-products alone (mainly retouched blade tools, $n < 5$ except in two cases for which $n = 8$, $n = 12$), or end-products (blade blanks exceeding tools) complemented by small bladelet cores in the same raw materials (n ranging between 12 and 87). These can be very reduced plain prismatic cores. They can also be made from various types of products, such as flakes, or formal tools that are now recognized as potential cores, for instance Raysse burins in the Gravettian (Klaric 2003) and carinated scrapers in the Aurignacian (Bon 2002). Although there is always evidence of some on-site reduction in the form of characteristic flakes (crests, core-rejuvenation flakes) and variable amounts of bladelets, the cores, whatever their nature, are rarely present in the assemblages, suggesting they were carried away when still serviceable as cores (or tools?). There is also evidence of retouch in the form of characteristic retouch flakes, implying that some, if not most, of the tools recovered were transformed on-site. Interpreting the patterns associated with low quantities requires that the problem of equifinality (Meltzer 1989) be

considered. If procurement during group movement is contemplated, the patterns point to the transport of portable personal gear in the context of provisioning individuals against encountered needs (Kuhn 1992). However, similar patterns may be produced by indirect procurement through social exchange, especially when transfers concern end-products alone. This will be argued more specifically for the early Aurignacian.

The Auvergne: seasonal direct mobility throughout the Upper Paleolithic – and earlier

The Upper Paleolithic settlement of the Auvergne goes back to the early Aurignacian, which is, however, very poorly represented, and only on the region's northern border (Le Châtelperron). Several settlements are known for the Gravettian and the Protomagdalenian (Perigordian VII), and many more for the Badegoulian and the Magdalenian (Table 3.2). They cluster in three main areas along the Loire and Allier rivers, and contrary to most Middle Paleolithic sites (see below) all contain significant amounts of northern (Cher and/or Grand-Pressigny) flint. The southernmost area, the Velay, is composed of volcanic plateaus with an average elevation of 1000 m, deeply incised by the upper Loire and Allier valleys contained between 450 and 850 m. It is bordered by mountainous higher altitude zones, which bore local glaciers during the second part of the Würm, when the area was isolated and could only be reached from the north by following the major watercourses (Bracco 1995). With the exception of Longetraye rock-shelter, on a high plateau, the other sites (Le Blot, Le Rond, Le Rond-du-Barry, La Roche à Tavernat) are all located in the valley bottoms where natural cavities are frequent. Le Blot on the Allier and Le Rond-du-Barry on the Loire are major sites and were repeatedly occupied, possibly because the first is a very large rock-shelter and the second is the only important and deep cavity in this area. During the Badegoulian and part of the Magdalenian, it is supposed that they functioned as main sites from which smaller satellite sites depended (Bracco 1996), a suggestion backed by the very large quantities of northern material, chiefly introduced as unworked nodules in view of the importance of cortical pieces (Masson 1981). Further north, the Limagne area includes the wider valley floor of the Allier (*ca*. 300 m elevation) and the adjacent higher tributary valleys. It is bordered to the east by the Forez Mounts and to the west by the Auvergne Mounts, which also bore local glaciers. The settlements include both rock-shelters (Enval-Durif, Blanzat, Thônes) and open air sites (Le Sire, Pont-de-Longues), clustering mainly along the Allier. Finally, to the northeast of the Limagne, on the left bank of the Loire, where the river ultimately flows out of the Massif central gorges at an elevation of 270 m, a series of open air sites were flooded by the Villerest dam; among these, the Gravettian site of La Vigne-Brun is well known for its six elaborate semi-sunken dwelling structures comparable to those documented in Eastern Central Europe for the same period.

Overall, owing to its pronounced relief and average elevation, and to the presence of local mountain glaciers, the Auvergne must have been a region of severe seasonal contrasts throughout the Upper Paleolithic, being particularly inhospitable during the winter months. That people deserted the Velay and the Limagne during the cold season (and possibly moved further north) seems to be backed by the faunal analyses carried out on a large number of Badegoulian and Magdalenian sites (Fontana 2000a). Reindeer, horse,

and ibex are the main hunted species irrespective of the time period, but hunting seemed to have been more specialized in the Limagne plain with a focus on horse and/or reindeer, and more balanced in the Velay's upper Loire and Allier valleys where the ibex is better represented. Most importantly, there is no evidence of hunting during the winter months in either region for any of these species. The extension of reindeer migrations during the Upper Paleolithic is a highly debated topic (see Fontana 1995 and 2000b for a comprehensive discussion), and in France at least this species may well have migrated over relatively short distances. However, the absence of any evidence of winter hunting does suggest that, whatever its relation to reindeer migrations, human occupation itself was seasonal in the Auvergne. Working on this assumption, it is proposed that the anticipated procurement of high-quality flint from the Cher and Grand-Pressigny areas was embedded in subsistence schedules (Binford 1979) and occurred in the context of planned seasonal moves, following natural routes between flint yielding regions and others known to be lacking suitable raw materials. Long-distance seasonal mobility (here over at least 180 km and generally ranging between 230 and 300 km, Figure 3.5) is precisely the pattern previously argued to obtain in Eastern Central Europe (Table 3.1). Although the Auvergne is part of Western Europe and therefore less "continental" in absolute terms, the environmental context was probably equally exacting owing to the physical characteristics of the region. That long-distance seasonal moves may in this case also be explained in terms of adaptive responses to environmental constraints is supported by the enduring northern origin of raw materials. This is documented not only throughout the Upper Paleolithic but even across the Middle-Upper Paleolithic divide; although in the former period the quantities introduced were far smaller and associated with the transport of portable personal gear. Four Middle Paleolithic assemblages from the Auvergne have yielded northern material transported over 230–300 km: Baume-Vallée and Saint-Anne I in the Velay (Masson 1981); Champ Grand and Vigne-Brun (Slimak 2004), two of the open air sites flooded by the Villerest dam and, like all the other sites of this area, thought to be connected with the seasonal exploitation of horses where the Loire opens out into the plain. There are very few other examples of such distances recorded for that period in Western Europe, and this case of continuity across the Middle–Upper Paleolithic divide is remarkably similar to one highlighted for Eastern Central Europe. In Moravia, beginning with the late Middle Paleolithic of Kůlna (Valoch 1986; Féblot-Augustins 1993), long-distance north–south trans-Carpathian instances of transfers following natural routes (the Moravian Gate) are systematically documented. And as in the Auvergne, it is only during the Upper Paleolithic that these transfers can be associated with a strategy of provisioning places.

The Pyrenees: expansion of cultural range during the middle Magdalenian

Given the hiatus in the occupation of the Pyrenees between the middle Magdalenian and the Badegoulian, represented only by its later phase at a very few sites in the Aude and one in the Ariège (Dachary 2002), interpretations derive from the research field of "landscape learning" (Rockman and Steele 2003). Arguments involve the comparison between contrasting procurement patterns documented for two assemblages of the central to eastern part of the Pyrenees, the earliest middle Magdalenian level of Enlène-Salle du Fond in the Ariège, and the late Magdalenian level of Troubat, some 80 km further west (Lacombe 1998).

Procurement patterns at Enlène are characterized by a severe imbalance between local and exotic stone, in favor of the latter. Over 63% of the analyzed assemblage (n = 5541 pieces >0.5 cm) is made of non-local raw materials. With the exception of some late Maastrichtian flint (0.4%), originating either from Chalosse, 160 km to the west, or from another closer source 50 km west of the site (Séronie-Vivien 2003), all non-local materials have a northern provenance. They are varied and include Senonian and early Maastrichtian flints from the Périgord, Turonian flint from Fumel, as well as some Santonian "grain de mil" from the western part of the Charente basin. The latter raw material, conveyed over 300 km, is only represented by a few tools (Figure 3.5). On the other hand, the bulk of the non-local northern flint, transported over 160 to about 200 km (Figure 3.5), arguably corresponds to the introduction of supplies in more or less advanced stages of processing, such as preformed bladelet cores and bladelet cores made from flakes, complemented by end-products; the amount of cortex varies depending on the raw material but is generally low. Local raw materials account for only 15%[1] and, among these, over 96% are strictly local, although other good sub-

local and regional sources were available. By contrast, at Troubat, where 5226 items were petrographically analyzed, there is a far more balanced exploitation of local to micro-regional (34%) and exotic stone (close to 28%)[2]. Moreover, the local to micro-regional stone is represented by several varieties, which were collected within a radius of 40–50 km, and the major component of exotic stone points to transport from the west and from the east along the Pyrenees rather than from the north. Only two types of northern raw material could be characterized with a measure of accuracy. There is some Senonian flint from the Périgord (0.6%) that is highly fragmented and for which the stages of introduction could not be reconstituted. A small amount of "grain de mil" (1.5%, n = 80) was transported over 270 km (Figure 3.5) mainly as blade blanks, some of which were transformed on-site into burins, as evidenced by a number of burin spalls; in addition, the presence of bladelets and of core rejuvenation flakes points to the introduction of a small bladelet core, which was not recovered.

The contrast in patterns documented for the middle Magdalenian of Enlène and the late Magdalenian of Troubat is consistent with the idea that the appreciation of raw material availability builds up through time, and that groups moving into unfamiliar territory transported supplies of stone over long distances as a hedge against the unpredictability of lithic resources (Kelly and Todd 1988). At Enlène, this would account for the large quantities of flint coming from various sources in the northern part of the Aquitaine basin, as well as for the near exclusive exploitation of local outcrops once on-site. As the area was settled, people became familiar with regional resources along the Pyrenees, but contacts would be maintained with the northern territories. This in turn would account for the more intensive and varied exploitation of micro-regional resources within a 40–50 km radius around Troubat, for the presence of a fair amount of western and eastern raw materials, and for the very small quantities of flints introduced from northern Aquitaine and Charente. These were possibly transported by people with a previous knowledge of the southern area. A similar scenario has been developed in the context of the Lateglacial-early Postglacial recolonization of the British Isles (Tolan-Smith 2003). It is, however, based on a larger number of sites. Clearly, other examples are required for the Pyrenees to support the assumption that the Enlène procurement pattern can be ascribed to an expansion of cultural range.

Direct versus indirect procurement

The patterns involving transfers of low quantities of either end-products alone or end-products plus small bladelet cores are contained between 110 and 380 km (Figure 3.5). While they overlap up to 270 km, beyond this value only the former pattern is represented. Not unexpectedly, the areas on the receiving end of transfers are not devoid of high-quality raw materials. For either pattern transfers are interregional, but connections do not follow natural routes, such as major watercourses; rather, they are mostly perpendicular to the valleys and entail crossing the hydrographic network. Because both patterns are associated with broadly similar types of spatial connections, and because their distributions overlap to a large extent, telling apart the scale of group mobility and the extent of interaction networks is not a straightforward process. What may, however, be suggested is that transfers exceeding 270 km, all associated with the transport of small numbers of retouched tools, can be ascribed with greater probability to indirect procurement through social exchange.

The patterns are documented for all periods – the early and late Aurignacian, the Gravettian, the Solutrean, the Badegoulian, and the Magdalenian[3] – but the data are particularly abundant for the early Aurignacian, accounting for two thirds of the transfers recorded in Figure 3.5. While this should be seen as a reflection of the state of current research and publication rather than as a mirror of the past, it does provide the opportunity to make some inferences about the nature of procurement behaviors.

The distinction between direct and indirect procurement has been addressed by Bordes *et al.* (2005) and Bon *et al.* (2005) for long-distance transfers associated with low quantities of end-products complemented by rare bladelet cores, a pattern identified in several assemblages from the Périgord and the Pyrenees (Le Piage, Roc de Combe, Brassempouy, Tuto de Camalhot) (Figure 3.6). The main arguments put forth in favor of direct procurement during group movement pertain to the broadly similar technological management of regional and more distant raw materials. This involves the anticipated manufacture of blade blanks for delayed use and the on-site *ad hoc* production of bladelets made from prismatic or carinated cores, introduced in more or less advanced stages of processing as a function of the distance from raw material sources. Extra-regional items corresponding to this pattern are

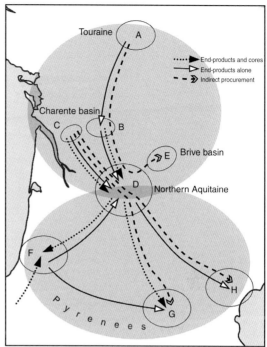

A   Grand-pressigny flint.
B   Turonian flints. *Les Vachons 2 c.1, Fontaury.*
C   "Grain de mil" flint.
D   Bergeracois, Fumel, Senonian flints. *Le Piage, Roc de Combe, Caminade, Castanet, Hui. Font-Yves, Font-Robert.*
E   Font-Yves, Font-Robert.
F   "Flysch" and Chalosse flints. *Brassempouy (Landes).*
G   *Tuto de Camalhot (Ariège).*
H   *Régismont-le-Haut (Hérault).*

Figure 3.6   Main interregional lithic connections in the French early Aurignacian south of the Loire River. Dotted lines with full arrowheads denote the transport of end-products (mainly blade blanks) complemented by small bladelet cores. Unbroken lines with open arrowheads denote the transport of end-products alone (mainly blade tools). Dashed lines with open arrowheads denote indirect procurement by down-the-line trade through social exchange. (Figure composed by Gérard Monthel (UMR 7055 du CNRS)).

therefore considered to represent the residual personal gear transported by the Aurignacians during long-distance moves. This interpretation can be extended to well documented Gravettian occurrences at Les Vachons and Arcy-sur-Cure (Klaric 2003; Primault 2003b) involving the transport of blade blanks and thick Raysse burins (i.e., bladelet "cores") in Grand-Pressigny flint; it can also be suggested for the "grain de mil" items recovered in the late Magdalenian level of Troubat (see above). Working on this assumption,

It is argued that transfers associated with the "end-products plus cores" pattern, and contained between 130 and 270 km (Figure 3.5), may be construed as proxies for the scale of group mobility.

It does not necessarily ensue that indirect procurement is the underlying mechanism for transfers involving only very small quantities of retouched tools, in particular when the connections and / or trajectories are the same as for the "end-products plus cores" pattern. For the early Aurignacian (Figure 3.6), examples of this include transfers of Turonian flints between the Charente and the Périgord (Le Piage, Roc de Combe, Castanet), of "grain de mil" flint between western Charente and the Périgord (Le Piage, Castanet), and of Chalosse and "flysch" flints between the Pyrenees and northern Aquitaine (Le Piage, Roc de Combe, Caminade, Castanet, Hui). Especially in the first two cases, where the trajectories are similar, the retouched tools may have been transported as part of larger personal gear also including blade blanks and a very few bladelet cores in other raw materials from the Charente basin. For instance at Le Piage, "grain de mil" and "Vachons" flints are represented by some retouched tools, "blond" Turonian by blanks plus bladelet cores; at Roc de Combe, the same holds for "blond" Turonian and "grain de mil" respectively. On the other hand, at Castanet "grain de mil" and "Vachons" flints, as well as Chalosse and "flysch" flints, are each represented by individual retouched tools, a feature indicative of indirect procurement for Pelegrin (n.d.).

Indirect procurement can more arguably be contemplated for the two early Aurignacian transfers that exceed 270 km (320 and 380 km) and relate to one item each of "grain de mil" flint conveyed from western Charente to the Ariège and the Hérault. Insofar as "grain de mil" flint occurs at several sites in the Périgord, and various northern Aquitaine flints (Bergeracois, Senonian, Fumel) have been identified at the Tuto de Camalhot and Régismont-le-Haut, the assumption is that the "grain de mil" was acquired through social exchange by down-the-line trade via Périgord "relay" sites. Indeed, the configuration of early Aurignacian lithic connections suggests that the north of the Aquitaine basin may have been the area where two mega territories intersected (Figure 3.6). Down-the-line trade can also be suggested in association with distances <270 km, thus substantiating the case for exchange also on a smaller scale, overlapping that of group mobility. This is illustrated by the presence of a few tools in Grand-Pressigny flint at Font-Yves

and Font-Robert in the Brive basin (200 km), which were possibly obtained through exchange with people from the south of the Charente, where this flint was identified at Les Vachons and Fontaury (Figure 3.6). In this specific case, the situation appears particularly complex, possibly associating exchange at some point on the Vézère with mobility along an east–west axis: various Turonian flints from the Charente were found at sites on the Vézère in association with Bergearcois flint, which also occurs in the Brive basin sites some 80 km east of the source.

It is of additional interest that the overall magnitude and configuration of lithic transfers for the early Aurignacian tally with the evidence of shell transfers (Taborin 1993; Blades 1999) from the Atlantic coastal plain and the Mediterranean. The possibility that in some cases shells were procured directly rather than exchanged should not be overlooked. Given the geographic origin of "grain de mil" flint, this can be argued for the Atlantic coastal species occurring in the Vézère sites, alongside this type of flint. Alternatively, they may have been acquired from groups from the south of the Charente, since they also occur at La Quina and Les Vachons, sites relatively close to the sources of "blond" Turonian and "silex des Vachons," types of flint identified at several Vézère sites. Other transfers can be interpreted in terms of indirect procurement with greater confidence, as in the case of the Atlantic coastal shells found at the Tuto de Camalhot, alongside with some "grain de mil" flint. Remembering that Régismont-le-Haut lies close to the Mediterranean coast and that indirect procurement via the Périgord was suggested for the "grain de mil" flint, this possibly also holds for the Mediterranean shells found at three Vézère sites: Blanchard, Castanet, and La Combe.

WESTERN CENTRAL EUROPE AND THE CONTINENTALITY GRADIENT

Located in the Ach and Lone valleys, small left-hand tributaries of the Danube, most of the Swabian Jura sites in Table 3.3 figure in the initial database. Throughout the Upper Paleolithic, long-distance transfers, associated with a small number of end-products, were found to be oriented along the Danube river, with maximum westward and eastward ranges of 180 and 160 km respectively. Given the recurrence of transfers along this natural route, the items were interpreted as personal gear transported during (seasonal?) group movement.

With the single exception of a north–south Baltic flint transfer to Hohlenstein-Stadel over 400 km in the Aurignacian (Burkert and Floss 2005), the new data do not dramatically alter previous transport patterns for this region. Rather, they emphasize the linear character of transfers along the Danube by extending both westward (220 km) and eastward (240 km) ranges. Again, the quantities involved are low and the items are end-products, except in the Aurignacian of Geissenklösterle where some debitage is also present. As previously, the overall transport pattern points to the provisioning of individuals as they go along.

An interesting point is that the magnitude of group mobility is roughly similar to that inferred from the "end-products plus cores" pattern for the Western European Upper Paleolithic. In this respect, the frequency distribution of MTDs for Western Central Europe (Figure 3.4) calls for comment. It is partly consistent with a location half-way along a continentality gradient, insofar as percentages for the [0–100] km procurement class are lower than in Western Europe but higher than in Eastern Central Europe, while percentages for the]100–200] km procurement class are very high, implying that most territories fall within this range. This suggests that although the magnitude of moves could be similar, longer moves were more frequent in Western Central Europe than in Western Europe, possibly as a result of greater continentality. Higher Western European percentages for the]200–300] km procurement class seem to contradict this assumption. It should, however, be noted that they are mainly due to transfers in the environmentally exacting Auvergne region, which account for 55% of the occurrences (17 out of 31) for this class.[4] When the Auvergne transfers are disregarded (Figure 3.7), the]200–300] km procurement class percentages for Western Europe and Western Central Europe are far more balanced (9% and 7% respectively); they are, in addition, both lower than the percentages for Eastern Central Europe (26%). This, as well as the very low percentages associated with the highest values (>300 km), draws Western Central Europe closer to Western Europe rather than to Eastern Central Europe.

There is indeed only one occurrence of transport over distances greater than 300 km, and this single end-product in Baltic flint suggests a particular mode of procurement, distinct from the provisioning of individuals. Baltic flint occurs in small quantities (<5%) at the Aurignacian site of Wildscheuer in the Rhineland,

some 140 km distant from the closest source (Floss 1994), and the tools recovered at Hohlenstein-Stadel were possibly conveyed by down-the-line trade to southwestern Germany.

CONCLUSIONS: SALVAGING THE CULTURAL ECOLOGICAL PARADIGM

In the examples discussed for Western Europe and Western Central Europe, the significance of distances greater than 100 km has been addressed in terms of the distinction between scale of group mobility and extent of interaction networks. Thus, planned seasonal moves over distances generally ranging between 230 and 300 km, and associated with a strategy of provisioning places, have been suggested for the Auvergne region throughout the Upper Paleolithic. A similar strategy was tentatively shown to operate in a different mobility context, related to the expansion of cultural range over 200 km in the Middle Magdalenian of the Pyrenees. Transfers up to 240–270 km and associated with a strategy of provisioning individuals have also been construed as proxies for the scale of group mobility for various regions at different periods of the Upper Paleolithic, in Western Central Europe as well as in Western Europe. Finally, the case has been made for indirect procurement through social exchange by down-the-line trade, in relation to the longest recorded distances for the Aurignacian of Western Europe (320, 380 km) and Western Central Europe (400 km), as well as in relation to shorter distances overlapping the scale of group mobility. The same mode of procurement is plausible for the two Solutrean shouldered points in Turonian flint from the Cher region found at Le Cuzoul (Lot) some 300 km from the source (Figure 3.5). Similar hunting implements in this type of flint are documented at Laugerie-Haute in the Périgord, in combination with Bergeracois flint, which also occurs at Le Cuzoul. Indirect procurement possibly also holds for the few tools in "grain de mil" found at Enlène (Middle Magdalenian), 300 km from the source (Figure 3.5), provided that ongoing research confirms the presence of this raw material in some Périgord sites (Pelegrin, pers. comm.).

As a result, the current propositions about the scale of group mobility and the extent of interaction networks in the European Middle and Upper Paleolithic (Table 3.5) are at variance with the initially formulated propositions

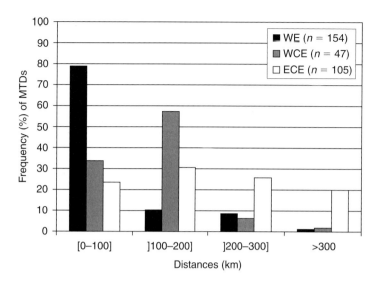

Figure 3.7 Current database minus the Auvergne sites. Frequency distribution of the maximum transport distances (MTD) per lithic assemblage for Western Europe (WE), Western Central Europe (WCE), and Eastern Central Europe (ECE). There is a single MTD per assemblage.

Table 3.5 Current propositions about the scale of group mobility and the extent of interaction networks in the European Middle and Upper Paleolithic, based on transfers of lithics alone. PI: provisioning individuals. PP: provisioning places.

	Western Europe	Western Central Europe	Eastern Central Europe
Middle Palaeolithic			
Group mobility	100–120 km ⇒ 230–300 km (exceptional)	110 km	100 km ⇒ 200–300 km (exceptional)
Interaction networks	?	?	200–300 km (exceptional)
Upper Palaeolithic			
Group mobility	130–270 km (PI) 180–300 km (PP)	180–240 (PI)	200–250 km (PP + PI) (common)
Expansion of cultural range	200 km (PP)	?	?
Interaction networks	• ≥300 km ⇒ 380 km • <300 km (probable)	400 km	• >300 km… 450 km… 700 km • <300 km (probable)

Same assemblages as in Table 3.1, with additional assemblages from Tables 3.2, 3.3, and 3.4.

(Table 3.1), significantly attenuating the differences between Western Europe and Eastern Central Europe. Concerning the scale of mobility, the evidence for the Middle and the Upper Paleolithic argues against a relationship between greater magnitude of moves and increasing continentality. Interestingly, extensive spatial networks of intergroup connections are now arguably revealed in Western Europe, not only though down-the-line trade in ornaments but also through indirect procurement of lithics. Moreover, whereas extensive shell networks seem more specific of the Magdalenian (Rensink *et al.* 1991), lithic networks are shown to operate much earlier, beginning as in Eastern Central Europe (Féblot-Augustins 1997a) with the initial stages of the Upper Paleolithic. Clearly, hunter-gatherers' dealings with space were far more independent of environmental constraints than previous data led to suppose.

But not wholly independent, however. While the relevance of the cultural ecological paradigm for addressing mobility and interaction related issues needs to be qualified, it can be argued that it is not totally invalidated. In the first place, attention should be paid to the weight of the Auvergne transfers. Once these are disregarded (Figure 3.7), MTDs for the three macro-regions are more coherently distributed along a continentality gradient. MTD percentages are highest for Western Europe in the [0–100] km procurement class, and for Western Central Europe in the]100–200] km class. They are also higher in the]200–300] km class for Eastern Central Europe than for the other two macro-regions. This suggests that when discussing group mobility in relation to environmental conditions, the frequency with which particular distance classes are covered should be considered in combination with the magnitude of transfers and is possibly more significant – as emphasized by Blades (2003) in connection with Aurignacian and Perigordian mobility in the Périgord. In the second place, and along the same lines, it should be remembered that the Auvergne is precisely the region where environmental constraints were shown to influence mobility strategies, leading to systematic and recurrent long-distance seasonal moves throughout the Upper Paleolithic and earlier – a case of continuity otherwise obtaining only in Eastern Central Europe. In the third place, both frequency and magnitude of transfers continue to set apart Eastern Central Europe from both Western Europe and Western Central Europe when distances of 300 km and beyond are involved. These transfers, construed as proxies for the extent of interaction networks, are far more frequent in Eastern Central Europe, and their range is also greater.

The sanguine view taken here of the cultural ecological paradigm as a still partly valid interpretive framework reflects the state of our knowledge in 2006. It could, however, be further weakened if, as a result of future research, the number of Western European sites with MTDs greater than 100 km were to increase. The required increase is considerable, since at present 70% of the recorded assemblages display shorter MTDs.

Be that as it may, it should be emphasized that while the recently collected data lead to toning down synchronic differences between western and eastern macro-regions, they reinforce previously stated diachronic differences. Transfers greater than 100 km remain exceptional in the European Middle Paleolithic, suggesting differences of a socio-economic nature. Moreover, the recorded occurrences remain associated

with a strategy of provisioning individuals, even when continuity in mobility patterns can be argued for the same region, as in the Auvergne and in Moravia. The fact that Middle Paleolithic people did not provision places with raw materials from afar is not necessarily indicative of specific cognitive capacities, as opposed to those of Upper Paleolithic people. Indeed, while the distances over which personal gear is transported may not reflect the scope of planning depth (Kuhn 1992), the nature of the items transported points to a higher degree of technological planning than previously thought. Current analyses insist on the flexible nature of such items: in the Quina Mousterian, for instance, large side scrapers and *limaces* are shown to be both tools and potential supplies of raw material in the form of long thin "retouch flakes" (Slimak 2004), and the same holds for MTA bifaces (Soressi 2002). Moreover, it has increasingly been found that in some assemblages (Delagnes *et al.* 2006; Park 2007) cores were worked from flakes, suggesting that flake blanks were transported both as potential tools (to be retouched) and as raw material supplies. This is strongly reminiscent of the nature of some of the Upper Paleolithic cores transported in the context of provisioning individuals. The possibility should be therefore considered that technical changes induced higher constraints on the quality of lithic materials in the Upper Paleolithic, thus justifying the higher cost of transport when the local stone proved unsatisfactory. The differences between Middle and Upper Paleolithic procurement patterns would thus highlight more specifically cultural facets of Paleolithic behavior related to the requirements of lithic technology, which are expressed irrespective of the environmental context.

ACKNOWLEDGMENTS

I am grateful to Brooke S. Blades for stimulating me to update my database and evaluate my former conclusions. Our numerous discussions about the problems of raw material interpretation proved both helpful and enjoyable.

ENDNOTES

1   Twenty per cent of the assemblage could not be petrographically characterized.

2   Thirty seven and a half per cent of the assemblage could
    not be characterized.
3   Transfers >100 km known to involve end-products plus
    small bladelet cores: early Aurignacian, 5 occurrences;
    Gravettian, 2 occurrences; Magdalenian, 1 occurrence.
    Transfers >100 km known to involve end-products alone:
    early Aurignacian, 18 occurrences; late Aurignacian,
    2 occurrences; Gravettian, 2 occurrences; Solutrean,
    4 occurrences; Badegoulian, 1 occurrence; Magdalenian,
    1 occurrence.
4   They also account for 11% of the occurrences (2 out of 18)
    for the [100–200] km procurement class.

REFERENCES

Aubry, Th. and Mangado-Llach, X. (2003) Interprétation
de l'approvisionnement en matières premières siliceuses
sur les sites du Paléolithique supérieur de la vallée du Coa
(Portugal). In: *Les matières premières lithiques en Préhistoire*.
Association Préhistoire du Sud-Ouest, Cressensac (France)
(*Préhistoire du Sud-Ouest, supplément* 5), pp. 27–40.

Binford, L.R. (1979) Organization and formation processes:
Looking at curated technologies. *Journal of Anthropological
Research* 35, 255–73.

Binford, L.R. (1980) Willow smoke and dogs' tails: Hunter-
gatherer settlement systems and archaeological site
formation. *American Antiquity* 45, 4–20.

Blades, B. (1999) Aurignacian lithic economy and early modern mobility: New perspectives from classic sites in the Vézère
valley of France. *Journal of Human Evolution* 37, 91–120.

Blades, B. (2003) End scraper reduction and hunter-gatherer
mobility. *American Antiquity* 68, 141–56.

Bon, F. (2002) *L'Aurignacien entre Mer et Océan: Réflexion
sur l'unité des phases anciennes de l'Aurignacien dans
le sud de la France*. SPF, Paris (*Mémoires de la Société
Préhistorique Française* XXIX).

Bon F., Simonnet, R., and Vézian, J. (2005) L'équipement
lithique des Aurignaciens à la Tuto de Camalhot (Saint-
Jean-de-Verges, Ariège), sa relation avec la mobilité
des groupes et la répartition de leurs activités dans un
territoire. In: Jaubert, J. and Barbaza, M. (eds) *Territoires,
déplacements, mobilité, échanges durant la Préhistoire*. CTHS,
Paris, pp. 173–84.

Bordes, J.-G. (2002) Les interstratifications Châtelperronien/
Aurignacien du Roc-de-Combe et du Piage (Lot, France):
Analyse taphonomique des industries lithiques:
Implications archéologiques. Thèse de Doctorat, Université
Bordeaux I, Bordeaux.

Bordes, J.-G., Bon, F., and Le Brun-Ricalens, F. (2005) Le
transport des matières premières lithiques à l'Aurignacien
entre le Nord et le Sud de l'Aquitaine: Faits attendus, faits
nouveaux. In: Jaubert, J. and Barbaza, M. (eds) *Territoires,
déplacements, mobilité, échanges durant la Préhistoire*. CTHS,
Paris, pp. 185–98.

Bracco, J.-P. (1994) Colonisation et peuplement en moyenne
vallée volcanique au Würm récent: Le campement badegoulien de la Roche à Tavernat (Massif Central). *Bulletin de
la Société Préhistorique Française* 91, 113–8.

Bracco, J.-P. (1995) Déplacements des groupes humains et
nature de l'occupation du sol en Velay (Massif central,
France) au Paléolithique supérieur: Intérêt de l'étude des
matières premières minérales. In: Chenorkian R. (ed.)
L'Homme Méditerranéen. Université d'Aix-en-Provence,
Aix-en-Provence, pp. 285–91.

Bracco, J.-P. (1996) Du site au territoire: L'occupation du sol dans
les hautes vallées de la Loire et de l'Allier au Paléolithique
supérieur (Massif central). *Gallia Préhistoire* 38, 43–67.

Burkert, W. and Floss, H. (2005) Lithic exploitation areas in
the Upper Palaeolithic of west and southwest Germany: A
comparative study. In: *Stone Age – Mining Age*: Proceedings
of the 8th international flint symposium, Bochum
(Germany), 1999. Deutschen Bergau-Museums, Bochum
(*Der Anschnitt, Beiheft* 19), pp. 35–49.

Chiotti, L., Leoz, L.E., Nespoulet, R., and Pottier, C. (2003)
Quelques exemples de stratégies d'approvisionnement dans
l'Aurignacien et le Gravettien à l'abri Pataud (Dordogne).
In: *Les matières premières lithiques en Préhistoire*. Association
Préhistoire du Sud-Ouest, Cressensac (France) (*Préhistoire
du Sud-Ouest, supplément* 5), pp. 115–22.

Dachary, M. (2002) Le Magdalénien des Pyrénées occiden-
tales. Thèse de Doctorat, Université de Paris X, Nanterre.

Delagnes, A., Féblot-Augustins, J., Meignen, L., and Park, S.-J.
(2006) L'exploitation des silex au Paléolithique moyen dans
le bassin de la Charente: Qu'est-ce qui circule, comment...
et pourquoi? *Bulletin de l'Association des Archéologues de
Poitou-Charente* 35, 15–24.

Demars, P.-Y. (1980) L'utilisation du silex au Paléolithique
supérieur: Choix, approvisionnement, circulation: L'exemple
du Bassin de Brive. Thèse de 3ᵉ cycle, Université
Bordeaux I, Bordeaux.

Digan, M. (2003) Les matières premières lithiques de l'unité
KL19 du site gravettien de la Vigne-Brun (Villerest, Loire):
Identification, modalité d'approvisionnement et diffusion.
In: *Les matières premières lithiques en Préhistoire*. Association
Préhistoire du Sud-Ouest, Cressensac (France) (*Préhistoire
du Sud-Ouest, supplément* 5), pp. 131–43.

Digan, M. (2006) *Le gisement gravettien de La Vigne-Brun
(Loire, France): Étude de l'industrie lithique de l'unité KL19*.
BAR International Series 1473, Oxford.

Féblot-Augustins, J. (1993) Mobility strategies in the late
Middle Palaeolithic of central Europe and western
Europe: Elements of stability and variability. *Journal of
Anthropological Archaeology* 12, 211–65.

Féblot-Augustins, J. (1997a) *La circulation des matières
premières au Paléolithique*. ERAUL 75, Liège.

Féblot-Augustins, J. (1997b) Middle and Upper Palaeolithic
raw material transfers in western and central Europe:
Assessing the pace of change. *Journal of Middle Atlantic
Archaeology* 13, 57–90.

Féblot-Augustins, J. (1999) La mobilité des groupes paléo-lithiques. *Bulletins et Mémoires de la Société d'Anthropologie de Paris* 11, 219–60.

Féblot-Augustins, J. (2005) *Survey of flint types from Bugey, France.* http://www.flintsource.net/flint/infF_bugey.html

Féblot-Augustins, J. and Perlès, C. (1992) Perspectives ethno-archéologiques sur les échanges à longue distance. In: Galley, A., Audouze, F., and Roux, V. (eds) *Ethnoarchéologie: Justification, Problèmes, Limites.* APDCA, Juan-les-Pins, pp. 195–209.

Féblot-Augustins, J., Park, S.-J., and Delagnes, A. (2005) *Lithothèque du bassin de la Charente en ligne.* http://www.alienor.org/ARTICLES/lithotheque/index.htm

Floss, H. (1994) *Rohmaterialversorgung im Paläolithikum des Mittelrheingebietes.* R. Habelt, Bonn.

Fontana, L. (1995) Chasseurs magdaléniens et rennes en bassin de l'Aude: Analyse préliminaire. *Anthropozoologica* 21, 147–56.

Fontana, L. (2000a) Stratégies de subsistance au Badegoulien et au Magdalénien en Auvergne: Nouvelles données. In: *Le Paléolithique supérieur récent: Nouvelles données sur le peuplement et l'environnement.* SPF, Paris (*Mémoires de la Société Préhistorique Française XXVIII*), pp. 59–65.

Fontana, L. (2000b) La chasse au renne au Paléolithique supérieur dans le sud-ouest de la France: Nouvelles hypothèses de travail. *Paléo* 12, 141–64.

Gamble, C. (1986) *The Palaeolithic Settlement of Europe.* Cambridge University Press, Cambridge.

Gamble, C. (1999) *The Palaeolithic Societies of Europe.* Cambridge University Press, Cambridge.

Geneste, J.-M. (1985) Analyse lithique d'industries moustériennes du Périgord: Une approche technologique du comportement des groupes humains au Paléolithique moyen. Thèse de Doctorat, Université Bordeaux I, Bordeaux.

Gould, R.A. (1980) *Living Archaeology.* Cambridge University Press, Cambridge.

Hahn, J. (1988) *Die Geissenklösterle-Höhle im Achtal bei Blaubeuren I.* Konrad Theiss Verlag, Stuttgart (*Forschungen und Berichte zur vor- und frühgeschichte in Baden-Württemburg* 26).

Keeley, L.H. (1988) Hunter-gatherer economic complexity and "population pressure": A cross-cultural analysis. *Journal of Anthropological Archaeology* 7, 373–411.

Kelly, R.L. (1983) Hunter-gatherer mobility strategies. *Journal of Archaeological Research* 39, 277–306.

Kelly, R.L. and Todd, L.C. (1988). Coming into the country: Early Paleoindian hunting and mobility. *American Antiquity* 53, 231–44.

Klaric, L. (2003) L'unité technique des industries à burins du Raysse dans leur contexte diachronique: Réflexions sur la variabilité culturelle au Gravettien à partir des exemples de la Picardie, d'Arcy-sur-Cure, de Brassempouy et du Cirque de la Patrie. Thèse de Doctorat, Université de Paris I, Paris.

Kuhn, S.L. (1992) On planning and curated technologies in the Middle Palaeolithic. *Journal of Anthropological Research* 48, 185–214.

Lacombe, S. (1998) Stratégies d'approvisionnement en silex au Tardiglaciaire: L'exemple des Pyrénées centrales françaises. *Bull. Soc. Préhist. Ariège-Pyrénées* LIII, 223–66.

Le Brun-Ricalens, F., and Séronie-Vivien, M.-R. (2004) Présence d'un silex d'origine nord-pyrénéenne (Chalosse?) dans l'Aurignacien du Piage (Lot, France) et implications. *Paléo* 16, 129–36.

Masson, A. (1981) Pétroarchéologie des roches siliceuses. Intérêt en Préhistoire. Thèse de 3ᵉ cycle, Université Claude Bernard-Lyon I, Lyon.

Meltzer, D.J. (1988) Late Pleistocene human adaptations in eastern North America. *Journal of World Prehistory* 2, 1–52.

Meltzer, D.J. (1989) Was stone exchanged among eastern North American Paleoindians? In: Lothrop, J. and Ellis, C. (eds) *Eastern Paleoindian Lithic Resource Procurement and Processing.* Westview Press, Boulder, pp. 11–39.

Morala, A., Turq, A., and Bordes, J.-G. (n.d.) Les ressources en matières premières lithiques autour de Castanet et réflexions sur l'approvisionnement. To appear in: *L'Aurignacien ancien de l'abri Castanet* (Coll.). Supplément à la revue *Paléo.*

Park, S.-J. (2007) Systèmes de production lithique et circulation des matières premières au Paléolithique moyen récent et final: Une approche techno-économique à partir des industries lithiques de La Quina (Charente). Thèse de Doctorat, Universite de Parix X, Nanterre.

Pelegrin, J. (n.d.) L'industrie lithique. To appear in: *L'Aurignacien ancien de l'abri Castanet* (Coll.). Supplément à la revue *Paléo.*

Piels-Desruisseaux, J.-L. (2004) *Outils préhistoriques: Du galet taillé au bistouri d'obsidienne,* 5ᵉ édition. Dunod, Paris.

Primault, J. (2003a) Exploitation et diffusion des silex de la région du Grand-Pressigny au Paléolithique. Thèse de Doctorat, Université de Paris X, Nanterre.

Primault, J. (2003b) Exploitation et diffusion des silex de la région du Grand-Pressigny au Paléolithique. In: *Les matières premières lithiques en Préhistoire.* Association Préhistoire du Sud-Ouest, Cressensac (France) (*Préhistoire du Sud-Ouest, supplément* 5), pp. 283–92.

Primault, J. (2006) *Collection pétrographique de référence.* http://www.archeosphere.com

Renard, C. (n.d.) L'organisation des productions lithiques solutréennes du Cuzoul (Vers, Lot): Implications techno-économiques. To appear in: Chalard, P., Clottes, J., and Giraud, J.-P. (eds) *The Cuzoul Monograph.* BAR International Series, Oxford.

Rensink, E., Kolen, J., and Spieksma, A. (1991) Patterns of raw material transport distribution in the Upper Pleistocene of northern and central Europe. In: Montet-White, A. and Holen, S. (eds) *Raw Material Economies among Prehistoric Hunter-Gatherers.* Lawrence, Kansas, University of Kansas (*Publications in Anthropology* 19), pp. 141–59.

Riche, C. (1998) Les ateliers de silex de Vassieux: Exploitation des gîtes et diffusion des produits. Thèse de Doctorat, Université de Paris X, Nanterre.

Rockman, M. and Steele, J. (eds) (2003) *Colonization of Unfamiliar Landscapes: The Archaeology of Adaptation*. Routledge, London and New York.

Séronie-Vivien, M. and Séronie-Vivien, M.-R. (1987) *Les silex du Mésozoïque nord-aquitain: Approche géologique de l'étude des silex pour servir à la recherche préhistorique*. Société Linnéenne de Bordeaux, Bordeaux (*Bulletin de la Société Linnéenne de Bordeaux*, supplément au tome XV).

Séronie-Vivien, M.-R. (1995) *La grotte de Pégourié (Caniac-du-Causse, Lot): Périgordien, Badegoulien, Azilien, Age du Bronze*. Association Préhistoire Quercinoise, [S.l.] (*Préhistoire Quercinoise, supplément 2*).

Séronie-Vivien, M.-R. (2003) Origine méridionale de silex recueillis dans le Paléolithique supérieur de la région Périgord-Quercy. In: *Les matières premières lithiques en Préhistoire*. Association Préhistoire du Sud-Ouest, Cressensac (France) (*Préhistoire du Sud-Ouest, supplément 5*), pp. 305–6.

Slimak, L. (2004) Les dernières expressions du Moustérien entre Loire et Rhône. Thèse de Doctorat, Université de Provence, Aix-en-Provence.

Soressi, M. (2002) Le Moustérien de tradition acheuléenne du sud-ouest de la France: Discussion sur la signification du faciès à partir de l'étude comparée de quatre sites: Pech-de-l'Azé I, Le Moustier, La Rochette et la Grotte XVI. Thèse de Doctorat, Université Bordeaux I, Bordeaux.

Surmely, F. (2000) Le peuplement magdalénien de l'Auvergne. In: *Le Paléolithique supérieur récent: Nouvelles données sur le peuplement et l'environnement*. SPF, Paris (*Mémoires de la Société Préhistorique Française XXVIII*), pp. 165–75.

Surmely, F. and Pasty, J.-F. (2003) L'importation de silex en Auvergne durant la Préhistoire. In: *Les matières premières lithiques en Préhistoire*. Association Préhistoire du Sud-Ouest, Cressensac (France) (*Préhistoire du Sud-Ouest, supplément 5*), pp. 327–35.

Surmely, F., Barrier, P., Bracco, J.-P., Charly, N, and Liabeuf, R. (1998). Caractérisation des silex par l'analyse des microfaciès et application au peuplement préhistorique de l'Auvergne (France). *Comptes rendus de l'Académie des sciences. Paris, Sciences de la terre et des planètes* 326, 595–601.

Taborin, Y. (1993) *La parure en coquillage au Paléolithique*. CNRS, Paris (*Gallia Préhistoire, supplément XXIX*).

Tolan-Smith, C. (2003) The social context of landscape learning and the Lateglacial-early Postglacial recolonizations of the British Isles. In: Rockman, M. and Steele, J. (eds) *Colonization of Unfamiliar Landscapes: The Archaeology of Adaptation*. Routledge, London and New York, pp. 116–29.

Torti, C. (1980) Recherches sur l'implantation en Limagne au Paléolithique moyen et supérieur. Thèse de 3e cycle, Université Bordeaux I, Bordeaux.

Valoch, K. (1986) Raw materials used in the Moravian Middle and Upper Palaeolithic. In: Takács-Biró, K. (ed.) *Proceedings of the First International Conference on Prehistoric Flint Mining and Lithic Raw Material Identification in the Carpathian Basin*, vol. 2. Magyar Nemzeti Múzeum, Budapest-Sümeg, pp. 263–7.

Whallon, R. (1989) Elements of cultural change in the Later Palaeolithic. In: Mellars, P. and Stringer, C. (eds) *The Human Revolution: Behavioural and Biological Perspectives in the Origins of Modern Humans*. Edinburgh University Press, Edinburgh, pp. 433–54.

Wiessner, P. (1982) Risk, reciprocity and social influences on! Kung San economics. In: Leacock, E. and Lee R.B. (eds) *Politics and History in Band Societies*. Cambridge University Press, Cambridge, pp. 61–84.

Chapter 4

SOURCING RAW MATERIALS FOR CHIPPED STONE ARTIFACTS: THE STATE-OF-THE-ART IN HUNGARY AND THE CARPATHIAN BASIN

Katalin T. Biró

Hungarian National Museum

ABSTRACT

The provenancing of chipped stone tool raw materials is an integral part of prehistoric research. Such studies include the necessary steps of field survey, fingerprinting, and characterization of geological outcrops, together with investigations of the archeological evidence. The state of such research, however, varies by country and region. It is argued here that joint projects involving international cooperation can advance the field, and promising examples will be presented from Central Europe with special focus on problems to be solved. An Internet based network of information systems is advocated to facilitate the wide dissemination of the results of lithic raw material studies.

INTRODUCTION

Lithic provenance analysis has a long tradition in Central Europe. The Carpathian Basin is a unique area within this region, surrounded by high mountains of the Alpean system (Alps, Carpathes, Dinarids) and currently comprising about eight independent political

Lithic Materials and Paleolithic Societies, 1st edition. Edited by B. Adams and B.S. Blades. ©2009 Blackwell Publishing. ISBN 978-1-4051-6837-3.

units (Figure 4.1). Due to the regional/national organization of research centers and projects, knowledge of lithic resources and their distribution is highly variable, even within the territory of a single country. Thematic conferences and joint projects have furthered the subject significantly, but we still lack the degree of practical geological knowledge our prehistoric ancestors possessed. Indeed, some source regions are still unknown to us or are imprecisely located, while others may not exist any longer due to prehistoric exploitation and/or modern destruction of the sources.

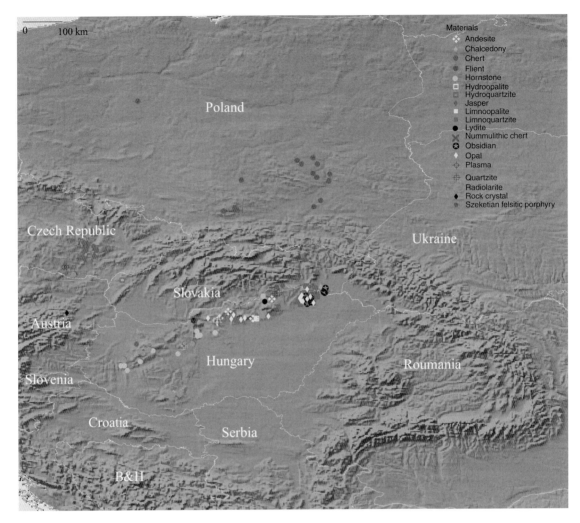

Figure 4.1   Relief map of the Carpathian Basin with location of important chipped stone raw material sources documented in the Lithotheca Collection of the HNM (see Figure 4.2 for key).

The practical application of source identification (i.e., fingerprinting raw materials by quantitative methods) and collection of distribution data from archeological sites also varies by country, region, and to some extent, chronological periods. Though lithic provenance studies as a rule are diachronic in the sense that the same raw material sources could be, and were, utilized over longer periods of time, there is substantial variation in accessibility, as well as recognizability of important raw materials.

This chapter will focus on lithic raw materials utilized during the Hungarian Paleolithic. It must be stressed, however, that much of this knowledge originates from the study of essentially younger prehistoric periods and wider geographical areas inside and outside the Carpathian Basin.

IDENTIFICATION OF LITHIC RAW MATERIALS

A necessary first phase in lithic raw material research is the identification of the raw material types. Simple as it may seem, even this phase is burdened with all the difficulties of interdisciplinary collaboration and analytical archeology. Macroscopic investigation of such material has its limitations despite the availability of large comparative collections, and was even more difficult before such collections came into existence. The geological/petrological/geochemical characterization of siliceous rocks is a young and complex field due to the relative chemical homogeneity and limited variation in mineral composition (Figure 4.2).

Tracing long-distance movements and postulated interactions between communities requires a systematic approach to the identification of raw material sources. Such an approach is represented by the geology–archeology oriented studies in a number of countries in Central Europe (Kozłowski 1972–73; Bárta 1979; Kaminska 1991; Floss 1994). In the late 1970s, lithic provenance studies were still rare in Hungary (Patay 1976; Lech 1981). In the early 1980s, a systematic collection of potential sources took place under the auspices of the Hungarian Geological Institute.

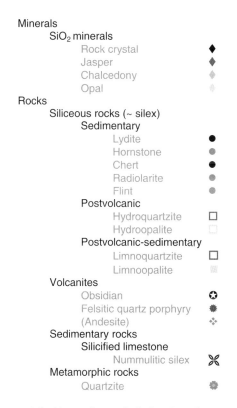

Figure 4.2 Nomenclature draft for chipped stone raw materials in the Carpathian Basin.

The results were presented in 1986 at a regional international conference in Sümeg, Hungary (Biró 1986; 1987) and the collected samples have been incorporated into the "Lithotheca" raw material collection housed in the Hungarian National Museum in Budapest. The Lithotheca contains geological specimens and provenance information, two catalogue volumes (Biró-Dobosi 1990; Biró-Dobosi-Schléder 2000) and a web-page with true color images of the most important raw material varieties have been published (www.ace.hu/litot). The collection continues to grow and has proved to be an excellent tool for provanance studies.

In addition to the expansion and continued documentation of the Lithotheca collection, the performance of analytical studies on the sample set continues in order to facilitate comparisons with archeological material, preferably by non-destructive methods. Recently, such analyses have focused on raw materials used for polished stone tools as part of the UNESCO-IGCP 442 project (Biró-Szakmány 2000), obsidian (Elekes *et al.* 2000; Kasztovszky-Biró 2004), grey flint (Kasztovszky-Biró 2002), radiolarite (Biró *et al.* 2002), and Szeletian felsitic porphyry (Markó *et al.* 2003). Current efforts also involve the analysis of other raw materials utilized in prehistory as part of the "Atlas of Prehistoric Raw Materials" project (www.ace.hu/atlas).

LITHIC RAW MATERIAL UTILIZATION IN THE PALEOLITHIC PERIOD

Provenancing Paleolithic material is a difficult task as artifacts are often heavily patinated, making macroscopic identification difficult, if not impossible. The availability of specific raw material sources also changes with time and the probability of the same resources being accessible throughout the Paleolithic period is less than in relatively modern times. Typically, effective provenance studies were primarily initiated on younger prehistoric assemblages with tightly controlled temporal affiliations. In such cases, settlement density, resource exploitation, supply patterns, and production on an industrial scale help facilitate archeological interpretations. In contrast, Paleolithic sites as **events** are dispersed over a vast period of time. It is difficult to identify the "trade partners" when the documented evidence is separated by at least several generations. It is also difficult to establish to what extent the distribution of goods (basically lithics, but also jewellery in the form of fossil shells and amber) reflects the actual movement of a particular community or its direct/indirect contact with other groups. A postulated sign of trade in the Upper Paleolithic is the establishment of workshop settlements around particular sources, as argued by Dobosi (1997). Recently, raw materials of apparent importance among the Paleolithic populations of the region were summarized at the Krosno conference on the Paleolithic of the Carpathian region (Biró 2002).

In the Neolithic the situation is more clear. Relatively dense settlement networks can be discerned, and the environs of lithic raw material sources tend to be uninhabited. Of course there are exceptions in certain periods, but these exceptions are associated with individual cultures, when particular communities appear to have exercised tight control over specific lithic resources.

To date, the most comprehensive sourcing studies of Paleolithic lithic inventories have been conducted on the earliest (Lower or Middle Paleolithic) assemblages.

Pioneering petroarcheological studies were restricted to the determination of mineral composition (quartz, chalcedony) and atypical fossils, often present in shapeless aggregates in the siliceous matrix (e.g., Vendl 1930). More accurate sourcing was accomplished in situations where the analyst had good regional knowledge on the site environs, primarily for Lower Paleolithic, Middle Paleolithic, and Early Upper Paleolithic sites. This was partly explained by proximity to raw material sources, usually woodlands and foot-hill regions, as an important factor in the selection of a settlement location. Examples of such situations include the sites of Tata (Végh-Viczián 1964), Érd (Dienes 1968), and Vértesszőlős (Varga-Máthé 1990) (Figure 4.3).

As Upper Paleolithic hunters moved from the environs of lithic sources to the open steppe regions they became detached from the raw material sources. Such groups appear to have been highly mobile, following herds of animals (typically reindeer) and covering

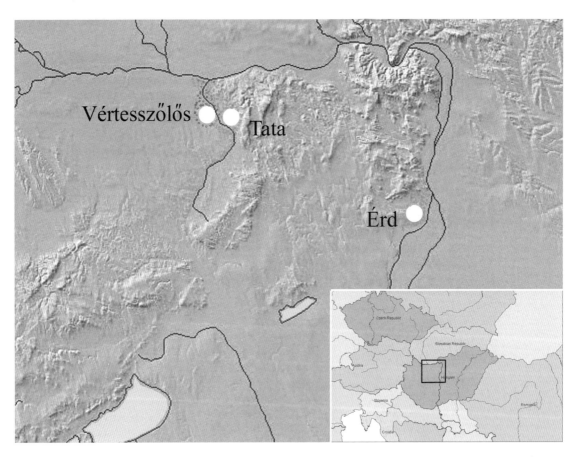

Figure 4.3   Raw material acquisition area for Lower and Middle Paleolithic sites in Transdanubia (Vértesszőlős, Tata, Érd).

considerable distances. The lithic raw materials used reflect such movements and, to a much lesser extent, the contacts of the community. The pattern of lithic raw material exploitation consisting of a centrally located archaeological site tied to possible resources in a "star-pattern" has been referred to as the **action radius model**. This model postulates that raw material sources were directly visited by the inhabitants of a settlement (Biró *et al.* 2001a,b). Figure 4.4 illustrates the action radius model based on data from the Upper Paleolithic Gravettian site of Bodrogkeresztúr in northeast Hungary. The same array of contacts is often proposed when assessing more recent (Neolithic or younger) sites, as demonstrated by Raczky *et al.* (2002: 850) for the Late Neolithic site of Polgár. In this case, however, we must be aware that we are dealing with the product of several factors and not necessarily the actual movement of people between a settlement and a raw material source.

Mapping the occurrences of source-specific or unique materials allows the reconstruction of broad temporal patterns of specific commodity utilization, and has been especially successful in Hungarian Paleolithic studies. Raw material sourcing projects supported by analytical methods have been successfully performed on several specific materials, including obsidian (Biró *et al.* 1986, 1988; Elekes *et al.* 2000; Kasztovszky-Biró 2004), Szeletian felsitic porphyry (Vértes-Tóth 1963; Simán 1986; Markó *et al.* 2003), rock crystal (Dobosi-Gatter 1996), and recently, nummulitic silex (Markó-Kázmér 2004). Source identification of these materials has been investigated by the application of different chemical and physical fingerprinting techniques. Obsidian has been of particular interest due to its easy recognition and limited source area compared to its wide archeological distribution (Vértes 1953; Gábori 1950; Biró 1984).

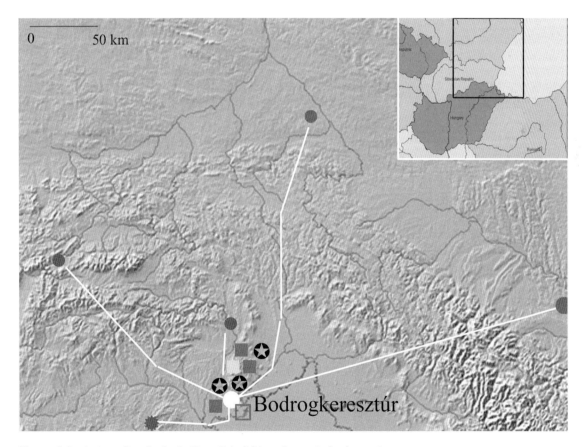

Figure 4.4   Action radius plot for the Upper Paleolithic settlement Bodrogkeresztúr.

The investigation of common raw materials that occur at several localities, such as radiolarite, limnic silicites, or hornstone, are more problematic. A particular problem is the frequent use of secondary sources of these materials, such as river gravels, during the Paleolithic. Such deposits exploited in the past do not necessarily correspond to the modern distribution of these same materials. Intensive patination can also be a problem. Chemical fingerprinting for these raw materials is still in an embryonic state (Elekes *et al.* 2000; Biró *et al.* 2002).

The diffusion radius of certain materials appears to be dependant on quality. Certain unique raw materials, such as obsidian or rock crystal, were moved over hundreds of kilometers, though the quantity of such materials decreases with distance. The regular supply zone seems to be confined to much smaller areas, where distance from the nearest source was an important limiting factor. As a result of both systematic surveys and diachronic petroarcheological investigation of archaeological lithic assemblages, it appears that all the important raw material sources currently identified were utilized during the Paleolithic period. It also appears that the use of regional and secondary resources (e.g., pebble deposits) was more important, than among subsequent Holocene groups still reliant on lithic raw materials.

TASKS FOR THE FUTURE

Elucidating the basic raw material stock of the Carpathian Basin was a major step in the analysis of Paleolithic raw material distribution and exchange networks in Hungary. Much work is needed, however, to refine our understanding of the lithic exploitation systems at this time. Important tasks include refining the precision of provenancing basic raw materials such as radiolarite, flint, or limnic quartzite. Also, the revision of lithic assemblage descriptions in relation to raw material utilization is important in light of our existing, as well as developing, knowledge. Provenancing techniques, mainly non-destructive geochemical methods, should be used more extensively both on archeological and comparative geological material and followed by comprehensive studies of the technological reduction chain of individual raw materials. By these steps, we can refine our insights into the movement, trade, and circulation of various lithic raw materials during the Paleolithic period.

REFERENCES

Bárta, J. (1979) K problematike proveniencie surovin na vyrobu stiepanej kamennej industrie v paleolite Slovenska. *Slovenska Archaeologia* 27, 5–15.

Biró, K.T. (1984) Distribution of obsidian from the Carpathian Sources on Central European Palaeolithic and Mesolithic sites. *Acta Archaeologica Carpathica Kraków* 23, 5–42.

Biró, K.T. (ed.) (1986) Őskori kovabányászat és kőeszköznyersanyag azonosítás a Kárpát medencében/International conference on prehistoric flint mining and lithic raw material identification in the Carpathian Basin – Sümeg Papers. *Budapest KMI Rota* 1, 1–342.

Biró, K.T. (ed.) (1987) Őskori kovabányászat és kőeszköznyersanyag azonosítás a Kárpát medencében/International conference on prehistoric flint mining and lithic raw material identification in the Carpathian Basin – Sümeg Proceedings. *Budapest KMI Rota* 2, 1–284.

Biró, K.T. (2002) Important lithic raw materials in the Carpathian region. In: Jan Ganczarski (ed.) Starsza i srodkowa epoka kamienia w Karpatach Polskich /The Older and Middle Stone Age in Polish Carpathians. *Krosno Mitel* 301–15.

Biró, K.T., Pozsgai, I., and Vladár, A. (1986) Electron beam microanalyses of obsidian samples from geological and archaeological sites. *Acta Arch Hung* 38, 257–78.

Biró, K.T., Pozsgai, I., and Vladár, A. (1988) Central European obsidian studies. State of affairs in 1987. *Archaeometrical Studies in Hungary* 1: Budapest KMI 119–30.

Biró, K.T. and Dobosi, V.T. (1991) Litotheca: Comparative raw material collection of the Hungarian National Museum. *Budapest Magyar Nemzeti Múzeum* 1–268.

Biró, K.T., Dobosi, V.T., and Schléder, Zs. (2000) Comparative raw material collection of the Hungarian National Museum II. *Budapest Magyar Nemzeti Múzeum* 1–320.

Biró, K.T., Bigazzi, G., and Oddone, M. (2001a) Instrumental analysis I. The Carpathian sources of raw material for obsidian tool-making: (Neutron activation and fission track analyses on the Bodrogkeresztúr-Henye Upper Palaeolithic artefacts. In: Dobosi V.T. (ed.) 2001 *Budapest, Magyar Nemzeti Múzeum* 221–40.

Biró, K.T., Elekes, Z., and Gratuze, B. (2001b) Instrumental analysis II. Ion beam analyses of artefacts from the Bodrogkeresztúr-Henye lithic assemblage. In: Dobosi, V.T. (ed.) Bodrogkeresztur-Henye (NE-Hungary) Upper Palaeolithic Site. *Budapest Magyar Nemzeti Múzeum* 241–45.

Biró, K.T., Elekes, Z., Kiss, Á. and Uzonyi, I. (2002) Radiolarit minták vizsgálata ionnyaláb analitikai módszerekkel/ Investigation of Radiolarite Samples by Ion-Beam Analytical Methods. *Archaeológiai Értesítő* 127, 103–34.

Biró, K.T. and György, S. (2000) Current state of research on Hungarian Neolithic polished stone artefacts. *Krystalinikum Brno* 26, 21–37.

Dienes, I. (1968) Examen petrographique de l'industrie. In: Gábori-Csánk, V., La station du Paléolithique moyen d'Érd, Hongrie. *Akadémiai Kiadó, Budapest* 111–14.

Dobosi, V.T. (1978) A pattintott kőeszközök nyersanyagáról Folia Archaeologica. *Budapest* 29, 7–19.

Dobosi, V.T. (1997) Raw material management of the upper palaeolithic (A case study of five new sites, Hungary). In: Schild, R. and Sulgostowska, Z. (eds) *Man and Flint*. Proceedings of the VIIth International Flint Symposium. *Warszawa* 189–95.

Dobosi, V.T. (ed.) (2001) Bodrogkeresztur-Henye (NE-Hungary) Upper Palaeolithic site, Hungarian National Museum. *Budapest* 1–245.

Dobosi, V.T. (1996) Palaeolithic tools made of rock crystal and their preliminary fluid inclusion investigation *Folia Archaeologica (Budapest)* 45, 31–50.

Elekes, Z., Biró, K.T., Rajta, I., Uzonyi, I., Gratuze, B., and Kiss Á.Z. (2000) Analyses of Obsidian and Radiolarite Samples by Ion Beam Techniques. Paper presented at 32th ISA Conference, Mexico City.

Floss, H. (1994) Rohmaterialversorgung im Paläolithikum des Mittelrheingebietes. *Monographien d. Römisch-Germanisches Zentralmuseum* 21, 1–407.

Gábori, M. (1950) Az Őskori obszidián – kereskedelem néhány problémája. *Archaeólogiai Értesítő, Budapest*, 77, 50–3.

Gáboriné Csánk, V. (1968) La station du Paléolithique moyen d'Érd, Hongrie. *Akadémiai Kiadó, Budapest*, 1968, 1–277.

Kaminska, L. (1991) Vyznam surovinej základne pre mladopaleoliticku spolocnost vo vychodokarpatskej oblasti/L'importance de la matiere pour les communantes du Paléolithique superieur dans l'espace des Carpathes orienta. *Slovenská Archeológia (Bratislava)* 39, 7–58.

Kasztovszky, Z. and Biró, K.T. (2006) Fingerprinting Carpathian Obsidians by PGAA: First Results on Geological and Archaeological Specimens. Proceedings of the 34th ISA Symposium, Zaragoza, E-book, http://www.dpz.es/ifc/libros/ebook2621.pdf – Zaragoza 2006, 301–8.

Kasztovszky, Z., Biró, K.T., and Dobosi, V.T. (2005) Investigation of gray flint samples with Prompt Gamma Activation Analysis. In: KARS–BURKE eds. Proceedings of the 33rd International Symposium on Archaeometry, 22–26 April 2112, Amsterdam GeoarchBioarch 2005, 3:79–82 – 1 saját 0.

Kozłowski, J.K. (1972–73) The Origin of Lithic Raw Materials Used in the Palaeolithic of the Carpathian Countries. *Acta Archaeologica Carpathica, Kraków* 13, 5–20.

Kretzoi, M. and Dobosi, V.T. (eds) (1990) Vértesszőlős – Site, man and culture. *Akadémiai Kiadó, Budapest*, pp. 1–552.

Lech, J. (1981) Flint mining among the early farming communities of Central Europe. *Przeglad Archaeologiczny, Wrocław* 28, 5–55.

Markó, A., Biró, K.T., and Kasztovszky, Z. (2003) Szeletian felsltic porphyry: Non-destructive analysis of a classical palaeolithic raw material. *Acta Archaeologica Hungarica, Budapest* 54, 297–314.

Markó, A. and Kázmér, M. (2004) The use of nummulithic chert in the Middle Palaeolithic in Hungary. In: Fülöp-Cseh (eds) *Die aktuellen Fragen des Mittelpaläolithikums. Tudományos Füzetek Tata*, vol. 12, pp. 53–64.

Patay, P. (1976) Les matières premières lithiques de l'âge du cuivre en Hongrie. *Acta Archaeologica Carpathica (Kraków)* 16, 229–38.

Raczky, P., Meier-Arendt, W., Anders, A., Hajdú, Z., Nagy, E., Kurucz, K., Domboróczki, L., Sebők, K., Sümegi, P., Magyari, E., Szántó, Z., Gulyás, S., Dobó, K., Bácskay, E., and Biró, K.T., and Polgár C. (1989–2000) Summary of the Hungarian-German Excavations on a Neolithic Settlement in Eastern Hungary. In: Aslan *et al.* (eds) Festschrift für Manfred Korfmann, Band 2. Remshalden-*Grunbach Verlag Bernhard Albert Greiner* pp. 833–59.

Simán, K. (1986) Felsitic quartz porphyry. *Sümeg Papers Budapest* 271–77.

Varga, I. (1991) Az Esztergom-Gyurgyalag lelőhelyen előkerült kőeszközök vizsgálata. *Acta Archaeologica Hungarica, Budapest* 43, 267–9.

Vargáné, M.K. (1990) Petrographic analysis of the lithic raw materials of the Vértesszőlős implements. In: Kretzoi-D. (ed.) *Vértesszőlős – Site, man and culture. Akadémiai Kiadó, Budapest* 287–99.

Vendl, A. (1930) A büdöspesti paleolitos szilánkok kőzettani vizsgálata. *Matematikai és Természettudományos Értesítő, Budapest* 47, 468–83.

Végh, A. and Viczián, I. (1964) Petrographische Untersuchungen an den Silexwerkzeugen. In: Vértes *et al.* (eds) *Archaeologia Hungarica, Budapest*, 43, 129–31.

Vértes, L. (1953) Az őskőkor társadalmának néhány kérdéséről (On some questions concerning palaeolithic society). *Archaeológiai Értesítő, Budapest* 80: 89–103.

Vértes, L. (1965) Az őskőkor és az átmeneti kőkor emlékei Magyarországon. A Magyar Régészet Kézikönyve. *Akadémiai Kiadó, Budapest* 1–385.

Vértes, L. *et al.* (1964) Tata, eine Mittelpalaeolithische Travertin-Siedlung in Ungarn. *Archaeologia Hungarica, Budapest* 43.

Vértes, L. and Tóth, L. (1963) Der Gebrauch des Glasigen Quartzporphyrs im Paläolithikum des Bükk-Gebirges. *Acta Archaeologica Hungarica, Budapest* 15, 3–10.

INTERNET RESOURCES

www.ace.hu/atlas
www.flintsource.net
www.ace.hu/litot

Chapter 5

UPPER PALEOLITHIC TOOLSTONE PROCUREMENT AND SELECTION ACROSS BERINGIA

Kelly E. Graf and Ted Goebel

Center for the Study of the First Americans, Texas A&M University

ABSTRACT

Two lithic techno-complexes characterize the terminal Pleistocene archeological record of Beringia, a non-microblade complex and a microblade complex. Archeologists working in the region have given two interpretations to explain this lithic variability. Some have argued that the complexes represent two distinct cultural groups, while others have suggested the variability represents two technological features of a single complex used by a single group of people. These interpretations are deeply rooted in descriptive analysis and culture history. In this chapter we employ a behavioral approach to explore these explanations by focusing on differences in toolstone procurement and selection represented in two sets of non-microblade and microblade assemblages from the Dry Creek site, Alaska (USA) and the Ushki-5 site, Kamchatka (Russia). Our results show that toolstone procurement and selection were both unpatterned and unplanned in non-microblade assemblages, but patterned and planned in microblade assemblages. The differences seen between the techno-complexes may have resulted from the varied ways people were provisioning and using the landscape.

INTRODUCTION

For two decades, some archeologists have argued that the earliest inhabitants of Beringia are represented by two separate technological complexes, one with and the

Lithic Materials and Paleolithic Societies, 1st edition. Edited by B. Adams and B.S. Blades. ©2009 Blackwell Publishing. ISBN 978-1-4051-6837-3.

other without microblades. Both complexes have been consistently found in well-dated, stratigraphic contexts in central Alaska and Kamchatka, Russia (Figure 5.1). In central Alaska, non-microblade assemblages make up the Nenana Complex, whereas microblade assemblages comprise the Denali Complex (Powers and Hoffecker 1989; Hoffecker *et al.* 1993). In Kamchatka, these assemblages have been referred to as the early Ushki and late Ushki complexes, respectively (Dikov 1979;

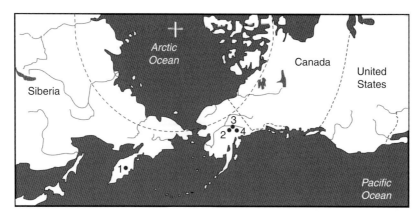

Figure 5.1 Map of Beringia with locations of archeological sites mentioned in text (1, Ushki sites; 2, Nenana Valley sites [Dry Creek, Moose Creek, Owl Ridge, Walker Road]; 3, Tanana River sites [Broken Mammoth, Swan Point]; and 4, Tangle Lakes sites [Phipps, Whitmore Ridge]).

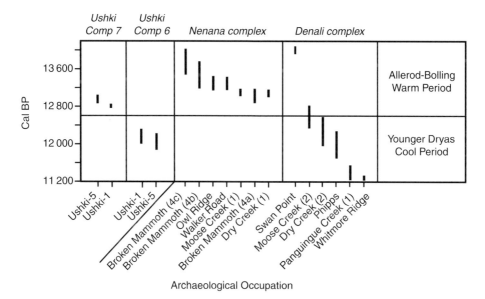

Figure 5.2 Comparison of calibrated AMS radiocarbon ages from Ushki Lake (components 7 and 6) and Alaskan archeological sites (Nenana and Denali complexes). (After Goebel *et al.* 2003.)

Goebel *et al.* 2003). The established radiocarbon record for these complexes suggests they are not coeval. Ages for the early non-microblade assemblages range from 13 800–12 200 cal BP, whereas earliest microblade assemblages span 12 200–11 100 cal BP (Hamilton and Goebel 1999; Hoffecker 2001; Goebel *et al.* 2003) (Figure 5.2).

There is, however, a significant exception to this non-microblade – microblade dichotomy, the Swan Point site in central Alaska, where C. Holmes and colleagues (Holmes *et al.* 1996; Crass and Holmes 2004) have uncovered a microblade industry associated with a buried paleosol dating to approximately 14 400 cal BP. The discovery of microblades at such an early age calls into question the notion that these complexes are temporally distinct, and has reopened arguments that they may represent behavioral facies of the same archaeological tradition (West 1996).

Clearly, resolution of the "non-microblade – microblade" question will require new approaches to studying the existing archaeological record. Previous investigations of Upper Paleolithic assemblages in Beringia have traditionally focused on describing and classifying them into complexes, phases, or traditions, to define culture histories. Typical questions asked include: Does variability exist? Were these complexes coeval or did one predate the other? What are the origins of these techno-complexes? Do they represent the same or different cultural groups (Dikov 1979; West 1981; Powers and Hoffecker 1989; Goebel et al. 1991, 2003; Hoffecker et al. 1993)? Even when explanations of variability call upon site activity instead of cultural differences (West 1983, 1996; Hoffecker 2001), studies have rarely provided evidence supporting said interpretations. At this point we need to move beyond description and cultural history and attempt to explain, from behavioral, ecological, and/or evolutionary perspectives, why there are two or more (Kunz and Reanier 1994) seemingly different techno-complexes in Beringia. We need to not only understand the origins of the first Beringians, but also the processes through which they colonized the New World's varied empty landscapes.

To begin to explain behaviorally the technological/typological patterns represented in the non-microblade and microblade techno-complexes of Beringia, research surrounding toolstone procurement and selection, technological organization, provisioning strategies, and foraging/land-use behavior needs to be undertaken. In this paper we take an initial step and address these questions by investigating toolstone procurement and selection at two multicomponent sites in Beringia that contain both industries. We focus on Dry Creek, Alaska and Ushki-5, Kamchatka (Figure 5.1). Each site and its associated lithic assemblages are described in detail below.

THE SITES: DRY CREEK AND USHKI-5

Dry Creek

Dry Creek is a multicomponent archeological site located about 5 km northwest of Healy, Alaska. It is situated upon a southeast-facing bluff that overlooks the braided floodplain of Dry Creek, a major tributary of the Nenana River (Figure 5.3). W. R. Powers directed

Figure 5.3   Aerial photograph of the Dry Creek site area (arrow points to site location). Note the braided floodplain of Dry Creek itself, containing numerous cobbles of potential toolstones. (Photograph by W. R. Powers.)

large-scale excavations of the site from 1973 to 1978 (Powers and Hamilton 1978; Powers et al. 1983); during this time an area of 347 m² was exposed (Powers et al. 1983). In 1992, N. Bigelow and Powers returned to the site to conduct geoarcheological investigations and collect additional carbon-14 samples (Bigelow and Powers 1994; Hoffecker et al. 1996).

These studies confirmed the presence of two stratigraphically separated Upper Paleolithic components dating to the terminal Pleistocene (Thorson and Hamilton 1977; Powers and Hoffecker 1989). Component I, the site's basal occupation, occurs in Loess 2 and has a single age of $11\,120 + 85$ ^{14}C BP (Powers and Hamilton 1978). Sealing component I is a thin sheet of wind-blown sand (Sand I). Above this sand is component II, which occurs in Loess 3 and has three ages ranging from $10\,060 + 75$ to $10\,090 + 250$ ^{14}C BP, averaging $10\,050 + 60$ ^{14}C BP (Bigelow and Powers 1994). Powers (1983) observed clear stratigraphic separation of the two Upper Paleolithic components across the entire excavation, although in places they became quite close. Powers and Hoffecker (1989) assigned components I and II to two separate cultural complexes, Nenana and Denali, respectively. Lithic assemblages from these components are the subject of our study.

The excavated lithic artifact assemblage for component I has been described by Powers (1983) and Goebel (1990). Goebel (1990) reported that the assemblage consists of 4524 articles, including 4461 debitage pieces, 7 cores, and 56 tools. These data are used in our study. Splinters and flake fragments comprise the majority of debitage pieces, with tertiary flakes, retouch chips, blades, blade-like flakes, cortical spalls, and cobbles/split cobbles making up the rest of the debitage assemblage. Microblades and microblade-related debitage are absent (Goebel 1990). Cores include a bipolar flake core, a monofrontal unidirectional prismatic blade core fragment, unidentifiable core fragments, and a possible platform rejuvenation spall from a flake core (Goebel 1990). The tool assemblage includes end scrapers, retouched blades, retouched flakes, cobble tools, side scrapers, bifaces, bifacial points, notches, gravers, and an undiagnostic tool preform. End scrapers dominate the tool assemblage; among them are simple end scrapers on blades and flakes, round end scrapers, carinated end scrapers, double end scrapers, and fragments. Among cobble tools are unifacial choppers and a scraper-plane. Bifaces include preforms of triangular-shaped points

and an undiagnostic biface fragment. Bifacial points include a complete triangular-shaped point and two basal fragments of triangular points. Among identifiable tool blanks, 50% are blades or blade-like flakes, and the rest are flakes and cobbles (Goebel 1990). In sum, the component I assemblage represents the "type-assemblage" for the central Alaskan Nenana complex – a blade-and-biface industry lacking any signs of microblade and burin technologies (Goebel et al. 1991; Hoffecker et al. 1993).

The component II lithic assemblage initially described by Powers (1983) consists of 28 881 lithic artifacts. Goebel (1990) analyzed complete sets of cores (109), core preforms (7), and tools (330), as well as a sample of 14 434 debitage pieces. The description that follows is based on Goebel's (1990) findings, and this assemblage is the subject of our study. While the vast majority of debitage is splinters and flake fragments, more expressive classes include tertiary flakes, microblades, microblade fragments, retouch chips, blades, blade-like flakes, cortical spalls, and whole or split cobbles. Cores consist primarily of wedge-shaped or "pseudo-wedge-shaped" microblade cores, "end" microblade cores, monofrontal subprismatic blade cores, and monofrontal unidirectional flake cores, as well as microblade core tablets, frontal rejuvenation spalls, undiagnostic core-like fragments, and several biface fragments interpreted to represent wedge-shaped core preforms. The tool assemblage includes burins, bifaces, retouched flakes, retouched microblades, side scrapers, retouched blades, cobble tools, bifacial points, gravers, notches, denticulates, and an end scraper. Burins dominate the tool assemblage; they consist of transverse burins, burins on snaps, dihedral burins, and angle burins. Bifaces occur in many recognizable shapes, including oval, spatulate, elliptical, ovate, lanceolate, oblong, triangular, discoidal, and deltoid forms, as well as in unexpressive or fragmentary states. Bifacial points, however, occur in just two forms – lanceolate and straight-to-convex-based or lanceolate and concave-based. Side scrapers are typically convergent or double-convex, but transverse, single-convex, single-straight, single-concave, and fragmented forms also occur. Cobble tools include bifacial chopping tools, unifacial choppers, hammerstones, and a pebble retoucher. Among identifiable tool blanks, 49% are made on flakes, while the rest are made on blades or blade-like flakes, microblades, and cobbles (Goebel 1990). The component II assemblage represents a comprehensive Denali complex industry, one characterized not just by the presence of wedge-shaped cores and microblades, but also burins, unhafted bifaces, lanceolate bifacial points, cobble chopping tools and choppers (called *chi-thos* in Alaska), and large side scrapers (Powers and Hoffecker 1989).

Faunal remains were poorly preserved at Dry Creek. Nonetheless, Guthrie (1983) identified teeth of three kinds of large mammal grazers: *Ovis* sp. and *Cervus* sp. (probably Dall sheep and wapiti) in component I, and *Ovis* sp. and *Bison* sp. (probably Dall sheep and steppe bison) in component II. Guthrie (1983) argued that both cultural components represented autumn–winter occupations. In this area today, high winds from the central Alaska Range are funneled through the Nenana River gorge, creating snow-free autumn and winter pastures for sheep in the Healy-Dry Creek area. From the overlook at the Dry Creek site, Paleolithic hunters could have easily spotted sheep, bison, and wapiti grazing nearby.

No archeological features were found at Dry Creek; however, artifact concentrations were recognizable and hence could be analyzed as toolkits or activity areas (Hoffecker 1983). Hoffecker (1983) delineated three such concentrations in component I and 14 in component II. Component-I concentrations are fairly homogeneous and appear to represent (i) butchering and/or hide-processing activities, and (ii) tool production. Component-II concentrations are more variable, with some representing specific activities like microblade and osseous point production, bifacial point production, butchering, or hide-working, while others represent combinations or the full-complement of these activities (Hoffecker 1983). Based on the overall structure of the site (including the lack of permanent features) and the compositions of identified activity areas, Hoffecker (1983) and Guthrie (1983: 286) concluded that the site functioned as a "hunting spike camp and processing station" during both Nenana and Denali times.

Ushki-5

The Ushki-5 site is one of several multicomponent sites located along the south shore of Ushki Lake, central Kamchatka. It is situated upon a low cape that juts out into the lake, about 200 m northwest of the famous Ushki-1 site (Figure 5.4). N. Dikov discovered Ushki-5 in 1964 and conducted test excavations there in 1974 (Dikov 1977). Among his excavations was

Figure 5.4 Photograph of Ushki-1 site taken from Ushki-5. Note Ushki Lake in foreground and Kliuchevskii volcano in background. (Photograph by T. Goebel.)

a stratigraphic control trench that cut through the heart of the site. In it Dikov found two stratigraphically separate cultural components he labeled 7 and 6, based on stratigraphic correlations with nearby Ushki-1 (Dikov 1977: 81–82; Dikov and Titov 1984: 73). From component 7, the basal cultural component, he described a stemmed bifacial point and core tablet of a small prismatic blade core. From above-lying component 6, he did not describe any artifacts (Dikov 1977). Although unable to date these components at Ushki-5, at Ushki-1 Dikov (1977) carbon-14 dated components 7 and 6 to about 14 400 and 10 000 ^{14}C BP, respectively. Dikov defined two distinct Upper Paleolithic complexes for central Kamchatka: (i) the early Ushki Paleolithic culture, characterized by small stemmed bifacial points, bifaces, chisels (or gravers), and end scrapers; and (ii) the late Ushki Paleolithic culture, characterized by wedge-shaped cores, microblades, small leaf-shaped bifacial points, and side scrapers (Dikov 1977, 1979, 1985; Dikov et al. 1983).

In 2000, a team led by M. Dikova, T. Goebel, and M. Waters returned to Ushki-5 to further investigate the late Upper Paleolithic components preserved there (Goebel et al. 2003). In a 20 m²-block excavation, they unearthed clear remains of cultural components 7 and 6. In a nearby test pit where the entire late Pleistocene-Holocene stratigraphic sequence was exposed (Goebel et al. 2003), component 7 occurred within a band of clay at a depth of nearly 250 cm below the modern surface, while component 6 occurred within a band of silty clay at a depth of around 200 cm. The two components were separated by about 30 cm of culturally sterile silt and clay (Goebel et al. 2003). Accelerator ¹carbon-14 dating of charcoal samples recovered during the 2000 excavations yielded ages of 11 110 ± 115 and 11 130 ± 50 ^{14}C BP for component 7 and 10 040 ± 75 and 10 060 ± 80 ^{14}C BP for component 6. Lithic materials from these excavations were cursorily described in Goebel et al. (2003) and are the subject of our study.

The lithic assemblage from component 7 at Ushki-5 includes 318 debitage pieces and 10 tools. Debitage is dominated by resharpening chips; however, small tertiary flakes, flake fragments, splinters, bipolar flakes, bladelets, bladelet fragments, and cortical spalls also characterize the debitage assemblage. The small sample of tools includes stemmed bifacial points, bifacial point fragments, a leaf-shaped biface, and non-diagnostic biface fragments (Goebel et al. 2003). In addition to these flake-stone artifacts, three tiny stone beads

and a bead preform were also recovered, similar to those Dikov (1977) found in component 7 at Ushki-1. Component 7 assemblages from Ushki-5 (described here) and Ushki-1 (Dikov 1977) clearly represent a non-microblade industry characterized by small bifacial points and knives and simple flake and blade tools.

The component 6 lithic assemblage at Ushki-5 consists of 601 articles, including 553 debitage pieces, 8 cores, and 38 tools. Debitage includes resharpening chips, microblades, microblade fragments, tertiary flakes, flake fragments, splinters, bladelets, bladelet fragments, blade-like flake fragments, and a cortical spall. Cores include wedge-shaped microblade cores, a subconical microblade core, and core tablets from wedge-shaped cores. Tools include retouched microblades, blades, and bladelets, burins, bifaces and biface fragments, cobble tools, side scrapers, an end scraper, bifacial point fragment, and smooth-backed knife (Goebel et al. 2003). In addition, one small stone bead was found. The component 6 assemblage from Ushki-5, like that from Ushki-1 (Dikov 1977), is a lithic industry characterized by microblade, blade, and bifacial technologies, similar to other microblade-rich late Upper Paleolithic complexes in Japan, Yakutia (Diuktai), and Alaska (Denali) (Goebel and Slobodin 1999; Mason et al. 2001).

The 2000 excavation at Ushki-5 revealed several archeological features. In component 7, two unlined hearth features were exposed (Goebel et al. 2003). These were roughly 1 m in diameter and consisted of charcoal, ash, and tiny burnt bone fragments. Lithic artifacts were dispersed around these hearths; however, the living floor itself was heavily bioturbated by rodent burrowing that appeared to be restricted to this stratigraphic layer. In component 6, Goebel et al. (2003) excavated a well-preserved semi-subterranean dwelling that was roughly 5 m in diameter and about 30 cm deep. In places it cut into the lower-lying component 7. The dwelling had a well-marked floor of charcoal and organic residue, a distinct shoulder traceable around the entire exposed perimeter of the dwelling, and a narrow "arctic-entry" passage that would have opened to the northwest. Near the center of the dwelling was a stone-lined hearth that contained evidence of multiple burning episodes, suggesting that the dwelling was occupied for an extended period of time. The hearth contained thousands of small fragments of burnt bone (Goebel et al. 2003). In sum, the permanent nature of the dwelling feature of component 6 suggests that the Ushki-5 site functioned as a habitation site. Although no recognizable dwelling features were found in component 7 at Ushki-5, Dikov (1977) reported the discovery of several dwelling features in component 7 at Ushki-1, indicating that during this time Ushki-5 also served as a habitation site.

RAW MATERIAL PROCUREMENT AND SELECTION

Our analyses focus on toolstone procurement and selection by human foragers during the late Pleistocene in Beringia. Admittedly, the current study is preliminary and data are limited; however, our hope is to take an initial step in understanding the technological organization and provisioning strategies of these humans. Dry Creek lithic data used in this paper are derived from those collected by the second author, Goebel (1990), and details on the local Dry Creek lithic landscape are based on a raw-material survey carried out by the authors in 2007. Analyses of the Ushki-5 assemblages are limited to those data collected during our 2000 field investigations; therefore numbers are small and in some cases statistically untestable. Despite these limitations, as we show below, several patterns are evident suggesting procurement and selection preferences.

Lithic variables used in our analysis are divided into two groups addressing two aspects of raw material use: procurement and selection. Variables that speak of procurement include toolstone availability, toolstone variability, presence/absence of cortex, and debitage size. Mapping availability of lithic raw materials on the landscape allows us to reconstruct the potential sphere of toolstone types available to foragers, and to consider these as a resource much as we do plant and animal foods (Blades 2001). Availability of toolstone resources clearly is a limiting factor on technology (Andrefsky 1994; Kuhn 1995). Characterizing toolstone variability by calculating actual frequencies of raw material types present in assemblages allows us to better understand the real sphere of toolstone procurement as well as relative distances traveled by foragers to acquire these resources. Frequencies in which cortex appears on different toolstone types can potentially help us infer which toolstones were locally (<5 km from sites) procured versus those that may have been procured extra-locally (≥5 km), especially in the context of our study where local toolstones are from secondary alluvial deposits. Variables addressing selection include

toolstone use in formal versus informal tool production and microblades from formally prepared cores versus flakes from informally prepared cores. In this study, formal tools include bifaces, side scrapers, end scrapers, and combination tools, while informal tools include marginally retouched blades and flakes, burins, denticulates, notches, and cobble tools. Some toolstones may have been better suited for a given set of functions, or may have been more durable or flexible and hence more likely used for formal tools manufactured in anticipation of future or long-term use (Andrefsky 1998: 213). Retouched microblades were omitted from the formal versus informal tool production variable because these artifacts do not necessarily represent a single tool, but a component of a highly formalized composite tool. Counting each microblade as a single tool, then, could artificially inflate real tool counts. As with the production of formal and informal tools, some specific toolstone types may have been more beneficial for the production and use of microblades. Therefore, retouched microblades and associated microblade technology are compared with non-microblade-related technologies to determine whether this was the case. Study of these lithic variables can, we think, lead to an increased understanding of toolstone procurement and selection behaviors of late Pleistocene foragers in Beringia, and can thus help explain documented technological differences between the microblade and non-microblade industries.

Dry Creek toolstone procurement

Toolstone availability

The lithic landscape within 2 km of the Dry Creek site consists of cobble-rich moraines, terraces, and alluvial fans dating to the Upper Pleistocene, as well as the broad, braided floodplain of Dry Creek itself, dating to the Holocene (Figure 5.3) (Wahrhaftig 1970; Ritter and Ten Brink 1986). All of these deposits are unconsolidated and poorly sorted, and composed of cobble-sized to silt-sized, well-rounded to sub-rounded clasts (Thorson and Hamilton 1977). Although about 80% of Dry Creek's bedload is made up of quartz-mica schist originating from bedrock of the northernmost ridge of the Alaska Range, the remaining 20% is characterized by well-rounded igneous and metasedimentary cobbles derived from Tertiary-aged Nenana Gravel (Thorson and Hamilton 1977). Nenana Gravel is the major bedrock formation in the Nenana valley north of Healy; it is a conglomerate of cobbles in a coarse-grained sand matrix (Wahrhaftig 1958; Ritter 1982). Lithic materials with properties amenable to controlled flaking from the Dry Creek floodplain, Nenana River terraces, and Nenana Gravel include "degraded quartzite" (Powers 1983), rhyolite, diabase, and a variety of cherts and chalcedonies. Although the dark-gray colored degraded quartzite is quite common in Dry Creek, the Nenana River, and other nearby creeks, the other raw materials are today quite rare and difficult to find.

East of the Nenana River, opposite the mouth of Dry Creek, local geomorphology is similar, but local lithology is very different. Lignite Creek, which empties into the river about 1.5 km north of the mouth of Dry Creek, flows through the same Nenana Gravel and Quaternary terrace formations as Dry Creek, has a bedload also consisting of quartz, quartzite, chert, and argillite originating from the Tertiary-aged Suntrana and Lignite Creek formations, the primary bedrock of the high foothills of the north Alaska Range east of the Nenana River (Wahrhaftig 1958). Our raw-material survey of Lignite Creek yielded many nodules of very hard but knappable quartzites, but surprisingly no knappable cherts or argillites. Panguingue Creek, located about 5 km north of Dry Creek, contains numerous cobble-sized clasts of knappable black chert. These can be found in the creek's floodplain, as well as in outwash-terrace alluvium of the Healy and Riley Creek glaciations.

Obsidian occurs in the Dry Creek component II assemblage; however, we do not think that this rock type was locally available. No natural occurrences of obsidian are known from the Dry Creek vicinity, but a small source does occur in the upper Teklanika River drainage in nearby Denali National Park, about 45 km southwest of Dry Creek. After this, the nearest sources of this valuable toolstone are Batza Tena and Wrangell Mountain, 300 km northwest and 350 km southeast of Dry Creek, respectively (Clark 1972; Cook 1995).

Thus, knappable materials locally available to the Upper Paleolithic occupants of the Dry Creek site include cobbles of cryptocrystalline silicates (CSS; cherts and chalcedonies) and dark-gray degraded quartzite, which are common, and rhyolite, diabase, and argillite, which are rare. These toolstones would have been retrievable from the surface of the braided floodplain of Dry Creek and nearby Nenana River, as well as from exposures of older glacio-fluvial deposits along the river and its side-valley streams.

Toolstone variability

The Dry Creek component I (DC I) assemblage analyzed here consists of 4491 lithic artifacts, including 48 (1.1%) manuports, 21 (0.5%) cores, 4366 (97.2%) pieces of debitage, and 56 (1.2%) bifacial and unifacial tools (Table 5.1). Artifact classes representing primary reduction, such as cores, cortical spalls, non-diagnostic debitage (i.e., shatter and flake fragments), flakes, and blades represent the overwhelming majority of artifacts (96%) in the DC I assemblage. Classes representing secondary reduction, such as resharpening chips and tools, are much less frequent (4%). Toolstone types utilized in DC I consist of degraded quartzite, nine varieties of CCS (black, gray, brown, green, tan, ferruginous, chalcedony, agate, and jasper), three varieties of fine-grained volcanics (FGV) (rhyolite, basalt, and dacite), and quartzite. Degraded quartzite and CCS clearly dominate the assemblage (about 95% of all artifacts are made of these two materials). Typical quartzite and FGV artifacts are relatively rare.

Comparing artifact class with toolstone type in DC I (Table 5.1), it is clear that nearly every artifact class has some elements made on degraded quartzite, brown CCS, rhyolite, and typical quartzite. Manuports and cores are made on six types of toolstones, and primary-reduction debitage pieces are made on 10 types of toolstones. Artifacts representing secondary reduction activities are characterized by a lot of the same raw materials; however, jasper CCS and basalt occur only in primary debitage, while green and agate CCS and dacite occur only in secondary debitage (Table 5.1).

The Dry Creek component II (DC II) assemblage analyzed here numbers 14 454 lithic artifacts, including 15 (0.1%) manuports, 126 (0.9%) cores, 14 483 (96.8%) pieces of debitage, and 330 (2.2%) bifacial and unifacial tools (Table 5.2). As with DC I, the majority of DC II artifacts are the result of primary reduction (with 90.5% of the assemblage consisting of cobbles, cores, and associated detached pieces), and only 9.5% of the assemblage being the result of secondary reduction (resharpening chips and tools). Unlike DC I, however, DC II contains microblade cores and associated microblades that represent a more formal level of tool production than the simple flake and blade core technologies of DC I. Toolstones utilized as artifacts include degraded quartzite, eight varieties of CCS (black, gray, brown, green, tan, chalcedony, jasper, and agate), and four varieties of FGV (rhyolite, diabase, basalt, and dacite), obsidian, quartzite, and argillite. Nearly 78%

Table 5.1 Dry Creek component I artifact class by toolstone type.

Toolstones					Artifact class				
	Blades	Flakes	Chips	Cortical Spalls	Shatter	Manuports	Cores	Tools	Total
Degraded quartzite	17	312	23	33	2433	0	6	9	2833 (63.1%)
CCS									1418 (31.6%)
Black	1	13	32	1	52	0	0	12	111 (2.5%)
Gray	4	38	31	13	192	0	0	12	290 (6.5%)
Brown	4	165	34	19	615	5	5	0	847 (18.9%)
Green	0	0	2	0	1	0	0	2	5 (0.1%)
Tan	0	1	0	3	8	0	0	2	14 (0.3%)
Ferruginous	2	30	4	2	80	0	0	9	127 (2.8%)
Chalcedony	0	0	0	0	12	1	1	4	18 (0.4%)
Jasper	0	0	0	0	0	2	3	0	5 (0.1%)
Agate	0	0	0	0	0	0	0	1	1 (0.02%)
FGV									89 (2%)
Rhyolite	0	13	8	2	55	2	1	2	83 (1.9%)
Basalt	0	1	0	0	3	1	0	0	5 (0.1%)
Dacite	0	0	0	0	0	0	0	1	1 (0.02%)
Quartzite	7	25	1	7	67	36	5	0	149 (3.3%)
Unidentifiable	0	0	0	0	0	0	0	2	2 (0.04%)
Total	35 (0.8%)	598 (13.3%)	135 (3.0%)	80 (1.8%)	3518 (78.3%)	48 (1.1%)	21 (0.5%)	56 (1.1%)	4491 (100.0%)

Table 5.2 Dry Creek component II artifact class by toolstone type.

Toolstones	Artifact class									
	Blades	Micro-blades	Flakes	Chips	Cortical Spalls	Shatter	Manuports	Cores	Tools	Total
Degraded quartzite	42	49	921	248	57	3796	0	1	25	5 139 (34.8%)
CCS										6357 (43.1%)
Black	3	9	36	36	5	232	0	1	21	343 (2.3%)
Gray	21	649	298	233	23	1112	0	33	70	2439 (16.5%)
Brown	7	18	25	26	4	73	0	3	39	195 (1.3%)
Green	0	0	2	1	0	1	0	3	0	7 (0.05%)
Tan	5	7	84	70	2	108	0	1	16	293 (2.0%)
Chalcedony	10	461	403	252	12	1774	0	50	54	3016 (20.4%)
Jasper	0	24	7	1	0	23	0	4	3	62 (0.4%)
Agate	0	0	0	0	0	1	0	1	0	2 (0.01%)
FGV										2578 (17.5%)
Rhyolite	5	274	201	145	16	1340	0	18	54	2053 (13.9%)
Diabase	9	9	89	31	2	323	0	1	3	467 (3.2%)
Basalt	3	2	15	0	0	29	0	1	5	55 (0.4%)
Dacite	0	0	0	0	0	2	0	0	1	3 (0.02%)
Obsidian	0	71	16	20	2	268	0	2	19	398 (2.7%)
Other										282 (1.9%)
Quartzite	5	1	38	5	9	160	15	7	19	259 (1.8%)
Argillite	0	13	3	2	0	4	0	0	1	23 (0.2%)
Total	110 (0.7%)	1587 (10.8%)	2138 (14.4%)	1070 (7.3%)	132 (0.9%)	9246 (62.7%)	15 (0.1%)	126 (0.9%)	330 (2.2%)	14754 (100.0%)

of the DCII assemblage is on degraded quartzite or CCS. FGV artifacts are also common (17.5%), while obsidian artifacts (2.7%) and other toolstones (1.9%) including quartzite and argillite are rare.

By comparing DCII artifact classes with toolstone types (Table 5.2), we see that degraded quartzite, black, gray, brown, tan and chalcedony CCS, rhyolite, diabase, obsidian, and quartzite are represented in nearly every artifact class. Although manuports are represented only by quartzite, cores are represented by 14 varieties of toolstone. Primary-reduction debitage pieces are manufactured on all of these same raw material types as well as on dacite but not argillite. Artifacts of secondary reduction activities are also represented by most of the same toolstones, with the exception of agate CCS occurring only in the form of a piece of shatter and flake core (Table 5.2).

Thus, DC I and DC II share a few similarities in terms of toolstone variability. In both assemblages, green CCS, agate CCS, and dacite are present in such low frequencies that they may represent relatively extra-local toolstones (Tables 5.1 and 5.2). A couple of other toolstone types, such as jasper CCS and basalt, also occur in low frequencies; however, these are represented by manuports and other primary reduction pieces in DC I and by both primary and secondary debitage pieces in DC II. Likely, these toolstones were locally procured.

There are also some differences in toolstone variability between the DC I and DC II assemblages. Although both assemblages share 13 of 17 toolstone types, ferruginous CCS appears only in DC I, while argillite, obsidian, and diabase appear only in DC II. The DC I assemblage also has some raw materials that were used as tools (e.g., green and agate CCS and dacite) that are not present in the primary-reduction debitage assemblage. These toolstones may represent relatively extra-local toolstones carried to the site as finished tools (all of them are end scrapers). Nearly all of the DC II toolstones were subjected to both primary and secondary reduction activities; all finished tools appear to have related cores and/or debitage (although as shown below some of the obsidian came from sources hundreds of km distant).

Differences in toolstone variability between DC I and DC II can be further seen in the diversity of frequently used toolstone types. The most frequently used types in the DC I assemblage include degraded quartzite and brown, gray, and ferruginous CCS, whereas in DC II frequently used toolstone types include degraded quartzite, chalcedony and gray CCS, rhyolite, and diabase.

While FGV types are more common in DC II, none of the FGV types are present in high frequencies in DC I. Chi-square analysis of toolstones by assemblage further underscores this pattern. Degraded quartzite appears more frequently than expected in DC I (63.1%) and less frequently than expected in DC II (34.8%). In contrast, CCS, FGV, and obsidian occur more frequently than expected in DC II (63.3%) and less frequently than expected in DC I (33.6%) (χ^2 value of 1524.441, 4 df, $p < 0.001$) (Tables 5.1 and 5.2). These results suggest that preferred toolstones are less diverse in DC I than in DC II. These data may suggest that DC I foragers were more discriminating users of a smaller set of toolstones than DC II foragers, or that DC I foragers were not as knowledgeable of the lithic landscape as DC II foragers. This is further explored below.

Presence of cortex and debitage size

In DC I cortex is present on 3.2% of the assemblage (Table 5.3). There is less degraded quartzite with cortex than expected and more CCS, FGV, and typical quartzite with cortex than expected (Table 5.3). These statistical results, though, should be treated cautiously since 25% of the cells fall below 5 in expected counts. Likely, they mean that degraded quartzite nodules were being worked more intensively than those of other toolstones. In DC II cortex is present on about 1.1% of the assemblage. In contrast to DC I, there were more cortical artifacts than expected on degraded quartzite and less cortical artifacts than expected on CCS, FGV, and obsidian (Table 5.3). These results suggest that in DC II some of the CCS, FVG, and obsidian toolstones were brought to the site from non-local sources in the form of finished or near-finished tools. In fact, 20 obsidian samples were sourced by J. Cook (pers. comm., 2004). Known sources include Batza Tena and Wrangell Mountain, which are located more than 300 km from Dry Creek, so that some of the obsidian was definitely procured from extremely distant sources and is therefore exotic. More than half of the obsidian artifacts sourced, however, came from unknown sources. Given the presence of cortex on several obsidian artifacts, perhaps some of the obsidian was procured in the Alaska Range relatively close to Dry Creek.

We also examined toolstone class by debitage size (Table 5.4). In DC I there seems to be a significantly higher than expected presence of large flakes made on degraded quartzite, and a higher than expected number of tiny resharpening chips made on CCS and FGV.

Table 5.3 Dry Creek presence of cortex by toolstone.

DC I[a]		Toolstone					
		Degraded quartzite	CCS	FGV	Obsidian	Quartzite	Total
Cortex	Count	34	53	6	0	49	142
	Expected count	90.4	44.1	2.8	0.0	4.8	142.0
	Percentage of total	0.8%	1.2%	0.1%	0.0%	1.1%	3.2%
No cortex	Count	2785	1321	80	0	100	4286
	Expected count	2728.6	1329.9	83.2	0.0	144.2	4286.0
	Percentage of total	62.9%	29.8%	1.8%	0.0%	2.3%	96.8%
Total	Count	2819	1374	86	0	149	4428
	Expected count	2819.0	1374.0	86.0	0.0	149.0	4428.0
	Percentage of total	63.7%	31.0%	1.9%	0.0%	3.4%	100.0%
DC II[b]		**Degraded quartzite**	**CCS**	**FGV**	**Obsidian**	**Other**	**Total**
Cortex	Count	57	50	19	2	29	157
	Expected count	56.1	66.5	27.4	4.1	2.9	157.0
	Percentage of total	0.4%	0.3%	0.1%	0.0%	0.2%	1.1%
No cortex	Count	5056	6012	2477	375	231	14151
	Expected count	5056.9	5995.5	2468.6	372.9	257.1	14151.0
	Percentage of total	35.3%	42.0%	17.3%	2.6%	1.6%	98.9%
Total	Count	5113	6062	2496	377	260	14308
	Expected count	5113.0	6062.0	2496.0	377.0	260.0	14308.0
	Percentage of total	35.7%	42.4%	17.4%	2.6%	1.8%	100.0%

[a]Chi-Square Test: Value 464.990[a], 3 df, $p < 0.001$. Note: 2 cells (25.0%) have an expected count less than 5. Minimum expected count is 2.76.
[b]Chi-Square Test: Value 250.169[a], 4 df, $p < 0.001$. Note: 2 cells (20.0%) have an expected count less than 5. Minimum expected count is 2.85.

These results suggest that most toolstones procured during the DC I occupation were being used for both primary and secondary reduction activities, but CCS and FGV tools may have been worked more intensively than degraded quartzite tools. Again, nearly all of these toolstones were probably procured locally, possibly from Dry Creek or Panguingue Creek alluvium or adjacent glacio-fluvial terrace deposits (as noted above, degraded quartzite and CCS are available today in cobble form along the creeks' stream beds). In DC II there is also a significant relationship between toolstone and debitage size. Specifically, there are more large debitage pieces made on degraded quartzite, quartzite, and argillite than expected, and more tiny resharpening chips made on CCS than expected, suggesting that degraded quartzite, quartzite, and argillite were primarily reduced on site, while other toolstones (especially some obsidian and CCS) were brought to the site in finished tool forms and resharpened there.

Dry Creek toolstone selection

Formal versus informal tool production

On investigating toolstone selection, no identifiable pattern is present in DC I (Table 5.5). All three classes of tools (bifaces, formal unifaces, and informal unifaces) were predominantly made on CCS (77.8%). The sample size was too small for statistical analysis, though it is clear that during the DC I occupation, selection was directed toward the use of CCS with no real preference of this toolstone in the manufacture of formal or informal tools. Degraded quartzite (16.7%) was predominantly used to manufacture unifacial tools (both formal and informal), while FGV was used only in the manufacture of formal unifaces.

Tool production in DC II is more directed, with degraded quartzite, FGV, and CCS being the toolstones of choice for manufacture and maintenance of

Table 5.4   Dry Creek debitage size by toolstone.

DC I[a]		Toolstone					
		Degraded quartzite	**CCS**	**FGV**	**Obsidian**	**Quartzite**	**Total**
Very small	Count	23	103	8	0	1	135
(<1 cm²)	Expected count	87.6	41.9	2.6	0.0	3.0	135.0
	Percentage of total	0.5%	2.4%	0.2%	0.0%	0.0%	3.2%
Small	Count	2645	1190	70	0	88	3993
(1–3 cm²)	Expected count	2595.1	1232.1	76.3	0.0	89.5	3993.0
	Percentage of total	62.4%	28.1%	1.7%	0.0%	2.1%	94.2%
Medium-large	Count	87	15	3	0	6	111
(>3 cm²)	Expected count	72.1	34.3	2.1	0.0	2.5	111.0
	Percentage of total	2.1%	0.4%	0.1%	0.0%	0.1%	2.6%
Total	Count	2755	1308	81	0	95	4239
	Expected count	2755.0	1308.0	81.0	0.0	95.0	4239.0
	Percentage of total	65.0%	30.9%	1.9%	0.0%	2.2%	100.0%
DC II[b]		**Degraded quartzite**	**CCS**	**FGV**	**Obsidian**	**Other**	**Total**
Very small	Count	248	619	176	20	7	1070
(<1 cm²)	Expected count	426.9	411.8	186.4	26.0	18.8	1070.0
	Percentage of total	2.0%	4.9%	1.4%	0.2%	0.1%	8.5%
Small	Count	4624	4207	1985	286	180	11 282
(1–3 cm²)	Expected count	4501.7	4342.1	1965.8	274.3	198.1	11 282.0
	Percentage of total	36.7%	33.4%	15.8%	2.3%	1.4%	89.6%
Medium-large	Count	150	18	32	0	34	234
(>3 cm²)	Expected count	93.4	90.1	40.8	5.7	4.1	234.0
	Percentage of total	1.2%	0.1%	0.3%	0.0%	0.3%	1.9%
Total	Count	5022	4844	2193	306	221	12 586
	Expected count	5022.0	4844.0	2193.0	306.0	221.0	12 586.0
	Percentage of total	39.9%	38.5%	17.4%	2.4%	1.8%	100.0%

[a]Chi-Square Test: Value 172.996[a], 6 *df*, *p* < 0.001. Note: 4 cells (33.3%) have an expected count less than 5. Minimum expected count is 2.12.
[b]Chi-Square Test: Value 515.515[a], 8 *df*, *p* < 0.001. Note: 1 cell (6.7%) has an expected count less than 5. Minimum expected count is 4.11.

formal tools (Table 5.5). Formal unifacial tools were most commonly manufactured on CCS, while formal bifaces tended to be manufactured on degraded quartzite and FGV. CCS, obsidian, and other toolstones were selected for the manufacture of expedient, informal tools. Clearly, in DC II there is a recognizable pattern in toolstones selected for formal versus informal tool production. Formal unifaces, such as scrapers, were selectively being manufactured on durable CCS, while bifaces were being manufactured on degraded quartzite and FGV toolstones.

Production of microblades

Considering toolstone by technological industry in DC II, a comparison of microblade-related, burin-related, and flake- and blade-related materials was made (Table 5.6). Microblades were manufactured more frequently than expected on CCS and obsidian, burins were manufactured more frequently than expected on CCS, and blades and flakes were manufactured more frequently than expected on degraded quartzite and other toolstones. These data show that local and extra-local, durable, high-quality toolstones were being

Table 5.5 Dry Creek formal versus informal tool production

DC I[a]		Toolstone					
		Degraded quartzite	CCS	FGV	Obsidian	Quartzite	Total
Bifaces	Count	0	7	0	0	0	7
	Expected count	1.2	5.4	0.4	0.0	0.0	7.0
	Percentage of total	0.0%	13.0%	0.0%	0.0%	0.0%	13.0%
Formal unifaces	Count	3	16	3	0	0	22
	Expected count	3.7	17.1	1.2	0.0	0.0	22.0
	Percentage of total	5.6%	29.6%	5.6%	0.0%	0.0%	40.7%
Informal unifaces	Count	6	19	0	0	0	25
	Expected count	4.2	19.4	1.4	0.0	0.0	25.0
	Percentage of total	13.2%	35.9%	0.0%	0.0%	0.0%	46.3%
Total	Count	9	42	3	0	0	54
	Expected count	9.0	42.0	3.0	0.0	0.0	54.0
	Percentage of total	16.7%	77.8%	5.6%	0.0%	0.0%	100.0%
DC II[b]		**Degraded quartzite**	**CCS**	**FGV**	**Obsidian**	**Other**	**Total**
Bifaces	Count	16	25	26	2	2	71
	Expected count	6.0	43.7	13.5	2.9	4.8	71.0
	Percentage of total	5.4%	8.5%	8.8%	0.7%	0.7%	24.1%
Formal unifaces	Count	3	83	17	4	4	111
	Expected count	9.4	68.3	21.1	4.5	7.6	111.0
	Percentage of total	1.0%	28.2%	5.8%	1.4%	1.4%	37.8%
Informal unifaces	Count	6	73	13	6	14	112
	Expected count	9.5	69.0	21.3	4.6	7.6	112.0
	Percentage of total	2.0%	24.8%	4.4%	2.0%	4.8%	38.1%
Total	Count	25	181	56	12	20	294
	Expected count	25.0	181.0	56.0	12.0	20.0	294.0
	Percentage of total	8.5%	61.6%	19.0%	4.1%	6.8%	100.0%

[a]Sample size too small for statistical analysis.
[b]Chi-Square Test: Value 58.565[a], 8 *df*, *p* < 0.001. Note: 4 cells (26.7%) have an expected count less than 5. Minimum expected count is 2.90.

selected for the production of microblades and burins. Again, data suggest during DC II times toolstone selection was patterned, unlike during DC I times.

Ushki-5 toolstone procurement

Toolstone availability

Our knowledge of the lithic landscape of Ushki-5 is based on several important references (Dikov and Titov 1984; Ivanov 1990), as well as observations of the local geology made while excavating there in 2000 (Goebel *et al.* 2003). Ushki Lake is situated along the Kamchatka River, the major watercourse draining the Central Kamchatka Depression. This tectonic depression is flanked on the east by the Kliuchevskii volcano group and on the west by the Sredinnyi mountain range. Bedrock in the vicinity of Ushki Lake is mainly volcanic and composed primarily of andesites and basalts (Dikov and Titov 1984); an outcropping of this bedrock occurs at the Ushki-1 site along the south shore of the lake (Figure 5.4). Other than these isolated volcanic bedrock exposures, surface geology is characterized by an unconsolidated mantle of fine-grained alluvial (sand, silt, and clay) and aeolian (loess and tephra) deposits reaching more than 7 m thick in

Table 5.6 Dry Creek component II technology types by toolstone.

		Toolstone					
		Degraded quartzite	CCS	FGV	Obsidian	Other	Total
Microblade-related	Count	49	1265	310	80	14	1718
materials	Expected count	598.4	740.2	300.2	46.3	32.8	1718.0
	Percentage of total	0.3%	8.6%	2.1%	0.6%	0.1%	11.7%
Burin-related	Count	0	75	5	4	0	84
materials	Expected count	29.3	36.2	14.7	2.3	1.6	84.0
	Percentage of total	0.0%	0.5%	0.0%	0.0%	0.0%	0.5%
Flake- and blade-	Count	5090	5017	2263	314	268	12952
related materials	Expected count	4511.3	5580.6	2263.1	349.4	247.6	12952.0
	Percentage of total	34.5%	34.0%	15.3%	2.1%	1.8%	87.8%
Total	Count	5139	6357	2578	398	282	14754
	Expected count	5139.0	6357.0	2578.0	398.0	282.0	14754.0
	Percentage of total	34.8%	43.1%	17.5%	2.7%	1.9%	100.0%

Chi-Square Test: Value 1128.602[a], 8 *df*, *p* < 0.001.

Note: 2 cells (13.3%) have an expected count less than 5. Minimum expected count is 1.61.

some exposures (Goebel *et al.* 2003). Cobble beds are exceedingly rare, even in modern point bars exposed along the Kamchatka River near Ushki Lake. Even when cobble deposits can be found, they do not appear to contain knappable materials. Our impression of the Ushki Lake vicinity is that it is a "toolstone desert", one largely devoid of lithic materials with properties amenable to controlled flaking. Our observations, however, are limited to the eastern, Kliuchevskii side of the Kamchatka River, and we know nothing about the lithic resources potentially available on the western, Sredinnyi side.

Toolstone variability

The Ushki-5 component 7 assemblage described here consists of 327 lithic artifacts, including 0 (0%) cores, 318 (97.2%) debitage pieces, and 9 (2.8%) bifacial and unifacial tools (Table 5.7). Artifact classes representing primary reduction, such as shatter, flakes, and blades characterize less than a quarter of the artifacts (23.2%). Classes representing secondary reduction, such as resharpening chips and tools, occur much more frequently (76.8%). Toolstone types utilized in this assemblage are numerous and consist of obsidian, 11 varieties of CCS (black, gray, brown, green, tan, white, jasper, and brown, gray, tan, and white chalcedony),

and basalt. CCS dominates the assemblage, making up approximately 94.2% of all toolstones utilized, while basalt (3.7%) and obsidian (2.1%) are much less common.

Comparing artifacts with toolstones in component 7, gray, green, and tan chalcedony CCS is represented by nearly every class of artifact. Debitage pieces characterizing primary reduction activities are manufactured predominantly on green, white, and tan chalcedony CCS, with basalt and jasper, white, brown chalcedony, tan and gray CCS appearing in lower frequencies. In contrast, gray chalcedony, black, and brown CCS, and obsidian are either not represented by these artifacts or they occur in very low frequencies. Artifacts associated with secondary reduction activities are characterized by the same toolstones, with the exception of brown chalcedony and tan CCS, which are absent, and gray chalcedony CCS, which only occurs as a single resharpening chip.

The Ushki-5, component 6 assemblage contains 621 lithic artifacts, including 8 (1.3%) cores, 574 (92.4%) pieces of debitage, and 39 (6.3%) tools (Table 5.8). Unlike component 7, most of the component 6 assemblage is the result of primary reduction: 57.3% of the assemblage consists of cores and associated detached pieces, while 42.7% consists of resharpening chips and tools resulting from secondary reduction activities.

Table 5.7   Ushki-5, component 7 artifact classes by toolstone type.

Toolstones	Artifact class						
	Blades	**Flakes**	**Chips**	**Cortical spalls**	**Angular shatter**	**Tools**	**Total**
Obsidian	0	3	4	0	0	0	7 (2.2%)
CCS							308 (94.2%)
Black	0	1	2	0	0	0	3 (0.9%)
Gray	1	15	136	0	5	3	160 (48.9%)
Brown	0	3	14	0	0	2	19 (5.8%)
Green	0	6	30	1	1	1	39 (11.9%)
Tan	0	1	0	0	1	0	2 (0.6%)
White	0	0	1	1	0	0	2 (0.6%)
Brown Chalcedony	0	1	0	0	0	0	1 (0.3%)
Gray Chalcedony	0	0	1	0	0	0	1 (0.3%)
Tan Chalcedony	1	22	40	1	0	1	65 (19.9%)
White Chalcedony	0	7	3	0	1	0	11 (3.4%)
Jasper	0	1	4	0	0	0	5 (1.5%)
Basalt	1	2	7	0	0	2	12 (3.7%)
Total	3 (0.9%)	62 (19.0%)	242 (74.0%)	3 (0.9%)	8 (2.4%)	9 (2.8%)	327 (100.0%)

Another difference is that component 6 contains microblade cores and microblade-related debitage. Toolstones include obsidian, 10 varieties of CCS (gray, brown, green, tan, white, brown chalcedony, gray chalcedony, green chalcedony, tan chalcedony, and white chalcedony), and basalt. CCS artifacts (72.1%) dominate the assemblage, while basalt (16.9%) and obsidian (10.8%) are less common.

Comparing artifact forms with toolstones in the component 6 assemblage, gray CCS and basalt are represented by nearly every artifact class (Table 5.8). Cores are represented by obsidian, gray CCS, and basalt. Primary reduction debitage is predominantly manufactured on gray and green CCS and basalt, and white CCS is the only toolstone not represented by these artifact classes. Secondary reduction debitage is dominated by gray and green CCS, obsidian, and basalt. Brown chalcedony CCS is the only toolstone not represented by secondary pieces.

Components 7 and 6 share a couple of similarities in toolstone variability. In both assemblages, tan, white, brown-chalcedony and gray-chalcedony CCS are present in such low frequencies that these may represent non-local toolstones (Tables 5.7 and 5.8). For both assemblages all of the materials occurring as tools are

also represented by both primary and secondary reduction activities. Though most of the toolstones in both assemblages are the same, black and jasper CCS occur only in component 7, whereas green-chalcedony CCS occurs only in component 6.

The range of toolstone variability in components 7 and 6 can further be seen in the diversity of frequently occurring toolstones. In component 7 these include gray, green and tan-chalcedony CCS. In component 6 these include gray CCS, basalt, and obsidian. Chi-square analysis of toolstones by assemblage further underscores this pattern: CCS appears more frequently than expected in component 7 (94.2%) and less frequently than expected in component 6 (72.1%). In contrast, obsidian and basalt occur more frequently than expected in component 6 (27.9%) and less frequently than expected in component 7 (5.9%) (χ^2 value of 64.487, 2 df, $p < 0.001$) (Tables 5.7 and 5.8). These results suggest that frequently occurring toolstones are less diverse in non-microblade component 7 than in microblade-rich component 6. These data may suggest that component 7 foragers were either more discriminating in their use of certain raw materials, or that they were not as knowledgeable of the lithic landscape as component 6 foragers.

Table 5.8 Ushki-5, component 6 artifact classes by toolstone type.

Toolstones	Artifact class								
	Blades	Microblades	Flakes	Chips	Cortical spalls	Angular shatter	Cores	Tools	Total
Obsidian	0	35	10	8	0	0	4	11	68 (11%)
CCS									448 (72.1%)
Gray	8	157	26	99	1	2	3	11	307 (49.4%)
Brown	0	3	4	11	0	4	0	0	22 (3.5%)
Green	1	6	11	30	0	0	0	4	52 (8.4%)
Tan	0	0	2	2	0	0	0	2	6 (1.0%)
White	0	0	0	2	0	0	0	0	2 (0.3%)
Brown Chalcedony	0	0	2	0	0	0	0	0	2 (0.3%)
Gray Chalcedony	0	0	2	1	0	0	0	1	4 (0.8%)
Green Chalcedony	0	0	2	1	0	0	0	0	3 (0.5%)
Tan Chalcedony	0	0	7	18	0	0	0	1	26 (4.2%)
White Chalcedony	0	0	8	13	0	0	0	3	24 (3.9%)
Basalt	3	0	50	41	0	4	1	6	105 (16.9%)
Total	12 (1.9%)	201 (32.4%)	124 (20.0%)	226 (36.4%)	1 (0.1%)	10 (1.6%)	8 (1.3%)	39 (6.3%)	621 (100.0%)

Presence of cortex and debitage size

In component 7, cortex is present on 0.6% of the assemblage. There does not appear to be a relationship between toolstone class and presence of cortex (Table 5.9), but the sample size is very small. Lack of cortex suggests that few if any raw materials were procured nearby the site; however, since all cortical pieces are of CCS, some of these toolstones may have been of local origin. We also examined toolstone class by debitage size; results suggest no relationship (Table 5.10). The preponderance of tiny resharpening chips and small flakes suggests that most toolstones were not being procured locally and that secondary reduction dominated reduction activities.

In component 6 cortex is present on four artifacts, 0.7% of the total assemblage. Therefore, only negligible primary reduction activities occurred in component 6. Further, there is no apparent relationship between toolstone and presence of cortex (Table 5.9); however, as with component 7, all pieces with cortex are of CCS.

Further, when examining debitage size according to toolstone, small-sized debitage is more commonly represented by CCS and obsidian (Table 5.10), while large-sized debitage is more commonly represented by basalt. Although these data suggest that some CCS and basalt were procured locally, the differential package size between basalt and CCS and obsidian may be the result of differential package size.

Ushki-5 toolstone selection

Formal versus informal tool production

In terms of toolstone selection, the component 7 tool assemblage, which consists of just nine bifaces, is too small to characterize statistically; however, of these tools seven were manufactured on CCS and two on basalt. In component 6 bifaces and informal unifaces are made on CCS, basalt, and obsidian, while formal unifaces are only made on CCS (Table 5.11). This may

Table 5.9 Ushki-5 presence of cortex by toolstone.

Component 7		Toolstone			
		Obsidian	CCS	Basalt	Total
Cortex	Count	0	3	0	3
	Expected count	0.0	2.8	0.1	3.0
	Percentage of total	0.0%	0.9%	0.0%	0.9%
No cortex	Count	7	298	10	315
	Expected count	6.9	298.2	9.9	315.0
	Percentage of total	2.2%	93.7%	3.1%	99.1%
Total	Count	7	301	10	318
	Expected count	7.0	301.0	10.0	318.0
	Percentage of total	2.2%	94.7%	3.1%	100.0%
Component 6		**Obsidian**	**CCS**	**Basalt**	**Total**
Cortex	Count	0	4	0	4
	Expected count	0.4	2.9	0.7	4.0
	Percentage of total	0.0%	0.7%	0.0%	0.7%
No cortex	Count	52	400	98	550
	Expected count	51.6	401.1	97.3	550.0
	Percentage of total	9.4%	72.2%	17.7%	99.3%
Total	Count	52	404	98	554
	Expected count	52.0	404.0	98.0	554.0
	Percentage of total	9.4%	72.9%	17.7%	100.0%

Sample sizes too small for statistical analysis.

Table 5.10 Ushki-5 debitage size by toolstone.

Component 7		Toolstone			
		Obsidian	**CCS**	**Basalt**	**Total**
Very small	Count	2	55	3	60
(<1 cm²)	Expected Count	2.8	54.4	2.8	60.0
	Percentage of total	2.4%	64.7%	3.5%	70.6%
Small	Count	2	19	1	22
(1–3 cm²)	Expected count	1.0	19.9	1.0	22.0
	Percentage of total	2.4%	22.4%	1.2%	25.9%
Medium-large	Count	0	3	0	3
(>3 cm²)	Expected count	0.1	2.7	0.1	3.0
	Percentage of total	0.0%	3.5%	0.0%	3.5%
Total	Count	4	77	4	85
	Expected count	4.0	77.0	4.0	85.0
	Percentage of total	4.7%	90.6%	4.7%	100.0%
Component 6		**Obsidian**	**CCS**	**Basalt**	**Total**
Very small	Count	2	46	9	57
(<1 cm²)	Expected count	2.0	46.1	13.9	57.0
	Percentage of total	2.3%	53.5%	10.5%	66.3%
Small	Count	1	15	8	24
(1–3 cm²)	Expected count	0.8	17.3	5.9	24.0
	Percentage of total	1.2%	17.4%	9.3%	27.9%
Medium-Large	Count	0	1	4	5
(>3 cm²)	Expected count	0.2	3.6	1.2	5.0
	Percentage of total	0.0%	1.2%	4.7%	5.8%
Total	Count	3	62	21	86
	Expected count	3.0	62.0	21.0	86.0
	Percentage of total	3.5%	72.1%	24.4%	100.0%

Sample sizes too small for statistical analysis.

Table 5.11 Ushki-5, component 6 formal versus informal tool production by toolstone.

		Toolstone			
		Obsidian	**CCS**	**Basalt**	**Total**
Bifaces	Count	1	2	3	6
	Expected count	1.6	3.1	1.3	6.0
	Percentage of total	3.7%	7.4%	11.1%	22.2%
Formal unifaces	Count	0	3	0	3
	Expected count	0.8	1.6	0.7	3.0
	Percentage of total	0.0%	11.1%	0.0%	11.1%
Informal unifaces	Count	6	9	3	18
	Expected count	4.6	9.3	4.0	18.0
	Percentage of total	22.2%	33.3%	11.1%	66.7%
Total	Count	7	14	6	27
	Expected count	7.0	14.0	6.0	27.0
	Percentage of total	25.9%	51.9%	22.2%	100.0%

Sample size too small for statistical analysis.

Table 5.12 Ushki-5, component 6 technology types by toolstone.

		Toolstone			
		Obsidian	CCS	Basalt	Total
Microblade-related materials	Count	43	177	0	220
	Expected count	23.9	159.0	37.1	220.0
	Percentage of total	7.0%	28.7%	0.0%	35.7%
Flake- and blade-related materials	Count	24	269	104	397
	Expected count	43.1	287.0	66.9	397.0
	Percentage of total	3.9%	43.6%	16.9%	64.3%
Total	Count	67	446	104	617
	Expected count	67.0	446.0	104.0	617.0
	Percentage of total	10.9%	72.3%	16.9%	100.0%

Chi-Square Test: Value 84.547, 2 df, $p < 0.001$. Note: 0 cells (0.0%) have an expected count less than 5. Minimum expected count is 23.89.

suggest CCS was intentionally selected for the manufacture of formal unifaces due to its durability.

Production of microblades

To consider toolstone by technological industry, microblade-, flake- and blade-related materials were compared. Only four artifacts resulting from burin manufacture were present in the assemblage, so these artifacts were omitted from the chi-square analysis. As shown in Table 5.12, during the component 6 occupation microblades were manufactured more frequently than expected on CCS and obsidian, while flakes and blades were manufactured on these toolstones and basalt. Basalt, however, was used for the manufacture of flakes and blades more frequently than expected. Thus, toolstone selection in the Ushki microblade complex appears to have been relatively patterned and planned.

DISCUSSION AND CONCLUSIONS

Our goal in this essay has been to determine whether differences in lithic raw material procurement and selection can be detected between non-microblade and microblade techno-complexes of Beringia, by examining in some detail the records from two Upper Paleolithic sites, Dry Creek, central Alaska and Ushki-5, central Kamchatka. These sites are unique in that they both contain two Upper Paleolithic complexes, so that

through our comparisons we can hold at least two things constant: site location and lithic landscape. Dry Creek represents a bluff-edge overlook site occurring in an environment with some locally available toolstones, while Ushki-5 represents a stream-side site occurring in a relatively toolstone-poor environment. Further, the original excavators of Dry Creek argued that both Upper Paleolithic components are the result of a series of short-term hunting occupations (probably during the fall-winter) (Guthrie 1983; Hoffecker 1983), and the original excavators of the Ushki sites argued that both components are the result of long-term habitations (Dikov 1977, 1979; Goebel et al. 2003). Obviously, Dry Creek and Ushki-5 are excellent places to investigate the non-microblade – microblade dichotomy in the Upper Paleolithic record of Beringia.

Toolstone procurement at Dry Creek. Lithic assemblages for both component I (non-microblade) and component II (microblade-rich) are dominated by local toolstones, primarily degraded quartzite and CCS. Few, if any, cores or tools can be clearly shown to have been transported in finished form from another location to the site. Just about the only difference that can be detected is that in component II FGV increases and obsidian and argillite appear for the first time. We do not think that obsidian or argillite were available along Dry Creek. Some of the obsidian came from more than 300 km from the site, while some of the argillite likely came from some other stream or terrace deposit in the Nenana Valley. Thus, toolstone procurement during the non-microblade occupation appears to have been locally oriented, while procurement during the microblade occupation appears to have been local and non-local. This may be an indication that the earliest occupants of the Dry Creek site did not know the lithic landscape of central Alaska as well as the site's later occupants, or it may reflect differences in provisioning strategies and logistical transport of toolstones. Perhaps settlement during the Nenana complex occupation of Dry Creek was more locally oriented, while settlement during the Denali complex occupation was more mobile and far-reaching.

Toolstone selection at Dry Creek. Our analyses suggest that the component I occupants of Dry Creek did not necessarily prefer any toolstone over another for production of formal or informal tools, indicating unpatterned selection of toolstones by the non-microblade occupants. The component II occupants, however, did inordinately manufacture side scrapers and burins on CCS, bifaces on degraded quartzite, and informal tools and microblades on CCS and obsidian, indicating patterned selection of toolstones by microblade-producing occupants. Possibly the early inhabitants of Dry Creek were place-oriented and had just begun to familiarize themselves with local resources, whereas later far-ranging, microblade-producing inhabitants were much more familiar with both local and non-local resources and thus able to more selectively use certain toolstones to meet specific needs

Toolstone procurement at Ushki-5. Both components 7 and 6 are dominated by what appear to be non-local toolstones, probably the result of a scarcity of knappable material in the vicinity of the site. Nevertheless, as at Dry Creek, the later microblade assemblage contains significantly more basalt and obsidian. This difference in toolstone procurement may indicate that initial, component 7 inhabitants of Ushki-5 were learning the lithic landscape, while later, component 6 inhabitants were more knowledgeable of the lithic landscape and made use of a wider array of regionally available toolstones. It is also possible that the differences reflect changes in provisioning strategies, with microblade-producing inhabitants being more mobile, effectively having more opportunities to access and transport a greater variety of non-local resources to the site.

Toolstone selection at Ushki-5. Unfortunately the tool assemblage from components 7 and 6 are really too small to say much about toolstone selection, but it is clear from our analysis that microblades were always made on CCS or obsidian and never basalt, possibly indicating a more selective behavior of component 6 inhabitants.

Conclusions. Even though raw materials were relatively plentiful at Dry Creek and relatively scarce at Ushki-5, toolstone procurement and selection was unpatterned and unplanned during the non-microblade occupations of both sites. In contrast, the results of this study suggest that toolstone procurement and selection was both patterned and planned during the microblade occupations of the two sites. Based on the results of this meager analysis, there are two, admittedly speculative, alternative explanations of the different archeological complexes of late Pleistocene Beringia. First, perhaps both assemblages represent different populations. In this case the non-microblade assemblages may represent initial inhabitants settling-in and actively learning the landscape, whereas microblade assemblages may represent later landscape experts, who had become familiar with the widespread distribution of lithic resources. Second, perhaps the two assemblages represent different ways of organizing technology that may have resulted from differing land-use strategies. In this case, the non-microblade assemblages may represent less mobile, place-oriented foragers, whereas the microblade assemblages may represent more logistically mobile, technology-oriented foragers. Finally, it is likely that both of these alternative explanations are correct. Early non-microblade complexes of Beringia may indeed represent hunter-gatherer groups who were conservatively and gradually expanding into new territories. Raw material procurement was local and technological organization was expedient, implying either infrequent residential moves or short-distance moves. As time passed, Beringia's early hunter-gatherers became more knowledgeable of the landscape, as evidenced by the presence of non-local toolstones, transport of these between sites, and preferential selection of certain toolstones for specific uses. This may imply a less conservative, far-reaching land-use strategy by microblade occupants. In this setting, microblade production may have become better suited as a technological choice.

We think our conclusions complement those offered to explain resource procurement, land-use, and landscape learning at the Broken Mammoth site in the Tanana Valley, central Alaska. Yesner *et al.* (2004)

suggest that hunter-gatherers at Broken Mammoth focused on local resources as they began to learn the local landscape. Near the end of the Late Glacial the Beringian landscape likely offered its initial inhabitants a complex set of ecological opportunities. Our study presented here, along with interpretations of Yesner *et al.* (2004), suggest that the non-microblade complexes of interior Beringia, especially Broken Mammoth, the Nenana Complex, and Ushki, represent a special adaptation during the Allerød interstadial, a period that Hoffecker and Elias (2007) argue was characterized by significant warming and the spread of the shrub-tundra biotic community. We argue that these non-microblade Beringian complexes represent a "settling-in" phase of interior Beringian colonization, as hunter-gatherers were learning the landscape and becoming more locally oriented. The early microblade occupation at Swan Point in the Tanana Valley, nearly 1000 years older than the non-microblade complexes analyzed here, may represent initial human exploration of central Alaska's unfamiliar landscapes. Further studies of the Swan Point lithic assemblage should help discern whether these first central Alaskans organized technology and procured lithic raw materials in a manner different from the later occupants of Broken Mammoth and the Nenana Complex sites. If Swan Point does indeed represent the initial exploration of the upper Tanana Valley, we predict that the technological strategies of its occupants would have been radically different from those of the later sites and would better match predictions of how initial explorers may have dispersed across unfamiliar landscapes of North America (Beaton 1991; Kelly 1996, 2003; Meltzer 2002, 2003, 2004)

Though the Dry Creek and Ushki data generally fit the model of Beringian colonization presented by Yesner *et al.* (2004), we think the non-microblade complexes of Beringia represent the settling-in phase following initial exploration and colonization. However, we stress that our findings require more rigorous testing with more detailed technological analyses of the Dry Creek, Ushki, and other early assemblages. We caution readers that our study of the Dry Creek assemblage was done nearly 20 years ago under a very different theoretical framework – one focusing on measurement of typological and technological similarities and differences to identify assemblage relationships (Goebel *et al.* 1991), not one focusing on reconstruction and explanation of raw-material provisioning and technological organization. Only through such organizational studies

will we be able to reconstruct the environmental and behavioral context of lithic assemblage variability in Paleolithic Beringia. This approach will ultimately help explain the apparent dichotomy between the non-microblade and microblade industries of Beringia and how they relate to the colonization of the New World.

REFERENCES

Andrefsky, W., Jr. (1994) Raw material availability and the organization of technology. *American Antiquity* 59, 21–35.

Andrefsky, W., Jr. (1998) *Lithics: Macroscopic Approaches to Analysis*. Cambridge University Press, Cambridge.

Beaton, J.M. (1991) Colonizing continents: Some problems for Australia and the Americas. In: Dillehay, T.D. and Meltzer, D.J. (eds) *The First Americans: Search and Research*, CRC Press, Boca Raton, pp. 209–30.

Bigelow, N. and Powers, W.R. (1994) New AMS dates from the Dry Creek Paleoindian Site, central Alaska. *Current Research in the Pleistocene* 11, 114–6.

Blades, B. (2001) *Aurignacian Lithic Economy: Ecological Perspectives from Southwestern France*. Kluwer Academic/Plenum Publishers, New York.

Clark, D. (1972) Archaeology of the Batza Tena obsidian source, west-central Alaska. *Anthropological Papers of the University of Alaska* 15(2), 1–22.

Cook, J. (1995) Characterization and distribution of obsidian in Alaska. *Arctic Anthropology* 32(1), 92–100.

Crass, B. and Holmes, C. (2004) Swan point: A case for land bridge migration in the peopling of North America. *Paper presented at 69th Annual Meeting*. Society for American Archaeology, Montreal.

Dikov, N. (1977) *Arkheologicheskie Pamiatniki Kamchatki, Chukotki i Verkhnei Kolymy*. Nauka, Moscow [in Russian].

Dikov, N. (1979) *Drevnie Kul'tury Severo-Vostochnoi Azii*. Nauka, Moscow [in Russian].

Dikov, N. (1985) The Paleolithic of northeastern Asia and its relations with the Paleolithic of America. *Inter-Nord* 17, 173–7.

Dikov, N., Brodianskii, D.L., and D'iakov, V.I. (1983) *Drevnie Kul'tury Tikhookeanskogo Poberezh'ia SSSR: Uchebnoe Posobie*. Izdatel'stvo Dal'nevostochnogo Universiteta, Vladivostok [in Russian].

Dikov, N. and Titov, E.E. (1984) Problems of the stratification and periodization of the Ushki sites. *Arctic Anthropology* 21(2), 69–80.

Goebel, T. (1990) *Early Paleoindian technology in Beringia: A lithic analysis of the Alaskan Nenana Complex*. Unpublished MA thesis, Department of Anthropology, University of Alaska, Fairbanks.

Goebel, T. and Slobodin, S.B. (1999) The colonization of western Beringia: Technology, ecology, and adaptations. In: Bonnichsen, R. and Turnmire, K. (eds) *Ice Age People of North America*, edited, Center for the Study of the First Americans, Oregon State University Press, Corvallis, pp. 104–54.

Goebel, T., Powers, W.R., and Bigelow, N. (1991) The Nenana Complex of Alaska and Clovis origins. In: Bonnichsen, R. and Turnmire, K. (eds) *Clovis Origins and Adaptations*, Center for the Study of the First Americans, Corvallis, pp. 49–79.

Goebel, T., Waters, M., and Dikova, M. (2003) The archaeology of Ushki Lake, Kamchatka, and the Pleistocene peopling of the Americas. *Science* 301, 501–5.

Guthrie, R.D. (1983) Paleoecology of the site and its implications for early humans. In: Powers, W.R., Guthrie, R.D., and Hoffecker J. (eds) *Dry Creek: Archeology and Paleoecology of a Late Pleistocene Alaskan Hunting Camp*. Unpublished report submitted to the National Park Service (Contract CX-9000-7-0047), pp. 209–87.

Hamilton, T. and Goebel, T. (1999) Late Pleistocene peopling of Alaska. In: Bonnichsen, R. and Turnmire, K. (eds) *Ice Age People of North America*, Center for the Study of the First Americans, Oregon State University Press, Corvallis, pp. 156–99.

Hoffecker, J. (1983) Human activity at the Dry Creek site: A synthesis of the artifactual, spatial, and environmental data. In: Powers, W.R., Guthrie, R.D., and Hoffecker, J. (eds) *Dry Creek: Archeology and Paleoecology of a Late Pleistocene Alaskan Hunting Camp*. Unpublished report submitted to the National Park Service (Contract CX-9000-7-0047), pp. 182–208.

Hoffecker, J. (2001) Late Pleistocene and early Holocene sites in the Nenana River valley, central Alaska. *Arctic Anthropology* 38(2), 139–53.

Hoffecker, J.F. and Elias, S.A. (2007) *Human Ecology of Beringia*. Columbia University Press, New York.

Hoffecker, J., Powers, W.R., and Bigelow, N. (1996) Dry Creek. In: West, F. (ed.) *American Beginnings: The Prehistory and Paleoecology of Beringia*, University of Chicago Press, Chicago, pp. 343–52.

Hoffecker, J., Powers, W.R., and Goebel, T. (1993) The colonization of Beringia and the peopling of the New World. *Science* 259, 46–53.

Holmes, C., VanderHoek, R., and Dilley, T. (1996) Swan Point. In: West, F. (ed.) *American Beginnings: The Prehistory and Paleoecology of Beringia*, University of Chicago Press, Chicago, pp. 319–23.

Ivanov, V. (1990) Problemy geomorfologii i Chetvertichnoi geologii v raione stoianki Ushki (dolina reki Kamchatki). In: *Drevnie Pamiatniki Severa Dal'nego Vostoka*, Akademiia Nauk SSSR, Dal'nevostochnoe Otdelenie, Severo-Vostochnyi Kompleksnyi Nauchno-Issledovatel'skii Institut, Magadan, pp. 161–70 [in Russian].

Kelly, R.L. (1996) Ethnographic Analogy and Migration to the Western Hemisphere. In: Akazawa, T. and Szathmary, E. (eds) *Prehistoric Mongoloid Dispersals*, Oxford University Press, Oxford, pp. 228–40.

Kelly, R.L. (2003) Colonization of new land by Hunter-Gatherers: Expectations and implications based on ethnographic Data. In: Rockman, M. and Steele J. (eds) *Colonization of Unfamiliar Landscapes: The Archaeology of Adaptation*, Routledge, New York, pp. 44–58.

Kuhn, S. (1995) *Mousterian Lithic Technology: An Ecological Perspective*. Princeton University Press, Princeton.

Kunz, M. and Reanier, R. (1994) Paleoindians in Beringia: Evidence from arctic Alaska. *Science* 263, 660–2.

Mason, O., Bowers, P., and Hopkins, D. (2001) The early Holocene Milankovitch thermal maximum and humans: Adverse conditions for the Denali Complex of eastern Beringia. *Quaternary Science Reviews* 20(1–3), 525–48.

Meltzer, D.J. (2002) What do you do when no one's been there before? Thoughts on the exploration and colonization of new lands. In Jablonski, N. (eds) *The First Americans: The Pleistocene Colonization of New Lands*, California Academy of Sciences, San Francisco, pp. 27–58.

Meltzer, D.J. (2003) Lessons in landscape learning. In: Rockman, M. and Steele, J. (eds) *Colonization of Unfamiliar Landscapes: The Archaeology of Adaptation*, Routledge, New York, pp. 222–41.

Meltzer, D.J. (2004) Modeling the initial colonization of the Americas: Issues of scale, demography, and landscape learning. In: Barton, C.M., Clark, G., Yesner, D., and Pearson, G. (eds) *The Settlement of the American Continents*, University of Arizona Press, Tucson, pp. 123–37.

Powers, W.R. (1983) Lithic technology of the Dry Creek site. In: Powers, W.R., Guthrie, R.D., and Hoffecker, J. (eds) *Dry Creek: Archeology and Paleoecology of a Late Pleistocene Alaskan Hunting Camp*. Unpublished report submitted to the National Park Service (Contract CX-9000-7-0047), pp. 62–181.

Powers, W.R. and Hamilton, T. (1978) Dry Creek: A late Pleistocene human occupation in central Alaska. In: Bryan, A. (ed.) *Early Man in America from a Circum-Pacific Perspective*, Archaeological Researches International, Edmonton, pp. 72–7.

Powers, W.R. and Hoffecker, J. (1989) Late Pleistocene settlement in the Nenana valley, central Alaska. *American Antiquity* 54(2), 263–87.

Powers, W.R., Guthrie, R.D., and Hoffecker, J. (1983) *Dry Creek: Archeology and Paleoecology of a Late Pleistocene Alaskan Hunting Camp*. Unpublished report submitted to the National Park Service (Contract CX-9000-7-0047).

Ritter, D. (1982) Complex river terrace development in the Nenana valley near Healy, Alaska. *Geological Society of America Bulletin* 93, 346–56.

Ritter, D. and Ten Brink, N. (1986) Alluvial fan development and the glacial-glaciofluvial cycle, Nenana valley, Alaska. *Journal of Geology* 94, 613–25.

Thorson, R. and Hamilton, T. (1977) Geology of the Dry Creek site: a stratified early man site in Interior Alaska. *Quaternary Research* 7, 149–76.

Wahrhaftig, C. (1958) Quaternary geology of the Nenana River valley and adjacent parts of the Alaska Range. In: *Quaternary and Engineering Geology in the Central Part of the Alaska Range*, U.S. Geological Survey Professional Paper 293, United States Government Printing Office, Washington DC, pp. 1–68.

Wahrhaftig, C. (1970) *Geologic Map of the Healy D-5 Quadrangle, Alaska*. Map GQ-807, US Geological Survey, Department of the Interior, Washington DC.

West, F. (1981) *The Archaeology of Beringia*. Columbia University Press, New York.

West, F. (1983) The antiquity of man in America. In: Wright, H., Jr. (ed.) *Late-Quaternary Environments of the United States*, Vol. 1, *The Late Pleistocene*, University of Minnesota Press, Minneapolis, pp. 374–82.

West, F. (1996) The archaeological evidence. In: West, F. (ed.) *American Beginnings: The Prehistory and Paleoecology of Beringia*. University of Chicago Press, Chicago, pp. 537–59.

Yesner, D., Barton, C.M., Clark, G., and Pearson G. (2004) Peopling of the Americas and continental colonization: A millennial perspective. In: Barton, C.M., Clark, G., Yesner, D., and Pearson, G. (eds) *The Settlement of the American Continents*, University of Arizona Press, Tucson, pp. 196–213.

Chapter 6

REDUCTION, RECYCLING, AND RAW MATERIAL PROCUREMENT IN WESTERN ARNHEM LAND, AUSTRALIA

Peter Hiscock

School of Archaeology and Anthropology, Australian National University

ABSTRACT

Complex lithic assemblage variation in Arnhem Land, Australia, was initially explained by suggestions that multiple cultural groups had co-existed or that a single group had used different toolkits as they moved seasonally to exploit temporary resources. These models have proved untenable and differences in assemblage composition across the landscape are now explained in terms of procurement economics: as knappers rationed, recycled, and substituted artifacts in response to the varying cost of obtaining replacement stone in each location. The economics of raw material use, and its articulation with technological strategies suited to different contexts of mobility and risk, provides mechanisms that can explain the persistence of geographical differences in artifact assemblages but which can also make sense of temporal subtle changes in technological activities. Unlike earlier archeological theories, we can now conclude that site function and identity of the local residence groups were factors of little significance in the production of assemblage differences.

Questions of why lithic assemblages display variation have been central to the explorations of Paleolithic societies on every continent. Increased understanding of lithic technology and more detailed paleoenvironmental reconstructions have, over recent decades,

Lithic Materials and Paleolithic Societies, 1st edition. Edited by B. Adams and B.S. Blades, ©2009 Blackwell Publishing. ISBN 978-1-4051-6837-3.

led to revision of many models about the character of assemblage variability and its articulation with past physical and social environments. In the context of Australia, where Paleolithic technology and economies existed until the last few centuries, reconceptualizing prehistoric ways of life involved abandoning linear models of cultural progression and recognizing that Paleolithic economic and technological systems were elaborate and dynamically changing (Hiscock 2008). These qualities are illustrated in new models of technological variability and the articulation of those ancient technologies with changing resources, altered levels of foraging risk, and modified social contexts. In many Australian regions stone working technology was reorganized at times of climatic and economic shifts in response to increased foraging risks and procurement costs (e.g., Cundy 1990; Hiscock 1994b, 1996, 1999, 2002, 2006; McNiven 1994, 2000; Clarkson 2002a, 2004, 2005, 2006; Attenbrow 2003; Hiscock and Attenbrow 2005a,b; Law 2005; MacKay 2005). This proposition constitutes a powerful explanatory model that is increasingly being applied to explicate variability in Paleolithic Australia, and in this chapter such factors are explored to make sense of the lithic assemblages in western Arnhem Land.

The Alligator Rivers region of Arnhem Land has taxed the imagination of several generations of archeologists. Fieldwork along the escarpment, and subsequently on the lowlands and wetlands to the north and west, gradually uncovered geographical differences in lithic assemblages. The perceived lithic variability was initially explained in traditional terms, involving the identity, territory, and seasonal use of territory by prehistoric foragers.

TRADITIONAL DEPICTIONS OF ASSEMBLAGE VARIATION IN ARNHEM LAND

Archeological investigations of western Arnhem Land have long recognized the existence of geographical variation in the typological composition of lithic assemblages. Initially the exploration of this phenomenon, by Schrire (e.g., White 1967a,b, 1971; White and Peterson 1969; Schrire 1972, 1982), focused on Late Pleistocene and Holocene assemblages observed at sites located in different environments along the East Alligator River corridor. These differences were expressed in terms of the position of five rock shelter deposits within the riverine corridors and their distance from the escarpment that visually dominates the landscape (see Figure 6.1). On the upper reaches of the riverine corridor Jimede I and II were overhangs located amidst the massive sandstone scarps of the dissected margins of the escarpment. Further downstream the Malangangerr shelter is a recess in a 10 m high sandstone residual block near the escarpment. Even further towards the coast two sites, Nawamoyn and Badi Badi, are located in small residual blocks of rock positioned on lowland plains. These five sites display differences in the abundance of tool types and Schrire attempted to make sense of them by distinguishing what she called "plateau sites" (Jimede I and II) from "plains sites" (Malangangerr, Nawamoyn and Badi Badi), and cataloguing the implement types she recognized at each.

Schrire employed a typological classification that defined a number of types found in many sites across this region (see Figure 6.2). The two dominant forms were points and scrapers. Points were flakes that had been retouched on the lateral margin(s) to produce a convergent plan shape (Figure 6.2a). They were retouched on either one or both faces, a pattern that allowed Schrire to differentiate between unifacial and bifacial points. The vast majority of points were broken and the convergence of the original margins was estimated from the shape of the remaining fragment. In Schrire's (1982) assemblages quartzite was the material from which most points were made, with chert and quartz employed to make others. Points were often shaped by invasive retouching accomplished by direct percussion. Scrapers were typically flakes which also had direct percussion retouch scars on their dorsal face but which had rounded rather than convergent plan shapes and retouched edges (Figure 6.2b). This was a morphologically varied class, and showed a different pattern of raw material usage, with most specimens being made from chert. Schrire also defined a number of types that were represented by low numbers of specimens but which she considered diagnostic of site use. Polished flakes had severe edge rounding and gloss on one or both faces (Figure 6.2c). Some specimens were retouched; others were not. Schrire also recognized what she called "bifacial ovals"; unifacially or bifacially retouched flakes with an ovate plan shape (Figure 6.2d). They often resembled points but had rounded rather than convergent forms. Another type was "fabricators," specimens that had been subjected to bipolar retouch, leaving scars which derived from opposing ends, features that Schrire referred to as "bruised ends" (Figure 6.2e). She argued that these were tools used as wedges or punches (White 1967a: 354–6; Schrire 1982: 41), although later Australian researchers concluded that such specimens were often bipolar cores (e.g., White 1968; Hiscock 1982, 1996, 2003). In addition to these flaked artifacts Schrire identified fragments of axes in many sites (Figure 6.2f). Identified by ground bevels creating the axe edge, these specimens were manufactured on igneous and metamorphic rocks by flaking and pecking prior to edge

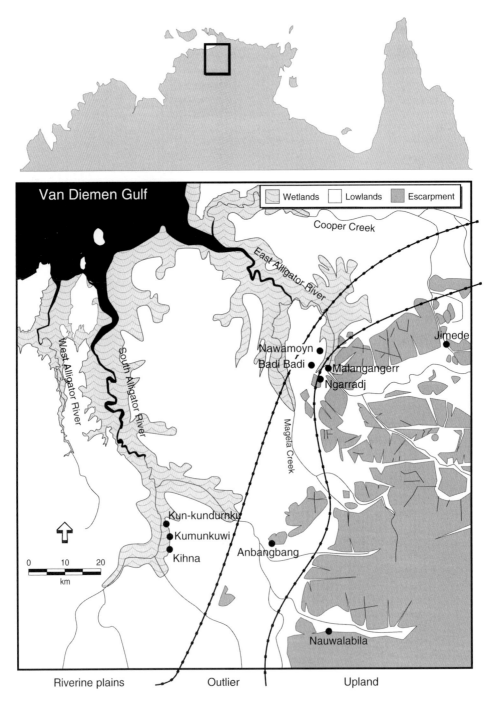

Figure 6.1 Location of discussed sites in western Arnhem Land, showing topographic distinctions and zones of raw material procurement.

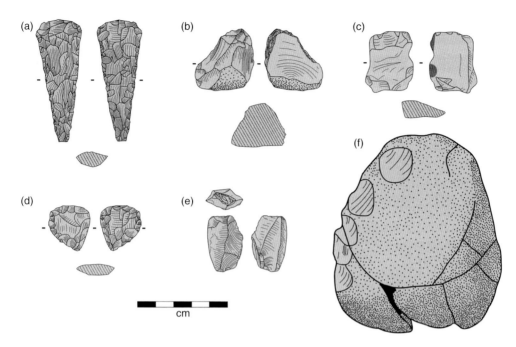

Figure 6.2 Illustration of the implement types recognized by Schrire in her study of assemblage variability in western Arnhem Land. (a) point, (b) scraper, (c) polished flake, (d) bifacial oval, (e) fabricator (bipolar core), and (f) axe. All specimens redrawn from Schrire (1982) and presented at the same scale.

grinding (White 1967c; Schrire 1982). Although later researchers have re-examined these types the basic divisions employed by Schrire have been maintained in many depictions of assemblage variation (e.g., Allen and Barton nd.; Jones and Johnson 1985; Meehan *et al.* 1985; Brockwell 1989, 1996). As a consequence many researchers have offered explanations for assemblage variation that echoed those proposed by Schrire, involving differences in the identity of resident groups and tool using activities that they undertook at each place.

Using these typological categories Schrire argued that assemblages on the plains differed from those on the plateau in the proportions of tool types (White 1967a,b, 1971; White and Peterson 1969; Schrire 1982). Sites along the escarpment contained large numbers of flakes, both unretouched and retouched; whereas on the plain her excavated sites contained few flakes. Schrire (1982: 244–5) characterized plains assemblage as having far fewer tools, proportionally fewer points and scrapers, but higher proportions of polished flakes, axes, and fabricators than in the plateau assemblages.

Initially White (1967b) hypothesized that assemblage differences were a consequence of the long-term co-existence of two cultural groups who possessed similar stone-tool traditions but emphasized different elements within them. She argued that prehistoric contrasts between plateau and plain were consistent with economic patterns observed in resident Aboriginal groups in the early decades of the twentieth century. She later rejected this hypothesis because the archeological pattern persisted for thousands of years and was not explicable in terms of the rapidly changing structure of land ownership and identity in the period following European contact (White and Peterson 1969). Furthermore "plateau sites" were actually located in gorges on the plateau margins and were accessible by people living on the plains; hence there was no topographic barrier to help maintain long-lived cultural differences.

Schrire subsequently offered an alternative interpretation, in which the landscape was used by a single group moving seasonally (White and Peterson 1969; White 1971). Her idea was that the focus of occupation shifted from the plains during the dry season to the plateau during the wet season, and assemblage composition in each sub-region reflected the seasonally different activities of foragers in each sub-region. Different tools were required to exploit the resources encountered in each place: on the plains foragers used bone-tipped spears and stone and shell scrapers to hunt and process fish and water-birds, while on the plateau land mammals such as kangaroos were captured and butchered using heavier stone-tipped spears and stone scrapers (Schrire 1982: 250). Knappers manufactured stone tools during their wet season occupation of the plateau, where high quality quartzite was available, and then, at the start of the dry season, they carried small numbers of stone tools onto the floodplains and emphasized organic tools such as bone points. Schrire's (1982: 251) model of seasonal mobility was therefore related to the economics of tool production. A seasonality model appeared to have the support of ethnographic information, which demonstrated that responses to seasonal inundation and variation in the abundance of plant and animal foods typically took the form of altered settlement localities and resource procurement activities (cf. White and Peterson 1969; also Spencer 1914; Thomson 1939).

Schrire's seasonality-based model has now been rejected. One reason was the discovery of further complexity in the archeological record, as new sites were reported and excavated. For example, Allen and Barton nd. reported a large, point-rich assemblage from Ngarradj Warde Djobkeng, a site situated at the base of a plateau near Malangangerr and the East Alligator River. Topographically Schrire considered this to be one of her "plains sites" but the assemblage was more like one of her "plateau sites," and it was considered an anomaly that signaled the inadequacy of models of seasonal transhumanence (Schrire 1982; Allen and Barton nd.). Further assemblage variation was identified when open sites on the South Alligator River were examined (Meehan *et al.* 1985). Studied in detail by Brockwell (1989, 1996), sites such as Kun-kundurnku, Kumunkuwi, and Kihna had few points, variable proportions of scrapers, but relatively high proportions of polished flakes and other forms of highly retouched flakes. While these assemblages resembled Schrire's "plains sites" more than her "plateau sites" they differed from both sets of assemblages. This discovery meant that a simple division of landscape and assemblages into two categories of "plain" and "plateau" was an inadequate

expression of environmental and archeological variability.

A second problem with the seasonality model was that it had limited capacity to explain chronological changes in assemblage composition, and as archeological research proceeded it became increasing clear that a number of sites displayed temporal change (see Allen 1989). It seemed likely that the seasonal contrasts had always existed throughout the Holocene, although they varied in magnitude, and several researchers were puzzled as to why implement frequencies would have altered over time if assemblage differences reflected seasonal climatic variation (e.g., Schrire 1982; Allen 1989; Allen and Barton nd.). Several researchers suggested that this enigma was connected to questions of environmental change.

The environment of western Arnhem Land underwent a series of radical changes, and the magnitude of landscape transformations represented a challenge to the seasonality model. Believing that the historical introduction of water-buffalo and pigs had degraded the riverine environments Schrire (1982) suggested that the plains had once been more productive and would have supported resident forager groups throughout the year. She concluded that there had not been a seasonal relocation of foragers between "plain" and "plateau." This proposition was adopted by Meehan *et al.* (1985) as the basis for interpreting South Alligator River sites as a closed system in which resident foragers remained annually and moved systematically between resource-rich swamp locations. Brockwell (1996: 99) hypothesized that differences between the South Alligator River assemblages reflected different site functions during the seasonal cycle of movement along the riverine plain; a seasonal model of assemblage variation and forager movement formulated at a much smaller scale than Schrire's. However even reconfigurations of seasonality models are challenged by the evidence for environmental transformations.

Landscape change in western Arnhem Land was thorough and severe, and it was not restricted to the period after European contact (see Hiscock and Kershaw 1992; Hiscock 1999, 2008). For instance, about 9000 years ago rising sea levels flooded the lower reaches of river valleys, creating mangrove fringed embayments which were then gradually filled with sediments and colonized by mangroves to form enormous swamps. By about 6000 years ago continued sedimentation had filled in the swamps to create grass covered flood plains. In parts of this new land high soil salinity then developed, making it impossible for vegetation to exist on the plains. Eventually salt leached from the profile and grasslands were again established. During the last millennia or two fresh water lagoons were formed at a number of places on the flood plain margin, and hence today's wetlands are only the final phase in a series of environmental changes. Away from the river valleys landscape changes also occurred throughout the Holocene. Along the coast progradation formed salt-rich mud flats during the mid-Holocene, extending the shores northwards and altering shoreline characteristics (Woodroffe *et al.* 1986). On the escarpment and adjacent plateaus erosion of sediments modified the landscape, filling gorges with sediments and altering vegetation communities (e.g., Hope *et al.* 1985). The consequences of this landscape evolution were locally varied, resulting in somewhat different environmental histories in each part of the region.

Repeated alterations to the environments of western Arnhem Land meant that resources were not constant for long periods; hence forager resource use in each locality and the functions of sites cannot have been constant. If patterns of implement variation reflected the use of available resources, as implied in all versions of the seasonality model, it would be expected that there would be frequent and substantial changes through time in assemblage variation. This is not what is found! Although there are subtle changes in assemblage patterning over time many of the characteristics of assemblage difference across the region are remarkably persistent while fundamental environmental changes occurred. The long-term persistence of geographical differences in lithic toolkits in a rapidly changing landscape is the central reason that models of seasonal movement cannot explain those spatial contrasts in assemblages.

This conclusion leaves archeologists again pondering the question of how to explain lithic variation in Arnhem Land. Although the pattern of assemblage variability is complex there is a geographical coherence to the assemblage differences that were known. If seasonal movement between the escarpment and the lowlands/wetlands cannot explain observed assemblage variation, does this imply that Schrire had originally been correct in hypothesizing that the region had been occupied by different cultural groups? One response to this question is that there are other possible causes of assemblage variation that have not been adequately considered in discussions of the region.

NEW DEPICTIONS OF ASSEMBLAGE VARIATION

Alternative explanations for assemblage variation require new depictions of the variation. Traditional characterizations of assemblage difference in western Arnhem Land were presented in terms of varying proportions of several classes of implement. The implement types represented in Figure 6.2 were treated as discrete end-products designed to carry out specific functions, and the difference between assemblages was considered to reflect the functional emphasis of prehistoric inhabitants. Hence lithic assemblages were thought of as expressions of seasonal differences in the foods foragers obtained.

Over the last decade many Australian researchers, like their colleagues around the world (e.g., Geneste 1985; Neeley and Barton 1994; Dibble 1995; Shott 1995), have abandoned the twin tenets of those typological analyses. The notion that different "types" of retouched flakes were the carefully crafted, finished form of distinctly different production activities has given way to consideration of different implements as mutable objects that can be transformed from one morphology to another, and the hypothesis that variation between "types" as well as within them may represent differences in the extent to which specimens were retouched rather than differences in design (e.g., Hiscock and Veth 1991; Hiscock 1994a, 2006; Hiscock and Allen 2000; Clarkson 2002a,b, 2004, 2005, 2006; Hiscock and Attenbrow 2002, 2003, 2005a). The notion that each type was functionally distinct has also been questioned, the alternative being that tools were often flexible and multifunctional (Hiscock 2006), and it has been pointed out that if a single specimen was morphologically transformed this would have functional implications (Hiscock and Attenbrow 2005b). In combination these propositions have facilitated studies of the relationships between economic/environmental contexts and the technological responses of ancient foragers, helping to re-express assemblage differences in terms of variation in the "intensity of reduction" and to thereby identify connections between procurement economics, mobility, and technology (e.g., McNiven 1994, 2000; Hiscock 1996, 2006; Clarkson 2002a, 2006; MacKay 2005). This perspective has proved to be a powerful way of understanding the dynamics of lithic variation, and its value is revealed by reinterpretations of typological variation in terms of intensity of reduction and recycling.

POINT REDUCTION AND RECYCLING TO EXTEND USE LIFE

One of the best examples of material conservation and recycling is the production and progressive modification of stone points. Specimens classified as points have converging, often straight, retouched lateral margins, and in Arnhem Land such points share many shape similarities, being elongate and symmetrical around their long axis, as well as similarities in production, being shaped by careful direct percussion retouch (Hiscock 2006: 76). However, points in this region are also extremely variable in the amount of retouch they have received. As noted above, some were retouched on only one face (unifacial points), while others have retouch on both faces (bifacial points). Furthermore, there is variation between points in the amount of their surface covered by retouch, the invasiveness of the scars, the proportion of the perimeter that was retouched, and their dimensions. Much of this variation is now understood to be related to differences in the intensity of retouching specimens have undergone and the extent to which knappers attempted to extend the use-life of points (Hiscock 1994a, 2006: 76–8). Differences in the treatment of points at each location in the landscape were responses to variation in the suitability of blanks and the cost of replacement material. This insight explains much of the typological variation between assemblages that puzzled Schrire and others, and it also provides a framework for exploring aspects of prehistoric settlement patterns in western Arnhem Land.

Paradoxically, prolonged reduction and recycling of Australian points has only recently been widely recognized. Many earlier researchers, such as Flood (1966; 1970), Schrire (1982), Flenniken and White (1985), and Allen and Barton (nd.) treated unifacial points, bifacial points, and other implement forms such as "bifacial ovals", as the independent and mutually exclusive end-products of different manufacturing processes. That conclusion was based on observations that each implement category contained different raw material proportions and dimensions. However Hiscock (1994a) pointed out that these patterns failed to evaluate whether each implement form was a distinct end-product or whether the classes of implement represented specimens undergoing a single manufacturing process but receiving different levels of reduction. For example, raw material differences between types of points would be created if specimens made on

some kinds of rocks (e.g., quartzite) were frequently transformed from one type to another after extended retouching while specimens made on other rocks were rarely modified. Size differences between the types would be created if small specimens were discarded after only small amounts of retouching and without being transformed whereas larger specimens were more intensively retouched and more frequently altered to such an extent they were classified as a different type. Such unequal treatment of blanks with different sizes, shapes, and materials now appears to be the best explanation of point variation.

Archeological evidence supports an interpretation in which some unretouched flakes were manufactured into unifacial points and some of those were in turn converted into bifacial points. Hiscock (1994a, 2006) has argued that in Arnhem Land and other regions of northern Australia the point reduction process generally began by retouching flakes along one lateral margin on the dorsal surface of a flake. Subsequently the second lateral margin might have been flaked in a similar fashion, and eventually one or both of these margins might have been made bifacial by the removal of flakes from the ventral surface. This sequence represents a reduction continuum from unifacial to bifacial points, as the former was gradually transformed into the latter, as shown in Figure 6.3. This sequence was not inevitable and alternatives are known, but Hiscock's (1994a) study of scar superimpositions on quartzite points from Jimede II indicates that more than nine out of ten bifacial points followed this one sequence. Studies of points in adjacent regions have also yielded evidence that unifacial and bifacial forms often represent specimens manufactured in the same way but reduced to different extents (e.g., Roddam 1997; Clarkson 2006).

Progressive reworking of some specimens proceeded even further, reducing the dimensions and shape of points, and altering some to such an extent that they no longer looked like points. For example, points were sometimes retouched until they no longer retained the pointed end and were instead oval or rectangular in shape. Schrire's "bifacial oval" is an example of an implement class that contains many highly reduced points (see Figure 6.3). Such specimens demonstrate that this process of retouching points to prolong their use-life was not focused exclusively on maintaining pointed artifacts but involved a variety of practices by which the usefulness of pieces of stone could be extended. Some of the retouching procedures may have acted as recycling, altering the form of the point to

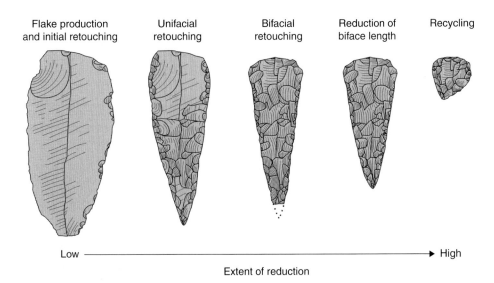

Low ——————————————————————————————————→ High

Extent of reduction

Figure 6.3 Morphological changes associated with increasing levels of retouching. All specimens oriented with their platform at the top. (Based on specimens depicted by Schrire 1982.)

an extent that it was employed for different functions. However the naivety of imagining that points, even when they were in their initial form, were only projectile heads has been discussed elsewhere (see Hiscock 2006). When points were used as tools they were not only small and transportable but were suited to multiple uses, providing stout, sharp margins as well as a pointed shape capable of penetration (Jeske 1989: 36). Use-wear evidence confirms that points from northern Australia were used on plant material, for woodworking, as well as projectile points. Furthermore, Australian unifacial and bifacial points are small and easily transported objects capable of producing large numbers of thin, sharp flakes. This means that points as well as cores could be carried around the landscape to supply the knapper with flakes as required. Hiscock (2006: 78) concluded that points could have been used in many ways, should not be considered to be simply "spearheads", and represent tools that were often transported, reworked, and recycled through long use-lives. The multifunctionality and capacity for extended reduction of points were properties particularly valuable in circumstances where the specimen had been carried to a locality in which replacement raw material was rare or costly, or where the access to replacement raw material was unpredictable. Extended reduction of points, signaled by morphological transformations,

such as the transition from unifacial to bifacial, is one indication that knappers prolonged their functionality in response to such contexts. The characteristics of point assemblages can therefore be used to explore geographical differences in forager economies. However there are other, even more extreme, processes for reducing artifacts in contexts of high material cost.

BIPOLAR REDUCTION TO GENERATE FLAKES

An entirely different mechanism that also extended the use life of stone artifacts was the conversion of specimens into bipolar cores. Bipolar knapping consists of placing a rock on an anvil and striking into it at 90°, in line with the point at which it is in contact with the anvil (Hiscock 1982). Because the specimen is immobilized by being held against the anvil, a heavy hammer can be used to apply large amounts of force and develop large compressive stresses (Cotterell and Kamminga 1987: 688–9; Andrefsky 1998: 26–8). Because bipolar techniques are uniquely suited to situations in which the worked piece has little weight, when low inertia presents a problem for continued reduction, they are often used towards the end of a reduction sequence. For this reason, in Arnhem Land,

signs of bipolar working are found only on small artifacts, and bipolar working was a technique applied to
artifacts, including points, which were already highly
reduced. In this sense bipolar knapping represents a
later stage of reduction than that discussed above for
point retouching and recycling.

The role of bipolar strategies in extending the life
of already highly reduced artifacts can be illustrated
by a description of bipolar reduction observed at sites
along the South Alligator River. Here a series of
research projects examined scatters of artifacts at the
edge of the riverine flood plain and in the woodland-
covered laterite slopes that border the flood plain (e.g.,
Meehan *et al.* 1985; Brockwell 1989, 1996; Hiscock
1996; Hiscock *et al.* 1992). A well studied example is
Kun-kundurnku 1, where several thousand artifacts
eroded from a low levee bank to form a large artifact
scatter (Figure 6.4). The assemblage is diverse,

Figure 6.4   Photographs of the surface artifacts visible at Kun-kundurnku 1 during the 1991–93 field seasons. (Photograph
by the author.)

containing manuports, axes, anvils, grindstones, hammers, and ground haematite in addition to flaked stone artifacts. Most of the artifacts are manufactured on the locally available quartz (71%), although tuff transported from the south and high quality quartzite transported from the east are also found. Many of the flaked artifacts made from good quality materials are extensively reduced. The high level of reduction is visible in a number of ways, including the high frequency of retouch on flakes made on imported rocks, the low weight (<10 g) of many of the cores and retouched flakes discarded at the site, and the abundance of cores (12% of all artifacts). Together these characteristics indicate that, as the central knapping strategy employed at Kun-kundurnku 1, bipolar knapping extended the usability of material brought to the site by employing techniques that enabled tools to be created from very small pieces of rock. In this site bipolar techniques were employed to reduce not only cores that had become too small to work with direct percussion, but also other tools which were converted into bipolar cores. An explication of this process is enlightening.

Hand-held cores, struck using direct percussion, were discarded when they were 20 to 50 g ($\bar{x} = 36.6$). Below 19 g problems of immobilizing the core made direct percussion untenable, and at about that weight knappers began to use bipolar techniques (Figure 6.5a).

All bipolar cores were lighter than 24 g, with the average ($\bar{x} = 5.4$) being much less than the smallest hand-held core. This demonstrates that there was a technical threshold in the process of core reduction; substantially above 20–25 g knappers could successfully remove flakes from cores simply by striking them with a hammer, but at or below that weight knappers often switched to bipolar knapping to enable them to continue reducing the cores. Since bipolar cores were more abundant than hand-held cores it is possible that many cores brought onto the site were extensively reduced to an extent that bipolar techniques were required.

Pressure to extend the usability of lithic material at the site meant that artifacts of all kinds were intensively reduced at Kun-kundurnku 1. Consequently hand-held cores were not the only kinds of stone artifacts subjected to bipolar knapping and transformed into bipolar cores. This is known because a number of specimens were discarded immediately after knappers began to reduce them with a bipolar technique; only a portion of their surface was covered with scars deriving from the bipolar platforms, leaving relic earlier surfaces intact. For such specimens, between 22 g and 10 g, it is often possible to determine what the artifact looked like before bipolar knapping began. While some of the specimens had been hand-held cores many of them had previously been retouched or unretouched

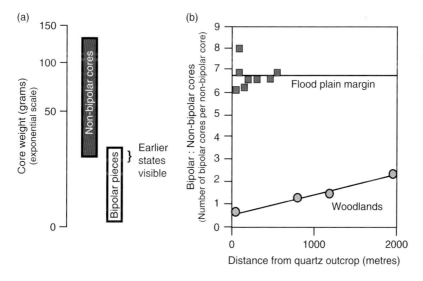

Figure 6.5 Patterns of bipolar working in sites on the edge of the South Alligator River flood plain. (a) Comparison of weights for bipolar and non-bipolar specimens at Kun-Kundurnku 1. (b) Relationship of bipolar abundance and raw material proximity near the South Alligator River.

flakes. At Kun-kundurnku 1 it was possible to identify the previous state of some specimens, and only 20% had originally been hand-held cores. Other specimens had been non-artifactual manuports, unretouched flakes, flakes with gloss, retouched flakes that would be classified as scrapers, and bifacial points (or "bifacial ovals"). One smaller bipolar piece without preserved relic features was made on dolerite and may originally have been an axe fragment. These data indicate that most of the classes of artifact brought into the site were extensively reduced, with many becoming recycled as bipolar cores.

Extensive reduction at sites on the South Alligator River flood plain, such as Kun-kundurnku 1, has multiple causes. While the low abundance, small size, and high reduction of artifacts such as points and scrapers, typically made on good quality, non-local quartzite, chert and tuff, can be partly attributed to the distance from those stone sources it is clear that distance to replacement stone is only one factor affecting the composition of assemblages. Hiscock (1996: 153–4) observed that sites positioned on the edge of the plains had high levels of quartz core reduction, shown by high ratios of bipolar: non-bipolar cores (Figure 6.5b), whereas artifact scatters in the woodlands 5–6 km away had much lower levels of quartz core reduction. These differences are not well correlated with distance to sources of quartz alone; some flood plain edge sites had intensive core reduction even though there were nearby rock outcrops. Hiscock argued that while the cost of stone procurement was a significant factor its affect was articulated with the pattern of forager settlement, and residential mobility was the key to these assemblage differences. At some sites on the floodplains' margins, especially those adjacent to lagoons, foragers lived for extended periods and provisioning of those sites with adequate amounts of suitable stone was potentially costly. Practices such as scavenging previously discarded artifacts and extending the life span of imported rock (even from local sources) were advantageous because they reduced the rate at which lithic material had to be carried to each site (Hiscock 1996: 152). In contrast, foragers exploiting the woodlands were more mobile; they encountered sources of stone frequently and transported it to any particular locality only at low rates, and consequently their need to conserve stone through intensive reduction was lower. This explanation of artifacts found near the South Alligator River assists in understanding the patterns of assemblage variation across the broader

region: assemblage variation was a consequence of the interaction between raw material costs, residential mobility, and forager risks.

RAW MATERIAL AVAILABILITY, FORAGER LAND-USE, AND ASSEMBLAGE REDUCTION

This conclusion indicates that a new regional synthesis of assemblage variation is possible. No doubt chronology and site use are implicated in the creation of differences in assemblage composition, but these factors fail to explain the enduring lithic variation in a context of regular environmental change. Persistence of a geographical pattern to assemblage composition can be understood as a reflection of the geological structure and history of the region. While traditional models depicted spatial differences between sites in terms of distinctions in biotic resources or topography, these did not provide a direct connection to broad-scale lithic patterns across the region. By instead thinking of this landscape in terms of the distribution of sources of rock from which people made their artifacts, it becomes possible to link environmental structure and assemblage variation.

A division of western Arnhem Land into three zones roughly describes raw material availability (Figure 6.1). The "uplands" zone contains the escarpment and nearby outliers less than 15 km from the scarp, and potential outcrops of good quality quartzite are close to sites in this area. The adjacent "outlier" zone contains small sandstone plateaus and boulders, and is 20–30 km from outcrops of quartzite in the escarpment country. "Riverine plains" are even further from quartzite outcrops found in the escarpment, but in the south the zone contains locations with outcrops of quartz and tuff. These zones are heuristic devices that simplify the geological patterns of the region, but they still offer potent descriptions of the environment. Even the low resolution description of raw material distances provided by this threefold division of the region reveals the connection of assemblage content with procurement costs.

These connections are apparent in the interrelationship of landscape position, raw material costs, and the extent of reduction that artifacts have undergone. Archaeological indicators of these factors are presented in Figure 6.6, which plots information for published assemblages of adequate size (Ngarradj levels I–III,

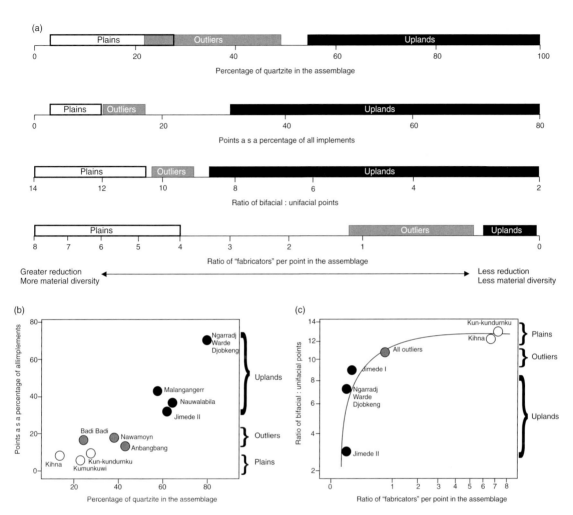

Figure 6.6 Illustrations of the assemblage differences between environmental "zones" in western Arnhem Land. (a) Univariate distributions of assemblage characteristics for each of the three zones: raw material composition (% quartzite) and relative point abundance (points as % of implements) in linear scales, and relative abundance of bifacial points (ratio of bifacial : unifacial points for assemblages with more than 20 specimens) and bipolar cores (number of "fabricators": number of points – exponential scale) in an exponential scale. (b) Scattergram showing the relationship of raw material composition (% quartzite) and relative point abundance (points as % of implements). (c) Scattergram showing the relationship of point bifacialness (ratio of bifacial : unifacial points) and relative abundance of bipolar cores (number of "fabricators": number of points – exponential scale). Data obtained from Allen and Barton (nd.), Jones and Johnson (1985), Schrire (1982), Meehan et al. (1985), and Brockwell (1996: 98) as well as personal observations.

Nauwalabila spits 2–15, Jimede II level I, Malangangerr level I, Anbangbang I, Nawamoyn level Ia, Badi Badi all levels, Kihna, Kumunkuwi, Kun-Kundurnku). These samples date to the same 3000-year period and can be interpreted as reflecting variation with contemporary

assemblages. Aspects of the patterning are summarized as follows (see Figure 6.6a):
• Quartzite is abundant in sites near the uplands where sources are found (55–85%), is much less abundant in the outlier zone (20–45%) and rare in assemblages on

the riverine plains (10–30%). This reflects both the decreased number of quartzite artifacts transported further than 30–40 km from the escarpment as well as the greater number of alternative material sources supplying stone to sites on the riverine plains.

• Points decline in absolute and relative abundance (as a proportion of the assemblage) from uplands to plains. In the uplands they are often the most common implement type, whereas at sites on the riverine plains less than a tenth of the implements were classified as points. There is a significant correlation between the abundance of quartzite and points in each assemblage (Figure 6.6b), a pattern that partly reflects the production of many points on quartzite but also reflects the altered levels of reduction that occurred with increased distance from the escarpment.

• One measure of reduction intensity presented in Figure 6.6 is the amount of retouching on points. Using the ratio of bifacial : unifacial points as an indication of the frequency of more reduced (bifacial) specimens it is clear that the points in the outlier zone tend to be more intensively retouched than in the uplands, and those found on plains sites are very intensively reduced. Heavy retouching in the outlier and riverine plains zones would have reduced many specimens to states in which they were no longer classified as points, contributing to low point abundance away from the uplands.

• Another measure of reduction intensity is provided by the frequency of bipolar cores, or "fabricators," an extremely reduced form. The ratio of "fabricators" : points in Figure 6.6 is an expression of bipolar abundance. High frequencies of bipolar cores were found on the riverine plains, not only because points were rare but also because many implements including points had been converted into bipolar cores. Both the bifacialness of points and the frequency of bipolar cores show the same pattern of stone reduction: outlier and riverine plains sites have more intensively reduced artifacts than sites in the uplands (Figure 6.6c).

These differences in assemblage composition across the landscape are explicable in terms of procurement economics: they emerged as knappers rationed, recycled, and substituted artifacts in response to the varying cost of obtaining replacement stone in each location. The processes involved may have been very simple. For instance, when knappers were far from replacement stone they changed their methods of manufacture and use to increase the return they obtained from material they already possessed, thereby reducing the cost and inconvenience of obtaining replacements. In western

Arnhem Land changed practices in response to material cost often took the form of extending the use-life of tools made on flakes by maintaining their edges through additional retouching. An example of this process is the re-working of points. Another strategy by which use-life was extended was to recycle objects, converting them from one usable form into another. An example of recycling was the reduction of exhausted stone tools as cores, to produce small flakes with usable edges, thereby transforming retouched flakes and other objects into bipolar cores. These two mechanisms, point retouching and bipolar recycling, are provided as illustrations of the kinds of changes to tool making and maintaining behaviors that are observable across Arnhem Land, but other ways of responding to high costs of replacement lithic material, such as employing locally available organic tools rather than costly, non-local stone ones, are also likely.

Technological behaviors sensitive to the economic contexts of artifact manufacture probably explain much of the geographical variation in assemblages, and their persistence over time. However, lithic variation in the region was not fixed; there were temporal changes in the extent of implement reduction. This is expected because the costs of material replacement are not merely a product of distance to stone sources; they are additionally a consequence of the settlement systems that were employed by ancient foragers. Two aspects of settlement have been widely discussed in the Australian context. The first is the level of sedentism of foragers, which has implications for the rates at which particular sites must be supplied with replacement stone. Mobile foragers employing a "provisioning of individuals" strategy (Kuhn 1995) would be advantaged by a portable toolkit that could be used for an extended period, thereby reducing rescheduling costs of frequent tool replacements (Hiscock 2006). The long-lived point reduction sequence employed in Arnhem Land is typical of extendable tools suited to those situations. Relatively sedentary foragers, such as the Late Holocene dwellers on the South Alligator River wetlands, may also have been under stress to extend tool usefulness as a cost reduction measure. A second feature of settlement affecting the extent of tool reduction was the level of foraging risk: the probability and cost of a failure to procure resources in a timely and cost-efficient manner (Bamforth and Bleed 1997). Maintaining a continuously functioning toolkit, including extending the use-life of tools, is a technological strategy that can provide benefits in situations with heightened risks (Hiscock 2006). In combination changes in mobility and foraging risk might be reflected in toolkit composition and extent of reduction.

These considerations provide an explanation for the chronological changes apparent in Arnhem Land assemblages. The onset of stronger El Niño conditions 3000–4000 years ago, created drier and more variable conditions that increased foraging risk and was probably associated with high levels of mobility. In response to these conditions there was intensive point production and intensive tool reduction creating bifacial forms which were more common in that period (Hiscock 2002, 2006). Point abundance and bifacialness declined during the Late Holocene, as levels of tool reduction diminished in line with lessened levels of mobility and risk. This parallels trends observed by Clarkson (2006: 104) in the region to the south, and he also concluded that the period of intensive implement reduction was a technological response in which foragers dealt with conditions of higher risk through extending the use-life of their tools. During the same period there were also alterations in forager mobility. In the Late Holocene riverine plains underwent a number of environmental transformations, culminating in the development of scattered wetlands and lagoons which provided reliable and abundant resources, and altered the economic context from one driven by foraging risk to one dominated by concerns for provisioning costs. The consequence was a shift from highly residentially mobile foraging strategies emphasizing point reduction as a solution to material costs, to relatively sedentary strategies emphasizing extended bipolar reduction of stone as a cost reduction measure. In a landscape where foragers constantly employed greater or lesser levels of reduction to balance tool using needs with provisioning costs, the temporal changes between mobile and more sedentary settlement patterns, and heightened and reduced foraging risk, was significant in creating chronological variations in persistent geographical patterns of lithic variation.

This model of lithic variability, focusing on the economics of raw material use and its articulation with technological strategies suited to different contexts of mobility and risk, provides mechanisms that can explain the persistence of geographical differences in artifact assemblages but which can also make sense of temporal subtle changes in technological activities. The complex lithic variation visible in Arnhem Land arose from the operation of multiple factors. However, unlike earlier archeological theories, we can now conclude that site function and identity of the local residence groups were factors of little significance in the production of assemblage differences. Instead, assemblage variation is largely a consequence of the interaction between raw material costs, residential mobility, and forager risks. That conclusion provides a basis with which to examine the changing adaptive strategies employed by prehistoric foragers living in northern Australia.

ACKNOWLEDGMENTS

I thank the Aboriginal owners of Kakadu, in particular the late Mick Alderson, for providing me access to the Kun-Kundurnku. ANPWS also provided permission for that study, although the arguments do not represent the opinions of ANPWS.

REFERENCES

Allen, H. (1989) Late Pleistocene and Holocene settlement patterns and environment, Kakadu, Northern Territory, Australia. *Indo-Pacific Prehistory Association Bulletin* 9, 92–117.

Allen, H. and Barton, G. nd. Ngarradj Warde Djobkeng. White Cockatoo Dreaming and the prehistory of Kakadu. *Oceania Monograph 37.*

Andrefsky, W. (1998) *Lithics: Macroscopic Approaches to Analysis.* Cambridge University Press, Cambridge.

Attenbrow, V. (2003) Habitation and land use patterns in the Upper Mangrove Creek catchment, New South Wales central coast, Australia. *Australian Archaeology* 57, 20–32.

Bamforth, D.B. and Bleed, P. (1997) Technology, flaked stone technology, and risk. In: Barton, C.M. and Clarke, G.A. (eds) *Rediscovering Darwin: Evolutionary Theory and Archaeological Explanation,* Archaeological papers of the American Anthropological Association No. 7, American Anthropological Association, Arlington, VA, pp. 109–39.

Brockwell, C.J. (1989) *Archaeological Investigations of the Kakadu Wetlands, Northern Australia.* Unpublished MA thesis, Australian National University.

Brockwell, S. (1996) Open sites of the South Alligator River Wetland, Kakadu. In: Veth, P. and Hiscock, P. (eds) *Archaeology of Northern Australia,* TEMPUS Volume 4. University of Queensland, Brisbane, pp. 90–105.

Clarkson, C. (2002a) Holocene scraper reduction, technological organization and landuse at Ingaladdi Rockshelter, Northern Australia. *Archaeology in Oceania* 37, 79–86.

Clarkson, C. (2002b) An Index of Invasiveness for the measurement of unifacial and bifacial retouch: A theoretical, experimental and archaeological verification. *Journal of Archaeological Science* 29, 65–75.

Clarkson, C. (2004) *Technological Provisioning and Assemblage Variation in the Eastern Victoria River Region, Northern Australia: A Darwinian perspective.* PhD thesis, Australian National University.

Clarkson, C. (2005) Tenuous types: "Scraper" reduction continuums in Wardaman Country, Northern Australia. In Clarkson, C. and Lamb, L. (eds) *Lithics "Down Under": Australian Perspectives on Stone Artefact Reduction, Use and Classification*, BAR International Series S1408, Archaeopress, Oxford, pp. 21–34.

Clarkson, C. (2006) Explaining point variability in the eastern Victoria River Region, Northern Territory. *Archaeology in Oceania* 41, 97–106.

Cotterell, B. and Kamminga, J. (1987) The formation of flakes. *American American*, 52(4), 675–708.

Cundy, B.J. (1990) *An Analysis of the Ingaladdi Assemblage: Critique of the Understanding of Lithic Technology.* Unpublished doctoral thesis, Australian National University.

Dibble, H.L. (1995) Middle Paleolithic scraper reduction: Background, clarification, and review of evidence to date. *Journal of Archaeological Method and Theory* 2, 299–368.

Flenniken, J.J. and White, J.P. (1985) Australian flaked stone tools: A technological perspective. *Records of the Australian Museum* 36, 131–51.

Flood, J. (1966) *Archaeology of Yarar Shelter.* Unpublished MA thesis, Australian National University.

Flood, J.M. (1970) A point assemblage from the Northern Territory. *Archaeology and Physical Anthropology in Oceania* 5, 27–52.

Geneste, J.-M. (1985) *Analyse Lithique d'Industries Moustériennes du Périgord: Une approche technologique du comportement des groups humains au Paléolithique moyen.* Doctoral dissertation, University of Bordeaux 1.

Hiscock, P. (1982) A technological analysis of quartz assemblages from the south coast, NSW. In: Bowdler, S. (ed.) *Coastal Archaeology in Eastern Australia*, Department of Prehistory, Australian National University, Canberra, pp.32–45.

Hiscock, P. (1994a) The end of points. In: Sullivan, M., Brockwell, S., and Webb, A. (eds) *Archaeology in the North*, North Australia Research Unit, Australian National University, Darwin, pp. 72–83.

Hiscock, P. (1994b) Technological responses to risk in Holocene Australia. *Journal of World Prehistory* 8(3), 267–92.

Hiscock, P. (1996) Mobility and technology in the Kakadu coastal wetlands. *Bulletin of the Indo-Pacific Prehistory Association* 15, 151–7.

Hiscock, P. (1999) Holocene coastal occupation of Western Arnhem Land. In: Hall, J. and McNiven, I. (eds) *Australian Coastal Archaeology*, ANH Publications, Department of Archaeology and Natural History, Australian National University, Canberra, pp. 91–103.

Hiscock, P. (2002) Pattern and context in the holocene proliferation of backed artefacts in Australia. In: Elston, R.G. and Kuhn, S.L. (eds) *Thinking Small: Global Perspectives on Microlithization*, Archaeological Papers of the American Anthropological Association (AP3A) number 12, Washington, DC, pp. 163–77.

Hiscock, P. (2003) Quantitative exploration of size variation and the extent of reduction in Sydney Basin assemblages: A tale from the Henry Lawson Drive Rockshelter. *Australian Archaeology* 57, 64–74.

Hiscock, P. (2006) Blunt and to the point: Changing technological strategies in Holocene Australia. In: Lilley, I. (ed.) *Archaeology in Oceania: Australia and the Pacific Islands*, Blackwell Publishing, Oxford, pp. 69–95.

Hiscock, P. (2008) *The Archaeology of Ancient Australia.* Routledge, London.

Hiscock, P. and Allen, H. (2000) Assemblage variability in the Willandra Lakes. *Archaeology in Oceania* 35, 97–103.

Hiscock, P. and Attenbrow, V. (2002) Reduction continuums in Eastern Australia: Measurement and implications at capertee 3. In: Ulm, S. (ed.) *Barriers, Borders, Boundaries*, Tempus volume 7, University of Queensland, St Lucia, pp. 167–74.

Hiscock, P. and Attenbrow, V. (2003) Early Australian implement variation: A reduction model. *Journal of Archaeological Science* 30, 239–49.

Hiscock, P. and Attenbrow, V. (2005a) *Australia's Eastern Regional Sequence revisited: Technology and change at Capertee 3.* British Archaeological Reports, International Monograph Series 1397, Archaeopress, Oxford.

Hiscock, P. and Attenbrow, V. (2005b) Reduction continuums and tool use. In: Clarkson, C. and Lamb, L. (eds) *Rocking the Boat: Recent Australian Approaches to Lithic Reduction, Use and Classification*, British Archaeological Reports, International Monograph Series 1408, Archaeopress, Oxford, pp. 1–151.

Hiscock, P. and Kershaw, P. (1992) Palaeoenvironments and prehistory of Australia's tropical top end. In: Dodson, J. (ed.) *The Naive Lands*, Longman Cheshire, Sydney, pp. 43–75.

Hiscock, P., Mowat, F., and Guse, D. (1992) Settlement patterns in the Kakadu Wetlands: Initial data on site size and shape. *Australian Aboriginal Studies* 2, 69–74.

Hiscock, P. and Veth, P. (1991) Change in the Australian Desert Culture: A reanalysis of tulas from Puntutjarpa. *World Archaeology* 22, 332–45.

Hope, G., Hughes, P.J., and Russell-Smith, J. (1985) Geomorphological fieldwork and the evolution of the landscape of Kakadu National Park. In: Jones, R. (ed.) *Archaeological Research in Kakadu National Park*, Australian National Parks and Wildlife Service Special Publication 13, Australian National University, Canberra, pp. 229–40.

Jeske, J. (1992) Energetic efficiency and lithic technology: an Upper Mississippian example. *American Antiquity*, 57, 467–81.

Jones, R. and Johnson, I. (1985) Deaf Adder Gorge: Lindner site, Nauwalabila 1. In: Jones, R. (ed.) *Archaeological Research in Kakadu National Park*, Australian National Parks and Wildlife Service Publication 13. Australian National University, Canberra, pp. 165–228.

Kuhn, S. (1995) *Mousterian Lithic Technology*. Princeton University Press, Princeton.

Law, W.B. (2005) Chipping away in the past, stone artefact reduction and mobility at Puritjarra Rockshelter. In: Clarkson, C. and Lamb, L. (eds) *Rocking the Boat, Recent Australian Approaches to Lithic Reduction, Use and Classification*, British Archaeological Reports, International Series 1408, Archaeopress, Oxford, pp. 81–93.

MacKay, A. (2005) Informal movements, changing mobility patterns at Ngarrabullgan, Cape York, Australia. In: Clarkson, C. and Lamb, L. (eds) *Rocking the Boat, Recent Australian Approaches to Lithic Reduction, Use and Classification*, British Archaeological Reports, International Series 1408, Archaeopress, Oxford, pp. 95–107.

Meehan, B., Brockwell, S., Allen, J., and Jones, R. (1985) The Wetlands sites. In R. Jones (ed.) *Archaeological Research in Kakadu National Park*, Australian National Parks and Wildlife Special Publication 13, Australian National University, Canberra, pp. 103–54.

McNiven, I.J. (1994) Technological organization and settlement in southwest Tasmania after the glacial maximum. *Antiquity* 68, 75–82.

McNiven, I.J. (2000) Backed to the Pleistocene. *Archaeology in Oceania* 35, 48–52.

Neeley, M.P. and Barton, C.M. (1994) A new approach to interpreting late Pleistocene microlith industries in southwest Asia. *Antiquity* 68, 275–88.

Roddam, W. (1997) *Like – But Oh How Different: Stone Point Variability in the Top End*. N.T. Unpublished Honours Thesis, Northern Territory University.

Schrire, C. (1972) Ethno-archaeology models and subsistence behaviour in Arnhem Land. In: Clarke, D.J. (ed.) *Models in Archaeology*, Methuen, London, pp. 653–69.

Schrire, C. (1982) *The Alligator Rivers: Prehistory and Ecology in Western Arnhem Land*, Terra Australia 7, Department of Prehistory, Research School of Pacific Studies, Australian National University, Canberra.

Shott, M.J. (1995) How much is a scraper? Curation, use rates, and the formation of scraper assemblages. *Lithic Technology* 20, 53–72.

Spencer, W.B. (1914) *Native Tribes of the Northern Territory of Australia*. Macmillan, London.

Thomson, D.F. (1939) The seasonal factor in human culture. *Proceedings of the Prehistoric Society*, 5, 209–21.

White, C. (1967a) *Plateau and Plain: Prehistoric Investigation in Arnhem Land*. Unpublished PhD thesis, ANU, Canberra.

White, C. (1967b) The prehistory of the Kakadu people. *Mankind* 6(9), 426–31.

White, C. (1967c) Early stone axes in Arnhem Land. *Antiquity* 41, 147–52.

White, C. (1971) Man and environment in Northwest Arnhem Land. In: Mulvaney, D.J. and Golson, J. (eds) *Aboriginal Man and Environment in Australia*, Australian National University Press, Canberra, pp. 141–57.

White, C. and Peterson, N. (1969) Ethnographic interpretations of the prehistory of western Arnhem Land. *Southwestern Journal of Anthropology* 26, 45–67.

White, J.P. (1968) Fabricators, outils écaillés or scalar cores? *Mankind* 6, 658–66.

Woodroffe, C.D., Chappell, J.M.A., Thom, B.G., and Wallensky, E. (1986) *Geomorphological Dynamics and Evolution of the South Alligator Tidal River and Plains, Northern Territory*. North Australia Research Unit, Mangrove Monograph No. 3, Australian National University, Darwin.

Part II

TECHNOLOGICAL AND ASSEMBLAGE VARIABILITY

Chapter 7

PALEOLITHIC EXPLOITATION OF ROUNDED AND SUB-ANGULAR QUARTZITES IN THE INDIAN SUBCONTINENT

Parth R. Chauhan

Stone Age Institute and CRAFT Research Center, Indiana University

ABSTRACT

The Indian subcontinent is geographically constrained by high altitude mountain ranges in the north (the Himalayas) and the Arabian Sea, the Indian Ocean, and the Bay of Bengal, to the west, south, and east, respectively. Within this large land-mass, there exists a diverse range of eco-zones which contain a host of raw material resources that were exploited in different ways by Pleistocene hominins. These sources occur in diverse environmental settings comprising fluvial, colluvial, and erosional deposits. Lithic raw material was a prime factor governing the dynamics of hominin ranging patterns and the majority of the Paleolithic evidence can be directly or indirectly linked to these sources, illustrating key changes and differences in patterns of raw material selection, acquisition, and processing. Overall, it is clear that there are three distinct morphological groups of raw material clasts that have been utilized throughout prehistory: (i) fluvially-rounded pebbles and cobbles, (ii) fluvially-rounded and sub-rounded cobbles and boulders, and (iii) tabular slabs and angular fragments. This chapter addresses the variable modes of exploitation of rounded and sub-angular quartzite in the Indian subcontinent in relation to morphological forms, size, and associated geological contexts. The behavioral evidence from select Lower, Middle, and Upper Paleolithic sites throughout the region suggests clast-size preference, multiple levels of technological acumen, and ecological adaptive strategies such as varied transport behaviors and site selection. Some sites discussed are: Riwat, Pabbi Hills, Toka, and Site 55 in the Siwalik Hills; Singi Talav in the Thar Desert; Kalpi in the Indo-Gangetic plains; Durkadi, Pilikarar, and Samnapur in the Narmada Basin; and Attirampakkam in the Kortallayar Basin.

Lithic Materials and Paleolithic Societies, 1st edition. Edited by B. Adams and B.S. Blades, ©2009 Blackwell Publishing. ISBN 978-1-4051-6837-3.

INTRODUCTION

The locations of most Paleolithic sites have been traditionally known to be governed by the geographic distribution of lithic raw materials along with water resources, topographic stability, and favorable ecological and climatic conditions. Most well-excavated sites throughout the Old World reflect variable patterns of raw material selection, transport, manipulation, and discard. Raw material (henceforth RM) sources are essentially represented by the diversity of ecological contexts such as fluvial, colluvial, bedrock outcrops, fan deposits, and occasionally, clastic material in cave contexts. The majority of these rock clasts occur in two distinct forms throughout the Old World, including the Indian subcontinent – in fluvially-deposited rounded to sub-rounded form and angular or tabular clasts or slabs. The former is usually represented by pebbles, cobbles, and boulders, derived from fluvial and colluvial sources and/or as weathered bedrock fragments; the latter is exposed as unweathered bedrock or basin-scale basement exposures. The majority of all sources is related to regional topographic properties and associated climatic and erosional processes. The long-term use of pebbles, cobbles, and boulders (the focus of this chapter), in contrast to the tabular/angular material, for Paleolithic toolkits was so dominant in some regions of the Old World, that some lithic assemblages and industries came to be referred to as parts of the "Chopper-Chopping Tool" or "Pebble Tool" tradition (Movius 1948). In most cases, such assemblages indicate the use of locations near water sources or paleochannel gravel/conglomerate deposits. Known to be comparatively resistant to natural fracture and its widespread availability, quartzite was the commonly-preferred RM during the prehistoric occupation of the Old World as well as North America (Seong 2004).

The Indian subcontinent or South Asia contains a continuous prehistoric archeological record dating back to, at least, the early Middle Pleistocene (Mishra 1994). Older archeological occurrences, from the Late Pliocene and Early Pleistocene, have also been reported from northern Pakistan and peninsular India, respectively (Paddayya *et al.* 2002; Dennell 2004); these, however, need further corroboration. South Asian Lower Paleolithic assemblages have generally been assigned to either the Acheulean or Soanian and Soanian-like traditions (Misra 1987), the latter of which may also represent early Middle Paleolithic assemblages (Gaillard and Mishra 2001; Chauhan

2003). The Movius Line (Schick 1994) traverses the region's northern reaches and, the region marks the easternmost domain of abundant and classic Mode 2 assemblages in the Old World (Petraglia 1998; Pappu 2001; Chauhan 2004, 2006). The Lower, Middle, and Upper Paleolithic sites are often found in spatial association with their respective RM sources, reflecting large-scale ecological adaptations and changing behavioral repertoires among South Asian hominin groups. There are many sites and low density scatters indicating intensive exploitation of rounded and sub-angular quartzite clasts in the subcontinent; however, only the most prominent examples have been included in this chapter. These sites are divided into two groups of quartzite clasts – pebble/cobble and cobble/boulder – and occur throughout the subcontinent (Figure 7.1; Table 7.1).

SOUTH ASIAN BIOGEOGRAPHY AND THE GEOGRAPHICAL DISTRIBUTION OF RAW MATERIALS

Located between the well-known paleoanthropological localities of East Africa, Georgia, and Java, the Indian subcontinent essentially comprises the regions of India, Pakistan, Nepal, Sri Lanka, Bhutan, and Bangladesh. To the west of peninsular India is the Arabian Sea, to the east, the Bay of Bengal, and to the south is the Indian Ocean. The entire region comprises a diverse spectrum of ecological and topographical zones combined with a complex geological history. The north is dominated by the Greater and Lesser Himalaya and the Siwalik Hills, ranges almost geographically parallel and temporally successive to each other. South of these mountain and hill ranges are the Indo-Gangetic plains, followed to the south by the great Thar Desert (in eastern Pakistan and northwestern India), and the Aravalli and Vindhyan range of hills. These hills are located north of the Deccan Plateau, which comprises the Western and Eastern Ghats (ranges of hills). The subcontinent is also interspersed with complex drainage systems and numerous associated ecological and geographical features such as deciduous woodlands, tropical evergreen forests, savanna landscapes, semi-arid and arid scrub lands, arid sand deserts, and periglacial loessic landforms (Korisettar and Rajaguru 2002); caves, canyons, rockshelters, lakes, pools, and springs are also found in high numbers. With the exception of the Narmada and Tapi Rivers, most rivers flow from east to

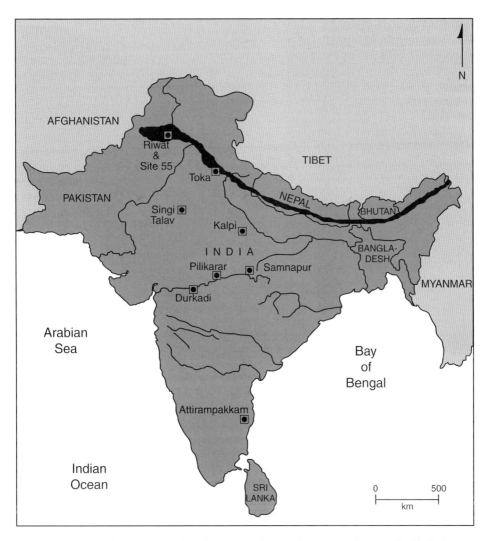

Figure 7.1   Locational map of sites associated with quartzite clasts as discussed in the text. The black shading across the northern part denotes the Siwalik Hills.

west and all exhibit unique fluvio-sedimentary regimes (Gupta 1995).

In the past as well as today, different ecozones provided access to different types and shapes of lithic RM, usually conditional upon factors of sedimentation and associated fluvial, tectonic, and erosional mechanisms. For example, for fluvial sources in the Siwalik region, rounded and sub-rounded quartzite and sandstone clasts dominated the landscape during various phases of hominin occupation. In parts of northcentral India, Acheulean assemblages were made on pink granite.

In the Western and Eastern Ghats, however, the primary RM was basalt and doleritic dykes, occurring as fragments from bedrock outcrops and water-worn clasts belonging to the Deccan volcanic Traps. Further south, in parts of Karnataka, limestone bedrock was the dominant RM type available in the form of tabular slabs. The possible lack of suitable/sizeable lithic RM in northeastern India may have limited the movements of hominin groups between South Asia and Southeast Asia – possibly explaining why Acheulean technology did not disperse eastwards from India. From the variety

Table 7.1 Key sites mentioned in the text and their associated age, classification, quartzite clast group, and significance. (Associated references are cited in the text.)

Site	Age	Classification	Clast group	Significance
Riwat	*ca.* 2 Ma	Pre-Acheulean Mode 1	Pebbles and cobbles	Possibly the oldest archeological evidence in Indian subcontinent;
Pabbi Hills	*ca.* 2–1 Ma	Pre- and non-Acheulean	Pebbles and cobbles	In fine-grained sediments and in stratigraphic association with numerous vertebrate fossils
Pilikarar	Mid-Pleistocene?	Early Acheulean	Cobbles and boulders	Oldest known Acheulean in Narmada Basin; possible quarry context; African affinities?
Singi Talav	>400 Ka	Early Acheulean	Cobbles and boulders	Oldest Acheulean in Thar Desert; 2nd known site in clay context after Attirampakkam
Attirampakkam	Pleistocene	Lower to Upper Paleolithic	Cobbles and boulders	Lengthy archeological sequence in clay context; preserved mammalian footprints; best-studied Acheulean site in recent times
Durkadi	Mid-Pleistocene?	Non-Acheulean Mode 1?	Cobbles and boulders	The only known Mode 1 occurrence in primary context in peninsular India
Toka	<200 Ka	Soanian (Mode 1/Mode 3)	Pebbles and cobbles	Richest-known Soanian site; partially stratified; differential clast preference for chopper production; rich typological diversity
Samnapur	Late Pleistocene	Middle Paleolithic	Pebbles and cobbles	Best-known Middle Paleolithic site in Narmada Basin but undated
Kalpi	*ca.* 45 Ka	Middle Paleolithic	Pebbles and cobbles	Rare Paleolithic occurrence in Ganga Basin; diminutive choppers and numerous bone tools
Site 55	*ca.* 45 Ka	Upper Paleolithic	Pebbles and cobbles	Rich site in loess context; use of stone platform

Figure 7.2   Quartzite pebbles and cobbles from Toka.

of RM exploited extensively by South Asian hominins, quartzite was one of the more dominant types used during the Lower, Middle, and Upper Paleolithic and at various types of sites contexts – quarry, factory, workshop, and occupational. It was available at numerous locations in the subcontinent and is generally associated with the extensive river systems of the subcontinent which contain variable amounts of clasts in (invariably) rounded form (Figure 7.2) reflecting extensive fluvial transport from their original geological sources. The biannual Indian monsoon system may have also played a major role in the possible seasonal availability of RMs in river or stream beds during the dry periods (also see Dennell 2007). Quartzites, in general, represent metamorphosed sandstone of varying levels of quartz-grain compactness, affecting conchoidal fracture patterns. The South Asian varieties are generally tan or brownish-yellow in color but other variations are also known – red, purple, brown, dark gray, and so forth. The cortex is usually a darker color than the interior which is often yellow or off-white.

Pebble-cobble sources and associated sites

As mentioned earlier, pebbles and cobbles of various types of RM (predominantly quartzite, sandstone, and basalt) are found as parts of fluvial/gravel deposits, paleochannel exposures, terrace sequences, and colluvial fan gravels. These forms of RM are distributed

throughout the Indian subcontinent and the frequencies and levels of clast sorting vary considerably, conditional upon the environment of deposition and associated post-depositional geomorphology. All types of Paleolithic sites are associated with this RM type, often with a prominent heavy-duty tool component (e.g., choppers, core-scrapers, single-platform cores on split cobbles). The prime large-scale example of occurrences in association with quartzite clasts is the Siwalik Hills region in the northern part of the subcontinent.

Sub-Himalayan zone

The Siwalik Hills or the Siwalik Foreland Basin consist of fluvial sediments deposited by hinterland rivers flowing southwards and southwestwards from the Lesser and Greater Himalaya (Acharryya 1994), when the region south of these mountains was originally a vast depression or basin (referred to as the foredeep) in the Miocene (Brozović and Burbank 2000). These sediments were later uplifted through ongoing tectonic processes that intensified in the Plio-Pleistocene when the hills, as a topo-geographical unit, relatively attained their present elevation. The range is less than 13 km wide in places with an average of 24 km, and it reaches an elevation between 900 m and 1200 m. The majority of the sediments are located within the political boundaries of Pakistan, India, and Nepal and have been divided stratigraphically into three Subgroups, which in turn are sub-divided into eight Formations: Kamlial (Lower Siwalik Subgroup); Chinji, Nagri, and Dhok Pathan (Middle Siwalik Subgroup); and Tatrot, Pinjore, and Boulder Conglomerate Formations (BCF) (Upper Siwalik Subgroup). According to Sangode and Kumar (2003), broad age estimates of sedimentation windows for the Upper Siwalik formations are 3.4–5.6 myr for Tatrot, 1.7–2.5 myr for Pinjore, and 0.7–1.7 myr for the BCF. Paleolithic sites on Siwalik slopes are situated on or above sediments belonging to all Siwalik Formations. The distribution of RM in the Siwalik region can be essentially viewed from two geographic perspectives, Primary and Secondary. The Primary sources of all quartzite clasts are represented exclusively by BCF deposits and Secondary sources by post-Siwalik or post-BCF deposits. Both types comprised sandstone and quartzite pebbles, cobbles, and boulders with a silty-sand matrix. Fine-grained material such as chert, chalcedony, and jasper, for example, were virtually absent and do not occur naturally in the Siwalik ecozone. Although the quartzite material here exhibits

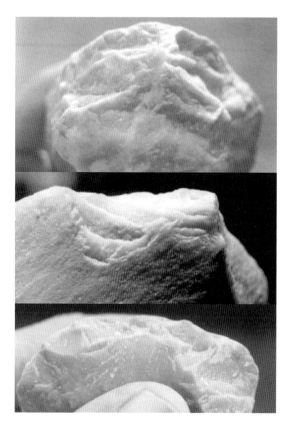

Figure 7.3 Variable degrees of use-wear on different grades of quartzite from Toka. Examples are of coarse-grained, medium-grained, and fine-grained quartzite (from top to bottom).

different levels of coarseness (Figure 7.3), there is very little indication of RM selectivity (e.g., Stout *et al.* 2005) within this subtle range of quality. This may be partially due to the homogenous nature of most quartzite clasts from the Indian subcontinent (though further detailed studies are clearly warranted).

Older Formations were not completely devoid of RM sources, but were not as abundant as the BCF and post-Siwalik exposures. In northern Pakistan, for example, Dennell and Rendell (1991: 95) mention the association of artifacts with "outcrops of quartzite-bearing Middle and Upper Siwalik conglomerates, or with spreads of quartzite pebbles and cobbles derived from these conglomerates." These early sources of RM were not geologically common prior to "BCF times," and appear to be isolated occurrences or as thin layers not more than few meters in length near the

main channel (Dennell 2007) rather than laterally extent components of the Upper Siwalik landscape. In fact, various exposures of the BCF visible today did not form a continuous and contemporary landscape during hominin occupation of the region. An important feature of these Primary sources, however, is that once deposited, they represented a relatively stable and long-term source of RM.

The unique period following the deposition of the BCF is chronologically referred to as "post-Siwalik" (Mukerji 1976; Stiles 1978), when antecedent Siwalik fluvial systems essentially modified their courses. Located across the Siwalik range of hills, the numerous post-Siwalik streams occur every 3–5 km at some places in the frontal zone. Many of these streams were pene-contemporaneous and originated in the BCF exposures to the north thus fluvially transporting conglomerates through narrow gorges within the hills before flowing out on to the plains as braided or meandering channels. The majority of Paleolithic sites are associated with this post-Siwalik or Secondary source of RM.

Late Pliocene-Early Pleistocene occurrences. The oldest evidence in the Siwalik Hills is the Paleolithic material from Riwat and the Pabbi Hills (Rendell *et al.* 1989; Dennell 2004). In fact, this Late Pliocene and Early Pleistocene evidence, respectively, is thought to represent, possibly, the earliest human occupation in the subcontinent. The site of Riwat is a part of the Soan syncline and comprises a gravel context chronologically equivalent to the Pinjore Formation (at a broad level) of the Upper Siwalik Subgroup. Out of the three quartzite specimens recognized as hominin-produced (Rendell *et al.* 1989; Dennell 2007), the most significant and diagnostic specimen is a multifaceted core-tool specimen on a large cobble with 8–9 flake scars. Unlike Riwat, however, the 607 specimens from the nearby Pabbi Hills (also representing typical Mode 1 technology) derive from fine-grained sediments (also of the Pinjore Formation) and are stratigraphically associated with a large number of vertebrate fossil remains (over 40,000 fossil specimens) across this landscape (Hurcombe 2004). Younger artifact types such as Acheulean bifaces generally do not occur in this area and the collected assemblage comprises varieties of flakes, flake-blades, debitage, angular fragments, cores, core fragments, and core tools manufactured on locally available quartzite cobbles. Additional types include discoidal, polyhedral, and sub-spheroid cores.

The cobble blanks were either flaked directly or split to provide additional striking platforms; subsequent flaking was either circumferential or bipolar. The majority of the specimens are distributed randomly across the Pinjore landscape and do not occur in spatially-discrete clusters (unlike the Soanian evidence described below). Such low concentrations of artifacts in this region may be partially attributable to the scarcity of quartzite clasts during Pinjore Formation times, which may explain the marginal occupation of hominins in the region (Dennell 2004, 2007).

Middle and Late Pleistocene occurrences. The most prominent exploitation of pebbles and cobbles is represented by the Soanian evidence, also found in the Siwalik region but in considerably younger contexts (i.e., post-Siwalik) than Riwat and the Pabbi Hills lithic material. The Soanian evidence is visibly rich, no doubt due to the preponderance of quartzite pebbles and cobbles in the Siwalik region since the Early Pleistocene in Pakistan and the Middle Pleistocene in India, both represented by the time-transgressive BCF. Soanian artifacts were manufactured on quartzite pebbles, cobbles, and occasionally on boulders, all derived from various fluvial sources on the Siwalik landscape. Such assemblages generally comprise varieties of choppers, discoids, scrapers, cores, and numerous flake types, all occurring in varying typo-technological frequencies at individual sites (Paterson and Drummond 1962; Chauhan 2005). Most Soanian assemblages appear to be contemporary with Late Acheulean and Middle Paleolithic assemblages as known from other parts of the subcontinent (Gaillard and Mishra 2001; Chauhan 2003). In addition to the geological evidence, this techno-chronological designation is also supported by the presence of advanced and diagnostic tool-types not generally associated with pre-Acheulean or Early Acheulean assemblages (e.g., Levallois flakes/cores, prepared cores, blades and blade-like flakes). Unlike the Soanian evidence, which is generally rich and occasionally stratified, the Mode 2 evidence in the Siwalik region is represented by isolated occurrences of Late Acheulean cleavers and handaxes in surface context or find-spots. The dominance of pebbles and cobbles and the low occurrence of boulders may explain this paucity of rich Acheulean assemblages, which required clasts sizeable enough to produce bifaces (discussed later).

Recent studies by the author demonstrate that Soanian sites on Siwalik frontal slopes between two major rivers vary considerably in their artifact quantities regardless of abundant RM sources (Chauhan 2008). Contextual investigations at the richest known Soanian site, Toka ($n = 4106$), reveal evidence of site formation and the acquisition of quartzite blanks from terrace and streambed contexts within a $1\,km^2$ area. General observations of tool-types, artifact quantity, and modes of lithic manufacture here indicate late Lower Paleolithic and early Middle Paleolithic levels of technology (Chauhan 2005). The most significant factor for intense occupation of this location was the post-Siwalik Tirlokpur Nadi. This streambed and its paleochannel, in the form of incised/uplifted terrace deposits, were the only available sources of quartzite clasts in the vicinity for stone tool manufacture. The next closest sources are the Run Nadi, another post-Siwalik stream $2\,km$ northwest and the antecedent Markanda River, $3-4\,km$ southeast. Over 46% of the Toka specimens are made on tan-colored quartzite (the predominant color in the region) and the remainder are burgundy, black, dark gray, white, brownish-tan, dark purple, and often combinations of these colors. No artifacts on Tatrot sandstone clasts were observed. Elsewhere in the Siwaliks, however, investigators have also documented non-biface artifacts on vein quartz, when quartzite clasts were unavailable or present in minimal quantities (Ganjoo *et al.* 1993–94), and also on "tuff" and chert (Corvinus 2002). A unique feature observed at Toka was the variation in preference of quartzite cobble size for the manufacture of side-choppers and end choppers; the former are more dimensionally homogenous than the latter (Figure 7.4).

Following the discovery of Toka in January 2001, it was anticipated that similarly rich scatters may also exist at analogous locations – near RM sources where post-Siwalik streams merged with the plains. To test this hypothesis, a survey was undertaken in the frontal slopes of the hills, as well as some interior zones between two major rivers. Newly-discovered surface occurrences were spatially correlated with nearby RM sources found in fluvial context, to understand variances in artifact quantities and general distances of clast transport. After Toka, the highest numbers of artifacts were found at Bhandariwale-Mirpur ($n = 279$) near the Run Nadi and at Karor Uparli ($n = 523$) near Jainti Majri Choe. Judging from such large numbers of artifacts, hominin occupation at such locations probably represents either short-term intensive activity or repeated visits over a longer period of time,

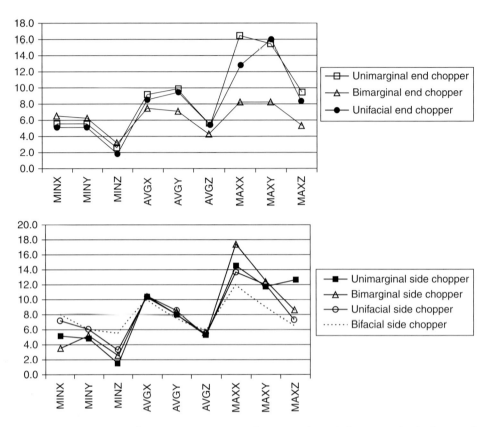

Figure 7.4 Minimum, average, and maximum comparative dimensions for end-choppers and side-choppers from Toka. (*Source:* Chauhan 2007.)

particularly at Toka (see Sullivan 1992). Although a broad "tethering" effect (Brantingham 2003) is archeologically visible, the distribution of sites and related frequencies of artifacts are not always proportionate to the amounts of RM available in the vicinity. For example, many equally ideal locations did not yield any artifacts, despite a visible abundance of quartzite clasts at these locations (Chauhan 2008). One potential explanation for the lack of rich sites (like Toka) at other similar locations may be that some post-Siwalik streams are time-transgressive. Certain post-Siwalik streams (i.e., deposition of RM) where sites are absent may have formed when hominin occupation ceased in the region. Another reason for the inconsistent site distribution in the frontal zones may be that some RM sources (particular streams or outcrops) visible today may not have been available in some streams during

hominin occupation. For example, not all post-Siwalik streams and rivers are connected to BCF exposures and thus contain variable quantities of quartzite. Therefore, the occupation of the Siwalik region during the Pleistocene would have required an in-depth ecological and geographical knowledge of the landscape which, in turn, may have fostered active social networks between major long-term habitation and workshop sites. The use of such interior locations away from the main sources of RM and water in the frontal zone also implies increased mobility and long-term planning.

The correlations made here between sites and conglomeratic RM sources are partly supported by similar evidence elsewhere in the Siwalik region. For example, most of the new occurrences of artifacts (some *in situ*) and surface scatters described by Stiles appear to be in association with either (Potwar) silts or conglomerates (both BCF and post-Siwalik gravels). In the Pinjore-Nalagarh dun, Karir (1985) has presupposed that all artifacts (including flakes) were made from locally-available pebbles. British Archaeological Mission to Pakistan (BAMP) located extensive lithic workshops of varying traditions and ages on the Lei (post-Siwalik) and Siwalik Conglomerates covered by loess deposits (Allchin 1995). For example, Rendell *et al.* (1989) describe a rich blade-dominated assemblage from these loess deposits at Site 55 dated to a minimum age of about 45 ka and unique features at this site include possible post-holes, a pit, and a wall-footing. Core preparation on the cobble blanks was minimal here and the majority of flakes may represent debitage from blade production. However, primary and secondary flakes were most common over other types. Some of the specimens possess variable amounts of edge retouch and a few may even be referred to as being microlithic (Rendell *et al.* 1989: 190).

Central and peninsular India

Similar examples of core-and-flake assemblages from peninsular India are less well known and occur in the form of sites, findspots, off-sites, and stray assemblages throughout the Indian subcontinent (e.g., Jayaswal 1982; Reddy *et al.* 1995; Chauhan in press). One recent example of a non-Soanian assemblage south of the Sub-Himalayan region is the site of Kalpi, a Middle Palaeolithic site also 45 kyr in age (as Site 55) from the Ganga Valley of northern India (Tewari *et al.* 2002). Here the investigators report a "pebble tool" component on quartzite, along with flakes, diminutive choppers, bone tools, and vertebrate fossil remains. Kalpi is located about 400 km south of the Siwalik Hills (the geographical domain of the Soanian industry).

In contrast, a Middle Paleolithic or Mode 3 assemblage/industry without a heavy-duty tool component was also produced from gravel/rubble clasts at Samnapur in the central Narmada Basin (Misra *et al.* 1990). Here, while chert was used for making the bulk of the tools, other rocks like quartzite and basalt were brought as cobbles from hill slopes and stream beds several kilometers away from the site. This site is situated on high alluvial flats on the right bank of the Narmada, but away from the main river channel, and at the foot of a "cherty quartzite" outcrop. The associated rubble horizon in the excavation is 0.5–0.7 m thick and formed of sub-rounded blocks of quartzite

with an average length of 15 cm. Over 3000 artifacts were collected from this excavation and the entire assemblage is flake-dominated, most specimens range from 5 to 20 cm in length. Blade-flakes with facetted platforms are common but very few Levallois flakes were recovered; a large number of cores of varying size and associated debitage were also documented. Among the shaped tools, there are single and double-sided scrapers with dorsal, ventral, and inverse retouch, end scrapers, denticulates, notched tools, borers, and retouched flakes. Late Acheulean bifaces or early Middle Paleolithic diminutive handaxes were not observed at this site.

Cobble-boulder sources and associated sites

Although pebbles, cobbles, and boulders generally occur in shared stratigraphic sequences, the level of clast sorting varies depending on the environment of deposition and local topographical gradients. For example, in the Siwalik region and most river valleys in peninsular India, gravel or pebbles/cobbles are most abundant and quartzite boulders are rarely found in such deposits. The boulders occur as parts of colluvial fans, occasionally high-energy fluvial sequences, and as sub-rounded clasts of weathered bedrock throughout peninsular India. As a result, several Paleolithic sites can be linked with these larger quartzite clasts in the Siwalik region as well as in peninsular India. Pebble-based tools (e.g., choppers) are also known from these cobble/boulder dominated assemblages, however they tended to occur in lower frequencies than in the assemblages described above.

Core-and-flake occurrences

In Nepal, in the Tui Valley, an industry of flakes and cores (bifaces are absent) was recovered from the basal alluvium of a quartzite cobble-boulder gravel, occurring below the stratified silts and clays of the Babai Formation at Brakhuti (Corvinus 2002). Similar specimens are found elsewhere in the Tui valley in high numbers, where the associated cobble-boulder gravel is exposed (above the bedrock and below the silt). The Brakhuti industry possibly constitutes a special requirement for heavy-duty tools in a forested habitat, implying considerable wood-work (Corvinus 1995). These assemblages from Nepal are of Late Pleistocene age (some may be even younger) and appear to share some typo-technological characters with similar industries from Southeast Asia (e.g., Van Tan 1997).

Durkadi. Further south in peninsular India, the most prominent evidence of cobble/boulder exploitation comes from the site of Durkadi and Mahadeo-Piparia in the central Narmada Basin (Khatri 1962; Armand 1983). Durkadi is better excavated than Mahadeo-Pipari, the latter of which represents a few isolated specimens from stratified sections along the river. Armand (1983) demonstrated the primary nature of Durkadi through meticulous excavations which yielded fresh Mode 1 artifacts on a conglomeratic bed – the primary source of the RM. Specimens from both these occurrences are larger in size than the average Soanian artifacts, probably due to the differential availability of rounded quartzite clasts in both areas (i.e., Siwalik Hills and the Narmada Basin). The majority of the specimens are produced on cobbles and occasionally from small boulders. Although specimens resembling "proto-bifaces" were identified, there are no definitive Acheulean bifaces at this site. Both occurrences comprise large quartzite clasts with large flake scars detached in either random or semi-centripetal patterns. Cores, choppers, core-scrapers, flakes (retouched and unretouched), and debitage are common in both assemblages and most specimens are fresh.

Mode 2 assemblages on cobbles/boulders

In addition to these core-and-flake occurrences on quartzite cobbles/boulders, numerous Mode 2 or Acheulean assemblages are known to have been produced from the same type of RM. The representative examples from northern, central, and peninsular India are Singi Talav in the Thar Desert (Misra 1995), Pilikarar in the Narmada Basin (Chauhan 2004b), and Attirampakkam in the Kortallyar Basin (Pappu et al. 2003; Pappu and Akilesh 2006).

Singi Talav. The oldest sediments in the Didwana area are represented by the Jayal Formation in the northern part of the Thar Desert, an extensive boulder conglomerate dated to the Early Quaternary (Misra 1987). Following its deposition, the gravel bed was subjected to uplift and covered by fine marly sediments, represented by the Amarpura Formation. Artifacts from this formation range from Early Acheulean to Late Middle

Palaeolithic–Early Upper Palaeolithic (Misra 1987: 102). Artifacts were recovered from several horizons from within a nearby stabilized sand dune and show evidence of hominin occupation in a combination of lake/playa/desert environments. Excavations at Singi Talav in the Amarpura Formation exposed an Early Acheulean industry on quartzite and quartz, and comprised handaxes, polyhedrons, spheroids, cores, flakes, and a few crudely-produced cleavers (Misra 1987: 104). Radiometric dating by Raghavan et al. (1989) places the lowermost cultural horizons at an age of more than 390 kyr (Mishra 1994). The RM was available as blocks or sub-angular weathered fragments from the nearby Aravalli outcrops and the Jayal Gravel Ridge and also gathered as clasts from then-extant streambeds (Misra 1987). The handaxes from Singi Talav are thick, large, and retain considerable amounts of cortex with biconvex cross-sections. Some artifacts are also produced on pebbles but have a marginal presence in the overall assemblage composition.

Pilikarar. The earliest depositional phase of the Narmada Quaternary sequence is represented by the Pilikarar Formation of Pleistocene age, which rests on the Cretaceous Deccan Traps (Tiwari and Bhai 1997). The exposed section at the type locality (Pilikarar) comprises four distinct strata (bottom to top): (i) a two-meter thick laterite bed; (ii) 0.5 m-thick spongy laterite; (iii) a boulder bed (0.3–3.0 m thick); and (iv) calcareous yellowish-brown soil with iron nodules of varying thickness. It is thought to underlie the Dhansi Formation in which the Geological Survey of India has possibly identified the Brunhes–Matuyama boundary (Rao et al. 1997). If the dating of both Dhansi and Pilikarar Formations is accurate, some of the Paleolithic evidence at Pilikarar may be among the oldest Early Acheulean assemblages in South Asia (Chauhan 2004; Patnaik et al. in press). Typologically, the bifaces broadly resemble other Early Acheulean assemblages in that they lack a sense of refinement and possess thick mid-sections and large flake scars. The Levallois technique is not clearly evident and younger artifact types are absent at the site. The Pilikarar assemblages come from two distinct stratigraphic contexts. The first assemblage appears to be in primary context from its location above a gravel/boulder deposit, representing a large colluvial fan. Here, the artifacts are in fresh condition and have been exposed recently by gully erosion. Some of the artifacts from above the gravel include handaxes, cleavers, discoidal cores, large

Figure 7.5 Main source of rounded and sub-angular quartzite clasts at Pilikarar at the base of the Vindhyan Hills.

retouched and un-retouched flakes, and debitage. The second assemblage was documented from within the conglomerate and is comprised of fresh and rolled cores, several choppers, and a unifacial cleaver-type artifact on a large side-struck flake. Both sets of artifacts are probably broadly contemporaneous and originate from the base of the nearby Vindhyan Hills where artifact concentrations are visibly higher and in association with the original quartzite clasts which form the source of the boulder bed of the Pilikarar Formation. These rounded to sub-angular clasts represent fluvially-transported or slightly-weathered quartzite blocks (Figure 7.5) eroding from the original bedrock from the Vindhyan Hills nearby. The sub-angular clasts found here (generally not found in fluvial channels) provided a greater number of natural striking platforms than completely rounded quartzites (another example is Isampur, described below).

Attirampakkam. The most significant Acheulean site in Tamil Nadu (southern India) is Attirampakkam, located in the Kortallayar valley (southeast India), and the adjoining regions have been intermittently studied for over a century by various workers (De Terra and Paterson 1939; Pappu 1996). More recently, S. Pappu has been meticulously investigating the site since 1999 (with the application of numerous modern techniques for the first time), and has contributed to revising previous geological interpretations. Techno-cultural levels at the site range from the Lower Paleolithic to the Upper Paleolithic, and the work has also resulted in the unique discovery of animal footprints within

banded clay sediments (Pappu *et al.* 2003). Acheulean artifacts, predominantly on quartzite, occur in ferricretes, ferricrete-gravels, and the underlying clay horizons and include choppers, discoids, sub-spheroids, bifaces (with minimum symmetry), cleavers, knives, and scrapers (Pappu 1996: 13). From geochemical studies and stratigraphical correlations, modes of tool use and discard have been preserved by episodic sedimentation at the site. Although the dating at this site is in progress, typological studies and geological observations at Attirampakkam point to a potentially early age of the lower Acheulean levels at the site – possibly between Late Lower and early Middle Pleistocene (S. Pappu, pers. comm.). Pappu and Akilesh (2006) report that RM quarrying, preliminary flake detachment, and basic shaping of large artifacts was done within a radius of 3–4 km, where cobbles and boulders were available as parts of fanglomerates of the Upper Gondwana Formation or the Allikulli Hills and associated outliers.

EXPLOITATION OF ANGULAR/ TABULAR RAW MATERIAL

The only known example of the exploitation of tabular or angular RM comes from the Hunsgi-Baichbal basins in southwestern India and is briefly mentioned here to show the contrast with rounded clasts. Systematic surveys and excavations have been conducted here since the late 1960s, exposing the presence of over 100 Acheulean occurrences in an erosional basin measuring 500 km² (Paddayya *et al.* 2002). The localities are found in a variety of depositional contexts: situated on the valley floor, along the plateau edge, in pediments, and along streams (Paddayya *et al.* 2002). Paddayya and Petraglia (1993) have observed that the artifact horizons in this region are close to the surface and overlie weathered bedrock (limestone) (Figure 7.6). Probably the most important Early Acheulean site from the Hunsgi complex is Isampur, representing the first known occurrence of artifacts in a quarry context in India (Paddayya and Petraglia 1997; Petraglia *et al.* 1999). Here, primary and buried limestone outcrops were exploited by Acheulean groups for stone-tool manufacture. Excavations at the locality displayed large slabs and cores of limestone that were flaked and fashioned into handaxes, cleavers, choppers, picks, knives, polyhedrons, scrapers, discoids, and unifaces. The debitage is present in abundance and shows variation in size, thus displaying all preserved stages of

Figure 7.6 Limestone slabs near Isampur in southern India. Note their acute angularity in contrast to the rounded quartzite clasts in Figure 7.2.

tool use – from the procurement stage to the retouch stage. Preliminary Electron spin resonance (ESR) dates at the locality of Isampur indicates a potentially early colonization of this region – at more than 1.27 myr ago (Paddayya *et al.* 2002).

DISCUSSION

Thus far, no Mode 1 assemblage or industry is known to have been produced exclusively on angular fragments or tabular clasts. The only Paleolithic evidence on such a form of RM is the Acheulean assemblages from Hunsgi-Baichbal valleys in southern India and the abundant younger lithic assemblages (i.e., Middle and Upper Paleolithic) elsewhere in the subcontinent which have been made from siliceous material. For example, in addition to the quartzite, basalt, granite, limestone, and sandstone clasts, fine-grained RM such as chert, jasper, chalcedony, and quartz also occurs as isolated clasts/fragments or small nodules often away from their original geological sources. In large riverbeds, such as the Narmada for example, such RM also occurs in fluvially-rounded form (signifying long-distance natural transport) or as sub-angular clasts (signifying short-distance natural transport of originally-tabular fragments). Such RM is almost always constrained in its overall dimension and morphology and is also found on (i) gravel beaches along rivers/streams, (ii) or occurring *away* from fluvial channels, and (iii) as veins (e.g., quartz) within other sedimentary bodies. With the exception of such areas as the Rohri Hills in southern Pakistan (Biagi and Cremaschi

1988), siliceous types such as chert were rarely utilized by Lower Paleolithic hominins in the Indian subcontinent. Such RM types were increasingly exploited only during the Middle and Upper Paleolithic phases (and later prehistoric times)[1]. At certain sites, however, such as at Attirampakkam and in the Siwalik Hills region, hard and coarse-grained RM (i.e., quartzite) continued to be utilized by "post-Lower Paleolithic" hominins when fine-grained RM was unavailable or occurred in minimal quantity.

Raw material transport behaviors

It can be presumed that South Asian hominins probably tested quartzite clasts in streambeds or bedrock outcrops for fracture quality, to avoid internal flaws, prior to blank transport. In older behavioral contexts such as the Oldowan, RM was transported from a few kilometers and sometimes up to thirteen kilometers from their original sources (Hay 1976; Stout *et al.* 2005). In South Asia, there is very little evidence of RM transport over distances greater than about 10 km. Indeed, the majority of Paleolithic sites are found near their respective sources of RM (Dennell 2004; Chauhan 2005). In the Siwalik region, it appears that long-distance transport of RM was practiced at a marginal level, presumably owing to the suitable quartzite clasts available at frequent lateral intervals (Mukerji 1976). Owing to this frequent availability of RM, the curation of finished tools rather than pebble/cobble blanks was more common. This is supported by the proximity of sites to major sources of RM, which do not seem to be further than 3–4 km in maximum distance. Sites are often located adjacent to such sources (e.g., Toka) or are a part of such scatters of clasts in the form of gravel/conglomerate outcrops (e.g., Karor Uparli) in the interior zones. Rarely are sites located in the plains to the south and interior hills or slopes, beyond 2–3 km from the Siwalik Frontal zone. At some locations, the relatively low numbers of artifacts was presumably explained by the low profile of quartzite clasts nearby. At locations with ample RM, however, reasons for the low number or virtual absence of artifacts are currently unknown, and may be related to such factors as chronological disparity of the post-Siwalik streams, topographic instability (Bhave and Deo 1997–98), or simply inconsistent patterns of land-use.

In the central Narmada Basin, it appears that Acheulean bifaces were produced near their sources of RM and utilized and discarded on the floodplains. For example, the pursuit of specific functions or subsistence behaviors have led to the frequent discard of Late Acheulean cleavers at Surajkund, where other tool types including debitage, cores, handaxes, scrapers, are generally absent (R. Patnaik *et al.* in press). The nearest source of RM for these occurrences is the extensive stretch of Vindhyan Hills at a distance of 0.5–3.0 km from the main channel. This evidence for very local procurement of stone is consistent with that from other Paleolithic studies, which suggest that stone was rarely carried more than 10 km, and usually less than 4 km from its source (Dennell 2004: 434). A rare example of RM transport over considerably longer distances in India comes from Yedurwadi where a quartzite spheroid may have been transported for more than 50 km (see Dennell 2007). Overall, it appears that quartzite was obtained in main river/stream channels and then transported and utilized within an average of a 5 km radius on the surrounding floodplains. Thus far, no such sites (i.e., using rounded quartzites) in lacustrine contexts are known.

Implications of dimensional and morphological constraints of the quartzite

Excepting the early evidence from Riwat and Pabbi Hills, it is apparent that most Lower Paleolithic sites and (Mode 2 and core-and-flake assemblages) in the Indian subcontinent were cobble-boulder dominated and most Middle and Upper Paleolithic sites were pebble-cobble dominated, undoubtedly related to artifact dimensional variation and morphological requirements in certain assemblage compositions. For the most part, artifact assemblages produced on pebble-cobbles show minimal cortex removal. However, even within Soanian assemblages, which are restricted in RM form and size, certain expedient artifacts such as choppers illustrate variable levels of cortex removal from the side(s) or face(s) of the clast (i.e., uni- and bimarginal verses uni- and bifacial choppers) (see Figure 7.4 in Chauhan 2003; Chauhan 2005, 2007). In addition, recent investigations by the author suggest that angular or amorphous cobble clasts at Toka may have been obtained by throwing these blanks on stationary anvils, to obtain striking platform(s) for further reduction of the otherwise unwieldy clasts. At the same time, the pattern of flake scars on some pebble specimens are indicative of the bipolar technique,

traditionally used to maximize flake output. Not surprisingly, the dimensional or morphological restriction of these rounded quartzites affected the assemblage composition of both Acheulean and non-Acheulean assemblages. For example, the Soanian industry rarely contains polyhedrons or spheroids, primarily due to the limited lateral thickness of the often flat quartzite cobbles utilized. Likewise, Acheulean assemblage composition and Acheulean settlement patterns were also directly affected by quartzite clast forms. At the same time, limited quartzite forms did not affect the morphological variability or typological diversity in the Soanian and similar industries.

According to Gaillard's (1995, 1996) observations on the Acheulean evidence from Singi Talav (Misra 1995) and the Soanian site of Dehra-Gopipur (Mohapatra 1990), these traditions have a strikingly similar processing sequence, mainly in terms of trimming, utilizing large cortical flakes, and utilizing semi or non-cortical flakes (the latter two receiving further retouching) (Figure 7.7a and b). This suggests that both groups of hominins (Acheulean and non-Acheulean) may have shared the same type of planning depth or foresight to produce pre-conceived tool types, regardless of their techno-morphological complexity (however, see Chauhan 2003). Overall however, the manufacturing process of Acheulean specimens from rounded quartzites was a much more complex process than producing core-and-flake assemblages, exercising a greater degree of clast manipulation (particularly boulders). From general observations on Acheulean flake morphologies and associated technological attributes, it is apparent that both hard and soft hammers were employed on the quartzite. Large and bold flake scars on Early Acheulean cores (Figure 7.8) indicate the detachment of primary and subsequent end-flakes or side-flakes for cleaver and flake-tool production. On such flake-based tools, the coarse-grained nature of quartzites rarely allow "precision retouch" or controlled detachment of numerous micro-flakes as is easily possible on siliceous RM. In the case of some cleavers produced on large primary flakes at Pilikarar, minimal subsequent trimming was required (Figure 7.9). For handaxe production, especially from boulders, preliminary flake detachment and basic shaping of artifacts probably took place at the quarry (e.g., Pappu and Akilesh 2006), followed by transport of the blanks to areas where they were shaped and trimmed further into the final product (in contrast to Isampur with tabular material). Unlike handaxes produced from boulders,

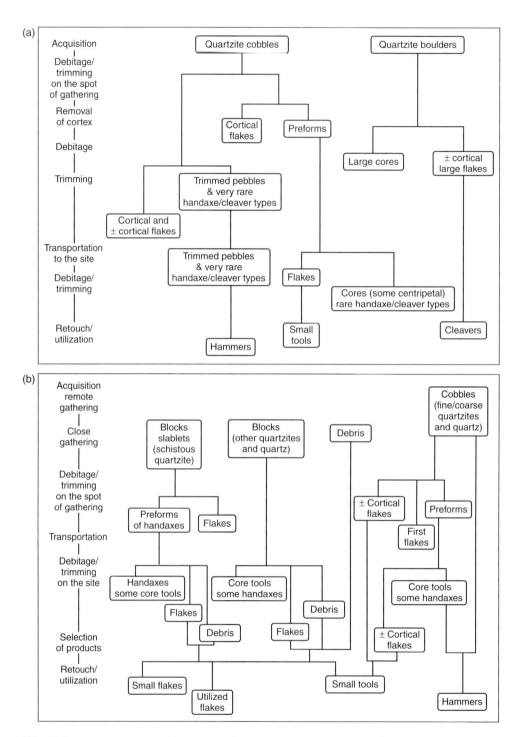

Figure 7.7 (a) Processing sequence at Dehra Gopipur. (b) Processing sequence at Singi Talav. (Redrawn from Gaillard 1995.)

Figure 7.8 Large core from Pilikarar with multiple primary flake scars. Geological hammer provides a relative scale. (Image courtesy: Stanley Ambrose.)

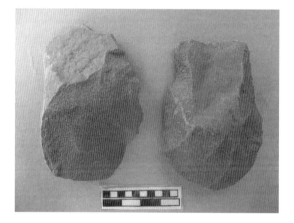

Figure 7.9 Large flake blank (left) and cleaver (right) from Pilikarar, both detached from quartzite boulder(s).

specimens made from cobbles often retain some cortex on the butts and one or more faces (e.g., Hou *et al.* 2000). This trend is more common in Early Acheulean types (at different sites) such as cleavers, handaxes, and picks, but not common in Late Acheulean assemblages where minimal cortex is retained and the "butts" are worked or shaped to varying levels. Trihedral picks appear to have been produced from core-fragments or from flakes thicker and larger than the primary/secondary blanks used for biface manufacture.

When manufacturing a biface (of reasonable size) from rounded RM, a relatively large blank is required since a greater amount of cortex needs to be removed in comparison with an angular/tabular blank. Quartzite is harder than most known RMs and its rounded forms appear to have required greater planning capacities and extra reduction strategies in comparison to limestone slabs. In rounded forms, one is forced to start chipping at the perimeter of the blank or clast and work one's way inwards to prepare the core/striking platform. This concept can also be applied to boulders when extracting a series of large flakes for biface manufacture, thus initially removing a higher amount of cortex more efficiently (see Figure 3 in Chauhan 2003). When the same is attempted on a cobble, the resulting cores and flakes are restrictive in size to allow further trimming for biface production. Therefore, from metrical analysis of bifaces (on rounded clasts) and an understanding of their reduction strategy through experimental flintknapping, one can (approximately) gauge the minimum size requirements of the original blanks. For instance, some Siwalik bifaces range between 9.9 and 15.2 cm in length (table 3 in Chauhan 2003). The minimum size of the blank would have to be proportionately larger in length and diameter, perhaps by at least 50–70%. This principle is explicit if the intention is to remove all or most of the cortex through hard-hammer percussion, unless they were made on large flakes derived from boulders. These differences in the size and reduction strategies, along with selective exploitation of the RM in the Siwalik region, have been previously studied by Mohapatra (1990) and recently highlighted by the author (Chauhan 2003). The correlation between raw-material size and form and the absence of bifaces has also been emphasized from studies on other lithic industries, such as the Clactonian in Britain (see White 2000).

CONCLUSIONS

From the preliminary assessments in this chapter, it is apparent that rounded and sub-angular quartzite clasts were abundant, and thus extensively utilized by Plio-Pleistocene hominins of South Asia. Not only were rounded quartzite clasts a widespread form of raw material, but they also proved to be functionally versatile throughout the entire Paleolithic period of South Asia. Most Paleolithic sites can be correlated with either pebble-cobble or cobble-boulder raw material groups, both of which occur as rounded or sub-angular forms near bedrock outcrops, stream channels, and paleo-channel deposits. Some sites belonging to the first group (pebbles/cobbles) are Riwat, Pabbi Hills, Toka, Site 55, Kalpi, and Samnapur; examples of those associated with the latter group (cobbles/boulders) are Singi Talav, Durkadi, Pilikarar, Attirampakkam, and several occurrences in Nepal. These different quartzite clasts forced hominins to make different decisions or choices for the most efficient manner of blank reduction. Indeed, the use of rounded quartzites reflects on hominin technological proficiency and associated cognitive levels which have major implications on their ability to reduce and shape unwieldy clasts or blanks in order to obtain suitable striking platforms. For example, to obtain simple choppers, cores, or flakes in Soanian and similar assemblages, pebbles and cobbles were specifically selected with one flattish face, allowing easier primary-flake detachment. In contrast, at Acheulean sites, boulders were probably split longitudinally and cobbles may have been thrown against stationary anvils to create suitable striking platforms for further flake detachment.

The distribution, accessibility, and morphology of rounded and sub-rounded quartzite clasts appear to have been the principal factors in determining settlement location, inter and intra-regional mobility, and associated assemblage compositions and subsistence behaviors. Locations with a high density of quartzite clasts occasioned intensive exploitation of such sources, possibly represented by multiple visits for clast procurement. High quartzite frequency probably also resulted in a greater flow of blanks and formal tools across different landscapes, including zones without prominent sources of raw material. The rich core-and-flake evidence belonging to the Soanian industry signifies inter-regional dispersal from peninsular India or northern Pakistan, leading to a novel environmental adaptation which spread quickly through hominin populations in the region within a relatively short time-span. Most of the assemblages suggest raw material transport distances of no more than 5 km from the localized sources. Hominins that produced Mode 1 or non-biface assemblages do not appear to have transported rounded clasts from their fluvial sources to considerable distances. The distributional patterns of such sites highlights the expedient use of such raw material forms and the most important factors for site location/selection appear to have been the presence of raw material combined with a stable topography.

Additional studies of raw material acquisition and processing in the Indian subcontinent are bound to yield new interpretations and models regarding early hominin home ranges and occupational histories of this region.

ACKNOWLEDGEMENTS

I am thankful to the editors, Brooke Blades and Brian Adams, for the invitation to contribute to this interesting volume and important topic. Some of the data discussed in this chapter was generated through two different projects funded by the Australian Research Council (PI: David Cameron), the National Geographic Society (PI: Rajeev Patnaik), and the Wenner-Gren Foundation for Anthropological Research (PI: the author), to all of whom I am very grateful. I also thank the Government of India for approving the Siwalik Paleolithic project for my doctoral research and for providing a research visa in affiliation with the Department of Geology, Panjab University (Chandigarh). I thank Stanley Ambrose for providing one of the photographs and K. Paddayya for permission to participate in excavations at Isampur during my post-graduate affiliation with Deccan College. Finally, I extend my thanks to Nicholas Toth and Kathy Schick, co-directors of the Stone Age Institute & CRAFT Research Center (Indiana University), for the postdoctoral position and the overall support of my research.

ENDNOTE

1   An exception to this is the episodic use of chert nodules for Oldowan tool production at Olduvai Gorge, where associated sources became available during fluctuations in the palaeo-lake levels (Stiles 1998).

REFERENCES

Acharyya, S.K. (1994) The Cenozoic Foreland basin and tectonics of the Eastern Sub-Himalaya: Problems and prospects. In: Kumar, R. and Ghosh S.K. (eds) *Siwalik Foreland Basin of Himalaya: Himalayan Geology* 15, Oxford & IBH Publishing, New Delhi, pp. 3–21.

Allchin, B. (1995) Early human occupation in the northern Punjab, Pakistan: An overview of the Potwar Project of the British Archaeological Mission to Pakistan (1981–1991). In: Wadia, S., Korisettar, R. and Kale, V.S. (eds) *Quaternary Environments and Geoarchaeology of India*, Geological Society of India, Bangalore, pp. 150–67.

Armand, J. (1983) *Archaeological Excavations in the Durkadi Nala: An Early Palaeolithic Pebble-Tool Workshop in Central India*. Munshiram Manoharlal Publishers Pvt. Ltd., Delhi.

Bhave, K. and Deo, S. (1997–98) Geomorphic perspective of archaeological sites in the Bhima Basin, Pune District, Maharashtra. *Puratattva* 28, 46–54.

Biagi, P. and Cremaschi, M. (1988) The early Palaeolithic sites of the Rohri Hills (Sind, Pakistan) and their environmental significance. *World Archaeology* 19(3), 421–33.

Brantingham, P.J. (2003) A neutral model of stone raw material procurement. *American Antiquity* 68(3), 487–509.

Brozović, N. and Burbank, D.W. (2000) Dynamic fluvial systems and gravel progradation in the Himalayan foreland. *Geological Society of America Bulletin* 112(3), 394–412.

Chauhan, P.R. (2003) An overview of the Siwalik Acheulian and reconsidering its chronological relationship with the Soanian – A theoretical perspective. *Assemblage* 7. http://www.assemblage.group.shef.ac.uk/issue7/chauhan.html

Chauhan, P.R. (2004) A review of the Early Acheulian evidence from South Asia. *Assemblage* 8. http://www.assemblage.group.shef.ac.uk/issue8/chauhan.html

Chauhan, P.R. (2005) The technological organization of the Soanian palaeolithic industry: A general "typo-qualitative" description of a large Core-and-Flake assemblage in surface context from the Siwalik Hills of northern India. In: Srivastava, V.K. and Singh, M.K. (eds) *Issues and Themes in Anthropology: A Festschrift in Honour of Prof. D.K. Bhattacharya*, Palaka Prakashan, Delhi, pp. 287–336.

Chauhan, P.R. (2006) Human origins studies in India: position, problems and prospects. *Assemblage* 9. http://www.assemblage.group.shef.ac.uk/issue9/chauhan.html

Chauhan, P.R. (2007) Soanian cores and core-tools from Toka, northern India: towards a new typo-technological organization. *Journal of Anthropological Archaeology* 26, 412–41.

Chauhan, P.R. (2008) Soanian lithic occurrences and raw material exploitation in the Siwalik Frontal zone, northern India: a geoarchaeological approach. *Journal of Human Evolution* 54(5), 591–614.

Chauhan, P.R. (in press) Core-and-flake assemblages of India. In: Norton, C.J. and Braun, D. (eds) *Asian Paleoanthropology: From Africa to China and Beyond* (Book Series: Vertebrate Paleobiology and Paleoanthropology). Kluwer-Academic Press.

Corvinus, G. (1995) Quaternary stratigraphy of the intermontane dun valleys of Dang-Deokhuri and associated prehistoric settlements in Western Nepal. In: Wadia, S., Korisettar, R., and Kale, V.S. (eds) *Quaternary Environments and Geoarchaeology of India*. Geological Society of India, Bangalore, pp. 124–49.

Corvinus, G. (2002) Arjun 3, a Middle Palaeolithic site in the Deokhuri Valley, Western Nepal. *Man and Environment* 27(2), 31–44.

Dennell, R.W. (2004) *Early Hominin Landscapes in Northern Pakistan: Investigations in the Pabbi Hills*. BAR International Series 1265. Archaeopress, Oxford.

Dennell, R. (2007) "Resource-rich, stone-poor": Early hominin land use in large river systems of northern India and Pakistan. In: Petraglia, M.D. and Allchin, B. (eds) *The Evolution and History of Human Populations in South Asia – Interdisciplinary studies in Archaeology, Biological Anthropology, Linguistics and Genetics*, Vertebrate Paleobiology and Paleoanthropology Series, Springer Press, New York, (in press), pp. 41–68.

Dennell, R.W. and Rendell, H.M. (1991) De Terra and Paterson and the Soan Flake Industry: A new perspective from the Soan Valley, Northern Pakistan. *Man and Environment* 16(2), 90–9.

De Terra, H. and Paterson, T.T. (1939) *Studies on the Ice Age in India and Associated Human Cultures*. Carnegie Institute Publication 493, Washington, DC.

Gaillard, C. (1995) An Early Soan assemblage from the Siwaliks: A comparison of processing sequences between this assemblage and an Acheulian assemblage from Rajasthan. In: Wadia, S., Korisettar, R., and Kale, V.S. (eds) *Quaternary Environments and Geoarchaeology of India*, Geological Society of India, Bangalore, pp. 231–45.

Gaillard, C. (1996) Processing sequences in the Indian Lower Palaeolithic: Examples from the Acheulian and the Soanian. *Indo-Pacific Prehistory Association Bulletin* Volume I, 14, 57–67.

Gaillard, C. and Mishra, S. (2001) The Lower Palaeolithic in South Asia. In: under the direction of Sémah, F., Falgueres, C., Grimaud-Herve, D., and Sémah, A. (eds) *Origin of Settlements and Chronology of the Palaeolithic Cultures in Southeast Asia*, Colloque International de la FONDATION SINGER POLIGNAC. Paris, June 3–5, 1998, Semenanjung, Paris, pp. 73–91.

Ganjoo, R.K., Safaya, S. and Lidoo, P. (1993–94) A report on the discovery of a Palaeolithic site near Katra, Udhampur District, Jammu. *Puratattva* 24, 33–4.

Gupta, A. (1995) Fluvial geomorphology and Indian rivers. In: Wadia, S., Korisettar, R., and, Kale, V.S. (eds) *Quaternary Environments and Geoarchaeology of India*, Geological Society of India, Bangalore, pp. 52–60.

Hay, R. (1976) *Geology of the Olduvai Gorge*. University of California Press, Berkeley.

Hou, Y.M., Potts, R., Baoyin, Y. *et al.* (2000) Mid-Pleistocene Acheulean-like stone technology of the Bose Basin, South China. *Science* 287, 1622.

Hurcombe, L. (2004) The stone artefacts from the Pabbi Hills. In: Dennell, R.W. Petraglia, M.D., and Allchin, B. (eds) *Early Hominin Landscapes in Northern Pakistan: Investigations in the Pabbi Hills*, British Archaeological Reports International Series 1265, Oxford, pp. 222–92.

Jayaswal, V. (1982) *Chopper-Chopping Component of Palaeolithic India*. Agam Kala Prakashan, Delhi.

Karir, B.S. (1985) *Geomorphology and Stone Age Culture of Northwestern India*. Sundeep Prakashan, Delhi.

Khatri, A.P. (1962) Mahadevian: An Oldowan pebble culture of India. *Asian Perspectives* 6(1), 186–97.

Korisettar, R. and Rajaguru, S.N. (2002) The monsoon background and the evolution of prehistoric cultures of India. In: Paddayya, K. (ed.) *Recent Studies in Indian Archaeology*, Munshiram Manoharlal Publishers Pvt. Ltd., New Delhi, pp. 316–48.

Mishra, S. (1994) The South Asian Lower Palaeolithic. *Man and Environment* 19(1–2), 57–72.

Misra, V.N. (1987) Middle Pleistocene adaptations in India. In: Soffer, O. (ed.) *The Pleistocene Old World-Regional Perspectives*, Plenum Press, New York, pp. 99–119.

Misra, V.N. (1995) Geoachaeology of the Thar Desert, Northwest India. In: Wadia, S., Korisettar, R., and Kale, V.S. (eds) *Quaternary Environments and Geoarchaeology of India – Essays in Honour of Professor S.N. Rajaguru*, Geological Society of India, Bangalore, pp. 210–30.

Misra, V.N., Rajaguru, S.N., Ganjoo, R.K. and Korisettar, R. (1990) Geoarchaeology of the palaeolithic site at Samnapur in the central Narmada Valley. *Man and Environment* 15(1), 107–16.

Mohapatra, G.C. (1990) Soanian-Acheulian relationship. *Bulletin of Deccan College Postgraduate and Research Institute* 49, 251–9.

Movius, H.L., Jr. (1948) The Lower Paleolithic cultures of southern and eastern Asia. *Transactions of the American Philosophical Society* 28, 329–420.

Mukerji, A.B. (1976) Geomorphological study of *choe* terraces of the Chandigarh Siwalik Hills, India. *Himalayan Geology* 5, 302–26.

Paddayya, K., Blackwell, B.A.B., Jhaldiyal, R. *et al.* (2002) Recent findings on the Acheulian of the Hunsgi and Baichbal valleys, Karnataka, with special reference to the Isampur excavation and its dating. *Current Science* 83(5), 641–7.

Paddayya, K. and Petraglia, M. (1993) Formation Processes of Acheulian localities in the Hunsgi and Baichbal Valleys, Peninsular India. In: Goldberg, P., Nash, D.T., and Petraglia, M.D. (eds) *Formation Processes in Archaeological Context*, Monographs in World Archaeology in World Archaeology No 17. Prehistory Press, Madison, pp. 61–82.

Paddayya, K. and Petraglia, M. (1997) Isampur: An Acheulian workshop in the Hunsgi Valley, Gulbarga District, Karnataka. *Man and Environment* 22, 95–100.

Pappu, R.S. (2001) *Acheulian Culture in Peninsular India*. D.K. Printworld (P) Ltd., New Delhi.

Pappu, S. (1996) Reinvestigation of the prehistoric archaeological record in the Kortallayar Basin, Tamil Nadu. *Man and Environment* 20(1), 1–23.

Pappu, S. and Akilesh, K. (2006) Preliminary observations on the Acheulian assemblages from Attirampakkam, Tamil Nadu. In: Goren-Inbar, N. and Sharon, G. (eds) *Axe Age: Acheulian Tool-Making from Quarry to Discard (Approaches to Anthropological Archaeology)*, Equinox Publishing, London, pp. 155–80.

Pappu, S., Gunnell, Y., Taieb, M., Brugal, J.P. and Touchard, Y. (2003) Excavations at the Palaeolithic site of Attirampakkam, South India: preliminary findings. *Current Anthropology* 44(4), 591–7.

Paterson, T.T. and Drummond, J.H.J. (1962) *Soan, the Palaeolithic of Pakistan*. Department of Archaeology, Government of Pakistan, Karachi.

Patnaik, R., Chauhan, P.R., Rao, M.R. *et al.* (in press). New geochronological, palaeoclimatological and Palaeolithic data from the Narmada Valley hominin locality, central India. *Journal of Human Evolution*.

Petraglia, M.D. (1998) The Lower Palaeolithic of India and its bearing on the Asian record. In: Petraglia, M.D. and Korisettar, R. (eds) *Early Human Behaviour in Global Context: The Rise and Diversity of the Lower Palaeolithic Record*, Routledge, London, pp. 343–90.

Petraglia, M.D., LaPorta, P., and Paddayya, K. (1999) The first Acheulian quarry in India: stone tool manufacture, biface morphology, and behaviors. *Journal of Anthropological Research* 55, 39–70.

Raghavan, H., Rajaguru, S.N. and Misra, V.N. (1989) Radiometric dating of a Quaternary dune section, Didwana, Rajasthan. *Man and Environment* 13, 19–22.

Rao, K.V., Chakraborti, S., Rao, K.J., Ramani, M.S.V., Marathe, S.D. and Borkar, B.T. (1997) Magnetostratigraphy of the Quaternary fluvial sediments and tephra of Narmada Valley, Central India. In: Udhoji, S.G. and Tiwari, M.P. (eds) *Quaternary Geology of the Narmada Valley – A Multidisciplinary Approach*. Special Publication No. 46 of the Geological Survey of India. Geological Survey of India, Calcutta, pp. 65–78.

Reddy, K.T., Prakash, P.V., Rath, A. and Rao, Ch. U.B. (1995) A Pebble tool assemblage on the Visakhapatnam Coast. *Man and Environment* 20(1), 113–18.

Rendell, H.M., Dennell, R.W. and Halim, M.A. (1989) Pleistocene and Palaeolithic investigations in the Soan Valley, Northern Pakistan. Oxford: *British Archaeological Report* (International Series) 544.

Sangode, S.J. and Kumar, R. (2003) Magnetostratigraphic correlation of the Late Cenozoic fluvial sequences from NW Himalaya, India. *Current Science* 84(8), 1014–24.

Schick, K.D. (1994) The Movius Line reconsidered – Perspectives on the earlier Paleolithic of Eastern Asia. In: Corruccini, R.S. and Ciochon, R.L. (eds) *Integrative Paths to the Past- Paleoanthropological Advances in Honor of F. Clark Howell*, Prentice Hall, Englewood Cliffs, New Jersey, pp. 569–96.

Seong, C. (2004) Quartzite and vein quartz as lithic raw materials reconsidered: A view from the Korean Paleolithic. *Asian Perspectives* 43(1), 73–91.

Stiles, D. (1978) Palaeolithic artefacts in Siwalik and Post-Siwalik deposits of Northern Pakistan. *Kroeber Anthropological Society Papers* (1) S3–S4, 129–48.

Stiles, D. (1998) Raw material as evidence for human behaviour in the Lower Pleistocene: The Olduvai case. In: Petraglia, M. D. and Korisettar, R. (eds) *Early Human Behaviour in Global Context: The Rise and Diversity of the Lower Paleolithic Period*, Routledge, London, pp. 133–50.

Stout, D., Quade, J., Semaw, S., Rogers, M.J. and Levin, N.E. (2005) Raw material selectivity of the earliest stone tool-makers at Gona, Afar, Ethiopia. *Journal of Human Evolution* 48, 365–80.

Sullivan, III, A.P. (1992) Investigating the archaeological consequences of short-duration occupations. *American Antiquity* 57(1), 99–115.

Tewari, R., Pant, P.C., Singh, I.B. *et al.* (2002) Middle Palaeolithic human activity and palaeoclimate at Kalpi in Yamuna Valley, Ganga Plain. *Man and Environment* 27(2), 1–13.

Tiwari, M.P. and Bhai, H.Y. (1997) Quaternary stratigraphy of the Narmada Valley. In: Udhoji, S.G. and Tiwari, M.P. (eds) *Quaternary Geology of the Narmada Valley – A Multidisciplinary Approach*, Special Publication No. 46 of the Geological Survey of India, Geological Survey of India, Calcutta, pp. 33–63.

Van Tan, H. (1997) The Hoabinhian and before. *Bulletin of the Indo-Pacific Prehistory Association* (Chiang Mai Papers, Volume 3) 16, 35–41.

White, M.J. (2000) The Clactonian question: On the interpretation of core-and-flake assemblages in the British Lower Palaeolithic. *Journal of World Prehistory* 14(1), 1–63.

Chapter 8

FILLING THE VOID: LITHIC RAW MATERIAL UTILIZATION DURING THE HUNGARIAN GRAVETTIAN

Viola Dobosi

Hungarian National Museum

ABSTRACT

The Gravettian period in the region of Hungary has been described by some as a "void" in the European Upper Paleolithic record. This characterization in part represents the failure of Hungarian Paleolithic specialists to disseminate the results of years of intensive research to the larger international scientific community. In this chapter, the results of lithic raw material analyses of assemblages from several Hungarian Gravettian sites are presented. A discussion of various local and non-local lithic raw materials utilized during the Hungarian Gravettian period, and how utilization patterns varied within this period, is presented. The value of the lithic raw material comparative collection, the *Lithotheca*, of the Hungarian National Museum in Budapest, is illustrated.

INTRODUCTION

As a result of intensive and opportune research over the past few decades in Hungary, a sketch of the Upper Paleolithic period has emerged that approximates the general European line of events. However, while some aspects of this sketch fit seemingly well, others are less consistent with the picture elsewhere in Europe, and certain issues still require investigation. Among these

Lithic Materials and Paleolithic Societies, 1st edition. Edited by B. Adams and B.S. Blades, ©2009 Blackwell Publishing. ISBN 978-1-4051-6837-3.

are: gaps in the absolute chronology of the Hungarian Upper Paleolithic; the contradictory chronology of the period surrounding the last peak of the Würm glacial; and the relationships, if any, among archeological cultures and the potential impact on terminology. Another critical area of research in Hungary is the on-going re-evaluation of key loess sequences. Research has demonstrated some loess sequences are older than previously thought resulting in a reconsideration of the age of fossil soil horizons and associated archeological layers. Important pieces of the picture are simply unavailable at the present time. Unfavorable preservation conditions result in a poor data set for the Late Paleolithic environment, and faunal and botanical (e.g., palynological) data are often missing at important Late Paleolithic settlements. Resolution of these important issues is on-going and beyond the scope of this chapter. Instead, the focus here will be to present our current understanding of lithic raw material utilization at several Hungarian Gravettian sites, and discuss what these data reveal about human adaptations at this time.

A static evaluation of Paleolithic sites can be performed using ethnoarcheological methods (e.g., Pincevent). Applicability of these methods requires, however, certain conditions that cannot be met in all cases. Temporal dynamics of cultural processes are recognized on the basis of conventional typology. This approach is rather subjective, but it has a strong tradition in the history of Paleolithic research. Spatial (and, partly, temporal) dynamics also can be identified by such objective data as raw material utilization and circulation. This latter perspective can be approached in two ways. The first is by tracing the geological sources of well-documented archeological finds. This is not always a successful endeavor, and can be especially difficult in the case of re-utilized materials. Sometimes it must be acknowledged that during the Upper Paleolithic, a much wider spectrum of natural resources was utilized than those known today. The second approach involves the mapping and characterization of the primary and secondary resources of raw materials used for the production of chipped stone implements we know today, such as natural and artificial outcrops, recent drifts, and alluvial sediments.

THE HUNGARIAN GRAVETTIAN

This chapter is concerned with raw material circulation as represented by Upper Paleolithic Gravettian sites, a cultural unit that requires some comments in view of generalizing remarks still encountered in the international technical literature. The Hungarian and Eastern Slovakian Upper Paleolithic, which share the same geological–geographical–ecological milieu and have a similar archeological heritage, is undoubtedly less rich than that of some surrounding countries. Despite this, we find the skepticism expressed in a recent major synthesis of the period unfounded:

> Les sites gravettiens sont déjá plus rares on Slovaquie orientale et en Hongrie. Dans le plaine de Pannonie, ou sur le versant oriental des Alpes, comme sur le versant nord des Alpes dinarique, les sites sont rares voire inexistants, et cette rareté fait conclure actuellement á un peuplement gravettian plus particuliérement localisé dans le partie septentrionale de l'Europe centrale. (Djindjan *et al.* 1999: 189)[1]

About one-third of the $92\,000\,km^2$ forming the territory of present-day Hungary is composed of alluvial cone sediments stemming from unique orographical conditions. In these areas, the likelihood of unearthing Paleolithic sites is greatest on insular, uplifted Pleistocene relict surfaces or sand and loess plateaux exposed by modern quarrying. In the remaining two-thirds of the country, however, over 60 sites to date have been confidently assigned to the period between the Interpleniglacial and the end of the Ice Age. Among these sites, 32 have been investigated by meticulous excavation. It is our belief that the "void" expressed by the above quote reflects the failure of Hungarian Paleolithic specialists to adequately disseminate the results of research in our country to the larger international scientific community. It is our goal to fill this "void" in order to present a more accurate and balanced picture of the Upper Paleolithic period in our region.

In Hungary, three Gravettian "phyla" are recognized. The Older Blade Industry, or Older Gravettian, dates to approximately 26–28 kyr and is possibly analogous to the neighboring Pavlovian circle. Settlements belonging to this unit include Megyaszó, a habitation settlement, and Püspökhatvan, a workshop site. The Ságvárian culture represents a reappearance of the pebble-working tradition and is dated to around 19–18 kyr. It is represented by the settlement at Mogyorósbánya. Finally, there is the Younger Blade Industry or Epigravettian phylum, probably descended from the Pavlovian circle. This phylum appeared after the Würm 3 cold maximum and lasted until the end of the Ice Age. Epigravettian sites discussed here are the settlements around Pilismarót, dated to approximately 16 kyr.

Within the Upper Paleolithic period, most sites with statistically relevant amounts of stone artifacts can be attributed to the Gravettian, both culturally as well as chronologically. In this chapter, sites are selected from all three phyla that demonstrate the raw material procurement strategy of the given community, the action radius, and the direction and intensity of contacts based on our current knowledge of geological source locations.

LITHIC RAW MATERIAL TYPES AND DISTRIBUTIONS

Raw material types are considered in relation to the distance between source and site based on principles presented in the Lithotheca catalogue volumes (see Biro, this volume, Chapter 4). Though these categories may initially appear subjective, they are in fact based on relatively objective criteria (Biró-Dobosi 1991: 8). Raw material types are divided into three general categories: local, regional, and long distance.

Local lithic materials are defined as those that did not require more than a single day's walking to procure. Based on our practical experience resulting from the study of workshop-settlements, local materials are conceivably collected from even smaller distances not exceeding a few hours walk; indeed, in some cases groups were practically "sitting" on raw material sources.

Regional lithic raw materials consist of loosely defined categories varying by age and culture, and denote raw materials used by groups inhabiting the general environs of the source.

With respect to Paleolithic assemblages, definition of this term will vary as the limits of the areas inhabited by particular cultural units are uncertain or simply unknown. Compared to local sources, regional sources can be defined as those within 1–2 days journey on foot, or perhaps the distance of maximal visibility of a particular source. Within the context of this chapter, typical examples of regional lithic raw materials exploited during the Hungarian Upper Paleolithic consist of the hydrothermal resources of the Mátra foothill region, 30–35 km distant from the sites on the northern margin of the Alföld in the Jászság area, or, in the case of Megyaszó, lithic sources of the prominent Tokaj Mountains lying within a distance of 20–25 km.

Long distance lithic raw materials are here defined as those transported over hundreds of miles/kilometers from a source area, and artifacts made from such raw materials may represent cross-cultural items. This category is highly suited for theories of interregional interaction and/or movements. Though in most cases specific long distance raw materials are fairly unique, making source identification straightforward, the particular route or pathway of an "exotic" artifact to a particular site can only be indirectly postulated. Distinctive "long distance" raw materials in the Hungarian Paleolithic include Northern flints, Upper

Silesian material from secondary contexts, Jurassic Krakow/Chocolate flint, Transcarpathian Prut River flint, and rock crystal.

Raw material varieties classified as "Northern flint" consist of a set of high-quality silices located to the north and northeast, beyond the Carpathian arch and outside the Carpathian Basin. Common types at Paleolithic localities are erratic flint, Jurassic flint from the environs of Krakow, and Swieciechów (possibly Volhynian) spotted hornstone. The petroarcheological terminology used for these materials is inconsistent. Some tend to use terms related to the mineralogical, physical-chemical component ratio (Pawlikowski 1989), while others prefer to use a topographical-chronological grouping (Prychistal 1989). Hungarian petroarcheological research has adopted both definitions, with an emphasis on formation type and age (Biró-Pálosi 1986). The difficulties in identifying such materials in archeological contexts is beyond the scope of this chapter (see Biró 1989; Simán 1989). The Upper Silesian secondary geological sources, yielding silex pebbles of Baltic origin transported by the continental ice sheet (moraines of Riss glaciation age) have been known since the 1920s. Due to the large extent of the source region and its variable composition, identification in archeological assemblages requires considerable experience using rather subjective methods based on macroscopic observations (e.g., characteristics of the pebble cortex or patina). However, the pebble cortex can be missing from the finished tools and patination can be misleading. It is easier to identify the sources and, consequently, estimate the distance between the source of the raw material and the place of utilization in the case of exotic raw materials with single isolated sources. Such is the case with Swieciechów spotted silex, located approximately 400 km away from sites in Hungary where it has been found. Due to its very unique appearance, this raw material has been widely discussed in the technical literature. In Hungary, its presence or absence has had an important influence on the cultural-chronological classification of certain archeological sites. The sources of Jurassic Krakow flint and Chocolate flint are located at comparable distances from Hungarian sites. In Paleolithic assemblages they are represented by a few isolated items and provide evidence of impressive long distance contacts. The very early and enigmatic presence of a considerable amount of flint from Transcarpathian sources is unexpected, as it is derived from the drift of the Prut River. At Esztergom-Gyurgyalag, *ca.* 600 km from the

Prut River sources, this material represents up to 80% of the total lithic assemblage (Varga 1991: 269). On other Hungarian Upper Paleolithic sites its presence is sporadic. Rock crystal is known to occur at various places within the Carpathian Basin, primarily in the northwestern part of the Tokaj-Prešov Mountains. Here, however, it is not of the size and quality suitable for the production of stone tools. On the basis of the Hungarian finds, rock crystal outcrops of Northern Transylvania (Maramures) located at a distance of 150–160 km, as well as pegmatites of the Eastern Alps, found at a distance of 400 km as the crow flies, should be considered as possible sources. The Bohemian/ Moravian occurrences of rock crystal (Rousmerov, Bobruvka) are even more distant (Valoch 1989: 74), and could be represent the source for the rock crystal item found at Sajóbábony-Méhésztető (Adams 2000: 175) or at Megyaszó-Szelestető (Béres 2001: 5–6). The provenance of rock crystal items can be investigated by the study of inclusions in the tools that are characteristic of the genetic types of the mineral. Another possible source of rock crystal tools found at Hungarian Upper Paleolithic sites (initially in the Pilismarót environs) are the central and eastern fringes of the Alps (Dobosi and Gatter 1996: 50).

The potential route of raw materials originating from the territories beyond the Carpathian range can be reconstructed based on geographical factors. The variable orographical endowments of the area had a considerable impact on interactions between Upper Paleolithic communities (Djindjan 1992). The Carpathian arch represented both an obstacle as well as a bridge. While the watershed ridge is often 60 km wide, the 1800–2000 m high range could be crossed at some passes and river valleys. These passes could be 40–50 km wide and less than 500–600 m above sea level, such as the Danube-Leitha valley in the northwest, a wide route between the Carpathians and the foothills of the Alps. In southeastern Transylvania there are some easily accessible passes, not higher than 900 m above sea level, leading outwards from the Carpathian Basin in the direction of the Prut and Dniestr rivers. These passes and river valleys could be used for the transport of both chipped stone raw materials and other commodities. Further, apart from the providing for the material needs of Paleolithic groups, these mountain passes undoubtedly facilitated the enrichment of the spiritual and social needs of the small, widely-dispersed communities within the basin.

EXAMPLES OF LITHIC RAW MATERIAL USAGE AT HUNGARIAN GRAVETTIAN SITES

Having summarized our current knowledge of the cultural and chronological aspects of the Hungarian Gravettian, as well as the nature and distribution of lithic raw material sources in the region, a presentation of the variable patterns of lithic raw material utilization during this time is presented.

MEGYASZÓ-SZELESTETŐ (OLDER GRAVETTIAN)

This flat, hilltop ("Szelestető" or "windy plateau") site is situated in the obsidian region between the Hernád River and the southwestern fringe of the Tokaj-Prešov mountain range, at an altitude of 225–230 m above sea level (Dobosi and Simán 1996). In the course of several field surveys and test excavations performed by the author and Katalin Simán (Dobosi and Simán 1996), an area measuring 50 × 100 m has been surveyed on the hilltop. The site has a C-14 date of 27 070 ± 300 BP (deb-5378).

Archeological finds were recovered from the surface, subsurface, and within two cultural layers. In accordance with previous observations, the cultural layer at the boundary of the recent soil and the Pleistocene sediments was disturbed and not *in situ*. The main cultural layer, located at a depth of 90–110 cm below the current surface, exhibited the characteristic features of Hungarian settlements occupied at about the time of the Paudorf interstadial period. Spatially, the settlement is widely dispersed, and while faunal material is sparse and features are restricted to ashy patches and grains of charcoal, the lithic industry is outstanding both in raw material selection and artifact workmanship (Table 8.1). Retouched tools are dominated by end scrapers (24.3%), followed by burins (21.1%), scrapers (7.8%), and shouldered tools (1.4%).

Lithic raw materials represented at Megyaszó-Szelestető are summarized in Table 8.2. As can be seen, the assemblage is heavily dominated by hidro- and limnoquartzites (63%), followed by obsidian (27.4%). Radiolarite, flint, silex, and Szeletian felsitic porphyry occur in very small amounts. Details of the various raw materials are summarized as follows:

1 Hidro- and limnoquartzites: Local outcrops of this material occur within 3–5 km to the site at the margin

Table 8.1 Basic composition of the Megyaszó-Szelestető lithic assemblage.

Tool		Blade		Core		Flake, others		Total
N	**%**	**N**	**%**	**N**	**%**	**N**	**%**	
560	6.8	415	5	352	4.3	6932	83.3	8263 pieces

Table 8.2 Distribution of raw materials at Megyaszó-Szelestető.

H/L		Obsidian		Radiolarite		Flint		Silex		SFP		Others		Total
N	**%**	**N**	**%**	**N**	**%**	**N**	**%**	**N**	**%**	**N**	**%**	**N**	**%**	
5208	63	2266	27.4	321	3.9	239	2.9	182	2.2	32	0.4	15	0.002	8263

H/L: Hidro- and limnoquartzites.
SFP: Szeletian felsitic porphyry.

of the neighboring hill. Limnic quartzite can be also collected from the drift of the nearby stream. The limnic quartzite outcrops lying at the western margin of the Tokaj-Prešov Mts. belong to the category of regional raw materials with respect to the Megyaszó site. All varieties of hydrothermal and limnic silices (jasper, limnic quartzite, silicified wood, etc.) occur in the lithic assemblage.

2 Obsidian: Obsidian occurs in relatively high amounts. The Hungarian (Carpathian 2) obsidian source is within visible distance, a few kilometers from the settlement in the southern part of the Tokaj-Prešov Mountains. Carpathian 1 obsidian, derived from the northern parts of the Tokaj Mountains falls into the regional range. (Regarding the discrimination of the obsidian types, see Biró *et al.* 1986.)

3 Szeletian felsitic porphyry (metarhyolite): This is a well-known and characterized raw material particularly associated with the Szeleta-culture. The source, consisting of secondary deposits, is located in the eastern part of the Bükk Mountains (Bükkszentlászló-Tatárárok). At Megyaszó, it can be classified as a typical regional raw material.

4 Radiolarite and other types of silex: These might have reached the settlement area in the form of fragments and pebbles selected from alluvial deposits of the Hernád River. Such materials originate in the catchment area of the Hernád River in Eastern Slovakia.

In summary, the Megyaszó-Szelestető lithic artifact assemblage is dominated by local raw materials derived from within 3–5 km of the site, which account for 63% of the recovered lithic artifacts. Obsidian (approximately 27%) is relatively abundant, and represents both a local and regional raw material. While long distance raw materials occur in the assemblage, they are not particularly abundant. Tool types are dominated by end scrapers and burins, and the abundance of cores and chipping debris is typical of sites located close to a raw material source.

PÜSPÖKHATVAN (OLDER GRAVETTIAN)

The Cserhát Mountains are located at the western edge of the North Hungarian Mid-Mountain range. The southern portion is bordered on the west by the wide valley of the Galga-stream. At the terrace margins, 80–100 cm thick natural exposures of hydroquartzite bench outcrops are exposed near the present surface. The site of Püspökhatvan, a special purpose (workshop) settlement of the Older Blade Industry (Pavlovian), is located to the southeast of the village of Püspökhatvan, on a steep terrace margin descending towards a road (i.e., along a wide stream valley). The site extends over a large area, and test excavations were performed at two artifact concentration zones (Diós and Öregszőlő locales). A total surface area of $120\,m^2$ was opened at an elevation of 265–270 m above sea level. At the Diós and Öregszőlő locations, dated to 27 700 ± 300 BP (deb 1901) a workshop settlement was exposed during three seasons of excavation (Csongrádi-Balogh and Dobosi 1995).

PÜSPÖKHATVAN-ÖREGSZŐLŐ

Among the artifacts described as tools, only a statistically insignificant quantity could be assigned to traditional tool types (Table 8.3). These are dominated by burins, various fragments with scraper retouch, or prepared cores exhibiting core base and/or rabot edge attributes.

The geological endowments of the site obviously influenced the choice of the Upper Paleolithic community in selecting this settlement location. At Püspökhatvan, the ratio of non-local raw materials is only 1.4%, and represents a minor deviation from the general raw material image, which is dominated by hydro- and limnoquartzites (99.6%) (Table 8.4).

The raw material and technological spectrum of the assemblage makes the character of the site obvious, and reflects the primary activity performed (i.e., quarrying/lithic reduction). The nature of this locality supports in all respects the observations of Simán (1983: 63–4) on the settlement strategy, faunal exploitation, and raw material use represented at northeastern Hungarian Paleolithic localities with regard to raw material use and settlement function:

> Based on the raw material distribution, a workshop-on-the-source or quarry site will have over 90% of local raw materials... In the case of workshop sites, apart from the local raw materials other raw materials of mixed type and different origin can occur in some, relatively small quantities

Among the non-local materials at Püspökhatvan, special mention must be made of "Szeletian" felsitic porphyry. While in the case of Megyaszó this easily recognizable and characteristic raw material was considered regional and predictable, at Püspökhatvan it represents a long-distance item due to the greater geographical distance to the source. The few items of "Szeletian" felsitic porphyry found to the west of the geological source region have been interpreted as evidence of a commercial/exchange route leading from east to west and, partly, south to north. The interpretation is probably correct as the only known source is located in the eastern part of the Bükk Mountains, within the territory of the Bábonyien and the Szeletian cultures, respectively. The archeological distribution of this raw material extends towards the northwest, to the Vág (Vah) valley and the Moravian Basin. Due to the foliated structure of the raw material, it is especially suitable for the production of bifacial tools. Because of this, the use of this type of raw material was formerly attributed to the Szeletian culture and

Table 8.3 Basic composition of the Püspökhatvan-Öregszőlő lithic assemblage.

Tools		Blades		Cores		Flake, others		Total
N	%	*N*	%	*N*	%	*N*	%	
215	7.2	65	2.2	1103	37.2	1583	53.3	2966

Table 8.4 Distribution of raw materials at Püspökhatvan-Öregszőlő.

Hidro- and limnoquartzites		Obsidian		Radiolarite		Flint		Silex		Szeletian felsitic porphyry	Others		Total
N	%	*N*	%	*N*	%	*N*	%	*N*	%	*N*	*N*	%	
2923	98.6	3	0.1	6	0.2	–		–		+	34	1.1	2966

the related lithic industries. Recent investigations have resulted in a reassessment of our understanding of the geographical and chronological settlement history of the Cserhát region. The excavations at Püspökhatvan served in a way as a prelude to this problem. Numerous surface collections, partly supported by excavations, have yielded several Paleolithic sites rich in this specific raw material, indicating that it is associated with a wide temporal range from Middle Paleolithic through the Neolithic period (Markó et al. 2002, 2003).

Occurrence of obsidian and radiolarite is not surprising at Püspökhatvan. Both raw materials were popular and widely distributed, and are considered here as long-distance elements. They are present at the site in a surprisingly low amount, although if the site functioned primarily as a lithic workshop associated with the exploitation of hidro- and limnoquartzite outcrops, then the paucity of other raw materials is understandable.

SÁGVÁRIAN CULTURE

Following the first population wave of the Gravettian culture in the Interpleniglacial period (Denekamp, Paudorf interstadial) there is a cultural–chronological gap lasting for several thousand years in the settlement history of the Hungarian Upper Paleolithic. The reason for this gap is not known yet. It is possible that the central areas of the Carpathian Basin became void of habitation; these changes, however, would not be necessary given the climatic conditions of the introductory phase of Pleniglacial B. It is also possible that the sites have yet to be found, or that the dating is incorrect.

It can be taken for certain, however, that about 20 000 years ago (20 kyr BP), a new population appeared in the Carpathian Basin and revitalized the tool-working tradition of a much older horizon (Dobosi 2001). Between the two blade-based industries, we find a population utilizing a much older tool working tradition based on the use of pebble-form raw materials. In the somewhat "fragmentary" tool assemblage, implements made from pebbles are quite frequent and the forms of traditional pebble tools appear again. The eponymous site of the culture is Ságvár while the stratotype is Mogyorósbánya-Újfalusi dombok. The climatic oscillation known as the Ságvár-Lascaux interstadial period (Gábori-Csánk 1978) and previously recognized on the basis of natural science research, acquired an archeological dimension with the discovery of the Mogyorósbánya Upper Paleolithic settlement.

MOGYORÓSBÁNYA

This site is located at the margin of a loess-covered terrace along the right bank of the Danube River and has been dated to approximately 19–18 kyr (19 000 ± 250 BP [deb-9673]; CalBC 21 050 – 20 140; 19 930 ± 300 BP [deb-1169]). It is situated over a valley leading from the Gerecse Mountains to the shallows of the Danube, and is an ideal location for a Paleolithic settlement. Our knowledge of the structure of the settlement indicates that a relatively large community settled here repeatedly within a relatively short time period. There are three settlement units separated by a 30–40 m wide expanse of empty space between each concentration. These units offer a unique possibility for the evaluation of the finds (Dobosi 2002). In Tables 8.5 and 8.6, data

Table 8.5   Technological tool type groups at Mogyorósbánya.

Tool		Blade		Core		Flake, others[a]		Total
N	%	N	%	N	%	N	%	
467	7.6	337	5.5	108	1.8	5196	85.1	6108

[a] "others" represents a wide range of artifacts that cannot be classed into the traditional Upper Paleolithic typological categories of Sonneville-Perrot. These consist of:
- Pebble tools, which are especially important in the Ságvárian industry and include geometrically broken pebbles, slices of segments, choppers and chopping tools (204 pieces);
- retouched flakes (74 pieces);
- production debris and raw material chunks (4918 pieces).

Table 8.6 Distribution of raw materials at Mogyorósbánya.

Hidro- and limnoquartzites		Obsidian		Radiolarite		Flint		Silex		Szeletian felsitic porphyry	Others		Total
N	%	N	%	N	%	N	%	N	%		N	%	
137	2.2	228	3.7	267	4.4	185	3.0	4719	77.2	–	38	0.7	6108

on material unearthed between 1984 and 1999 are presented. As Table 8.5 indicates, the Mogyorósbánya assemblage is dominated by the unique category of "pebble tools", which are represented by 204 pieces. The retouched tool assemblage is dominated by burins (31.7%), followed by end scrapers (11%), scrapers (5.8%), and shouldered tools (1.7%).

Various lithic raw materials classified as "silex" dominate the raw materials represented at this site, accounting for approximately 77% of the assemblage (Table 8.6). Among these are lydite and an increasing amount of nummulithic chert (A. Markó, pers. comm.), both representative of the pebble fraction of the assemblage. Both can be collected in the drift of the Danube River, a few hundred meters from the site. Among the radiolarites, the characteristic dark brown colour variant with greenish marbly pattern typical of the White Carpathes (Vlara Valley) has been identified, and does not necessarily indicate a distant raw material. The occurrence of this raw material close to the site is likely due to secondary deposition by the Vág (Vah) and the Danube rivers in the form of pebbles. Obsidian occurring at Mogyorósbánya is undoubtedly a long distance raw material in this case. Its ratio of occurrence decreases as a function of distance from sources. Its abundance at Mogyorósbánya possibly reflects the preferred pebble technology, as the obsidian nodules can be worked in a fashion similar to pebbles. Flint here refers to erratic flint, identified by macroscopic examination. Typical attributes of flints include a chalky-limy, smooth cortex (if preserved) and very fine-textured, homogeneous, and silky non-cortical surfaces. In addition, most of the flint pieces are covered with an intense bluish-white patina. There is a small amount of raw material of hydrothermal origin that can be classified as long distance items here. Though there are some inferior quality hydroquartzite occurrences 20–40 km distant from the site,

their use in archeological contexts is not yet adequately documented (Biró 1984: 45).

From the provenance data derived from the lithic raw materials, direct long distance contact can be hypothesized with the hydroquartzite sources in the northeast and obsidian sources further to the east. The most probable origin of the pebble material, the most abundant raw material represented at the site, does not indicate distant contacts but only the catchment area of the Danube River (Dobosi 1997). One of the most enigmatic raw materials in the Hungarian Paleolithic is amber. One of the two pieces known so far was found at Mogyorósbánya. Infra-red spectral analyses on both specimens have been performed by M. Földvári (1992) of the Hungarian Geological Institute. Comparison to other archeological sites and modern amber used for jewellery resulted in inconclusive results (Földvári 1992: 17).

EPIGRAVETTIAN (YOUNGER GRAVETTIAN)

The Epigravettian culture belongs to the last significant settlement wave of the Würm and, at the same time, the end of the Pleistocene period. Several small temporary campsites with small lithic assemblages have been identified from the last few thousand years of the Ice Age. As opposed to the settlement sites of the classical phase of the Gravettian culture, these sites can be found on small, low-lying Pleistocene relict elevations in the Alföld (Great Hungarian Plain).

The most intensively studied and best-known chain of settlements dated to this period is the sequence of campsites around Pilismarót. Here, the Danube River makes a sharp turn from its east–west course to a north–south course. The triangle formed in this bend is bordered along an imaginary hypotenuse by loess terraces at a height of 150–160 m above sea level. On a plateau dissected by dry valleys and streamlets, dispersed hunters' camps are located within visual sight of each other. These represent the remains of seasonal hunting camps or perhaps sites occupied by a nuclear family for a few weeks only. The thickness of the cultural layer is variable beneath the sandy loess and is associated with an embryonic soil dated to the Ságvár-Lascaux interstadial period. In the Hungarian loess sequences, the age of this well marked Late Würm level is dated around 16 kyr BP (Dobosi *et al.* 1982: 310).

On the approximately 5 km long terrace margin, five settlements areas have been excavated. For preliminary evaluation, the artifacts from these areas have been condensed into one set, for two reasons. First, the series of settlements are coeval from a cultural-archeological point of view, such that the settlement strategy, raw-material management, and stone tool typology can be considered identical. Second, only a condensed set of artifacts produces a statistically valid sample from the little assemblages. The material from the site is currently under evaluation, and while numerical values may change slightly when analysis is completed, general tendencies of the assemblage are already apparent. The basic composition of the lithic artifact assemblage is presented in Table 8.7. Among retouched artifacts, burins predominate (33%), followed by end scrapers (23%), scrapers (4.3%), and shouldered tools (1.0%).

Analysis of 1840 artifacts indicates that the raw material use of this group is the most varied among the communities studied (Table 8.8). It appears that during this period there was easy access to multiple types of raw material, and different procurement strategies may be represented in the assemblage. For example, there is evidence of direct contact with, or access to, the geological source region. Alternatively, indirect contact by exchange or through a chain of mediators is possible. Finally, the basic raw material needs may have been embedded in the general round of activities conducted by the ever-moving communities. None of these scenarios can be excluded as of yet.

The most abundant raw materials represented at the site consist of radiolarites, flints, and silex, which account for about 95% of the analyzed sample. Among the silex types, erratic and Prut flint were obtained from distances of several hundreds of kilometres, from beyond the Carpathian Basin.

The remainder of the sample consists of a varied mix of raw material types. Andesite as a possible raw material for chipped stone tools was integrated into Paleolithic research in Hungary by the activity of private collectors, A. Péntek and S. Béres in the Cserhát Mountains. The possible use of this material has been documented at Korolevo in Transcarpathian Ukraine, where in the context of the Middle Paleolithic bifacial industry (Koulakovska 2001: 209) and the Early Upper Paleolithic material (Gladilin and Demidenko

Table 8.7 Composition of the lithic material from Pilismarót.

Tool		Blade		Core		Flake, others		Total
N	%	N	%	N	%	N	%	
438	10.8	533	13.2	101	2.5	2973	73.5	4045

Table 8.8 Lithic raw material proportions from Pilismarót[a].

Hidro- and limnoquartzites		Obsidian		Radiolarite, Flint (erratic, Prut), Silex		Szeletian felsitic porphyry		Others Sandstone, rock crystal, andesite, quartzite		Total
N	%	N	%	N	%	N	%	N	%	
19	1.03	14	0.8	1754	95.3	2	+	51	2.8	1840

[a] These are preliminary figures only; total count of sourced artifacts likely to change upon completion of analysis.

1989: 146) more than 90% of the total raw material used was local andesite. Recent test excavations by the author have confirmed the suspected use of andesite for chipped stone tools at the Aurignacian site of Acsa. Rock crystal was found at Pilismarót in a surprisingly large quantity, and fluid inclusion analysis by Gatter and Dobosi (Dobosi and Gatter 1996) indicate Alpine pegmatitic sources.

CONCLUSION

The lithic raw material data from the Hungarian Gravettian presented above permit the following conclusions. First, throughout this period, a wide range of lithic raw material types were utilized, derived from primary and secondary sources. The precise modes of acquisition likely varied, both diachronically and synchronically, and more research is needed to refine our knowledge of this topic. Second, based on current data, it appears that utilization of local raw materials was more common among Older Gravettian and Sagvarian groups, between about 28–18 kyr. Among Hungarian Epigravettian groups, there is an apparent dramatic shift in the utilization of long distance raw materials. The evidence suggests an increased utilization of long distance raw materials as climatic conditions deteriorated between the Interpleniglacial and the latter part of the Würm, between about 28–16 kyr. It is likely that this pattern represents a combination of factors, including population movements and changing subsistence/settlement systems associated with an increasingly colder and drier environment. More research is needed to determine if there was in fact an association between the use of long distance lithic materials and the changing social and environmental conditions of the later Würm. It is hoped that the data presented above will help us begin to fill the void in the archeological literature that has existed for the Hungarian Gravettian.

ENDNOTE

1   The Gravettian sites are less frequent in Eastern Slovakia and in Hungary. On the Pannonian plain, east of the Alps and to the north of the Dinarides, the sites are rare or non-existing. This scarcity let us conclude that the Gravettian population can be specifically localised to the Northern area of Central Europe. (Djindjan et al. 1999: 189)

REFERENCES

Adams, B. (2000) Archaeological Investigations at two open-air sites in the Bükk Mountain region of Northeast Hungary. In: Orschiedt, J. and Weniger, G.-C. (eds) *Neanderthals and Modern Human – Discussing the Transition: Central and Eastern Europe from 50.000–30.000BP*. Neanderthal Museum, Mettmann, 169–82.

Béres, S. (2001) Újabb hegyikristály lelet Megyaszó-Szelestető felső paleolit telepóről. *Ősrégészeti Levelek* 3, 5–6.

Biró, K. (1984) Őskőkoriés őskori pattintott kőeszközeink nyersanyagának forrásai. *Archaeologiai Értesitő* 111, 42–52.

Biró, K.T. (1989) Northern flint in Hungary. In: Kozlowski, J.K. (ed.) *"Northern" (Erratic and Jurassic) Flint of South Polish Origin in the Upper Palaeolithic of Central Europe*. Institute of archaeology, jagellonian University, Cracow, 75–85.

Biró, K.T. and Dobosi, V.T. (1991) Lithotheca Comparative Raw Material Collection of the Hungarian National Museum. National Museum, Budapest.

Biró, K.T. and Pálosi, M. (1986) *A pattintott kőeszközök nyersanyagának forrásai Magyarországon. A M.Áll.Földtani Intézet évi jelentése az 1983. évről*, Budapest, 407–35.

Biró, K.T., Pozsgai, I. and Vladár, A. (1986). Electron beam microanalyses of obsidian samples from geological and archaeological sites. *Acta Archaeologica ASH* 38, 257–78.

Csongrádi-Balogh, É. and Dobosi, V.T. (1995) Palaeolithic Settlement traces near Püspökhatvan. *Folia Archaeologica* 44, 37–59.

Djindjan, F., Kozlowski, J. and Otte, M. (1999) *Le paléolithique supérieur en Europe*. Armand Colin, Paris.

Djindjan, F. (1992) L'influence des frontiers naturelles dans les déplacements des chasseurs-cueilleurs au würm recent. *Preistoria Alpina* 28, 7–28.

Dobosi, V.T. (1997) Raw material management of the Upper Palaeolithic. (A case study of five new sites). In: Schild, R. and Sulgostowska, Z. (eds) *Man and Flint*, Institute of Archaeology and Ethnology Polish Academy of Sciences, Warszawa. pp. 189–93.

Dobosi, V.T. (2001) About Ságvárian: Chronological-cultural sketch of the Upper Palaeolithic in Hungary. In: Ginter, B., Drobniewicz, B., Kazior, B. et al. (eds) *Problems of the Stone Age in the Old World*, Uniwersytet Jagelloński, Krakow, pp. 195–201.

Dobosi, V.T. (2002) Mogyorósbánya-Újfalusi dombok. Upper Palaeolithic Site. *Archaeological Investigations in Hungary* 1999, 5–14.

Dobosi, V. and Gatter, I. (1996) Palaeolithic tools made of rock crystal and their preliminary fluid inclusion investigation. *Folia Archaeologica* 45, 31–50.

Dobosi, V.T. and Simán, K. (1996) New Upper Palaeolithic site at Megyaszó. *Communcationes Archaeologicae Hungariae* 1996, 5–22.

Dobosi, V.T., Vörös, I., Krolopp, E., Szabó, J., Ringer, Á. and Schweitzer, F. (1982) Upper Palaeolithic settlement in Pilismarót-Pálrét. *Acta Archaeologica Hungarica* 35, 287–311.

Földvári, M. (1992) Analysis of the amber from Mogyorósbánya. *Communcationes Archaeologicae Hungariae* 1992, 16–7.

Gábori-Csánk, V. (1978) Une oscillation climatique a la fin du Würm en Hongrie. *Acta Archaeologica Hungarica* 30, 4–11.

Gladilin, V.N. and Demidenko, Y.E. (1989) Upper Palaeolithic stone tool complex from Korolevo. *Anthropologie* 27(2–3), 143–78.

Koulakovska, L. (2001) Le complexe bifacial á Korolevo (Couche II-A). Actes de la table-ronde internationale organisée a Caen. 1999. *ERAUL* 98, 208–11.

Markó, A., Biró, T.K. and Kasztovszky, ZS. (2003) Szeletian felsitic porphyry: Non destructive analyses of a classical palaeolithic raw material. *Acta Archaeologica Hungarica* 54, 297–314.

Markó, A., Péntek, A. and Béres, S. (2002) Chipped stone assemblages from the environs of Galgagyörk (Northern Hungary). *Praehistoria* 3, 245–57.

Pawlikowski, M. (1989) On the necessity of standardization of petrological investigation in Archaeology. In: Kozlowski, J.K. (ed.) *"Northern" (Erratic and Jurassic) Flint of South Polish Origin in the Upper Palaeolithic of Central Europe*, Institute of archaeology, jagellonian University, Cracow, pp. 7–15.

Prychistal, A. (1989) A survey of Moravian Raw Mterials used for chipped Artifacts in teh Palaeolkithic. In: Kozlowski, J.K. (ed.) *"Northern" (Erratic and Jurassic) Flint of South Polish Origin in the Upper Palaeolithic of Central Europe*, Institute of archaeology, jagellonian University, Cracow, pp. 63–70.

Simán, K. (1983) Palaeolithic settlement pattern in county Borsod-Abauj-Zemplén. *Herman Ottó Múzeum Évkönyve* XXBV–XXVI, 55–67.

Simán, K. (1989) Northern flint in the Hungarian Palaeolithic. In: Kozlowski, J.K. (ed.) *"Northern" (Erratic and Jurassic) Flint of South Polish Origin in the Upper Palaeolithic of Central Europe*, Institute of archaeology, jagellonian University, Cracow, pp. 87–94.

Valoch, K. (1989) Flint and rock crystal in the Moravian Palaeolithic. In: Kozlowski, J.K. (ed.) *"Northern" (Erratic and Jurassic) Flint of South Polish Origin in the Upper Palaeolithic of Central Europe*, Institute of archaeology, jagellonian University, Cracow, pp. 71–4.

Varga, I. (1991) Mineralogical analysis of the lithic material from the Paleolithic site of Esztergom-Gyurgyalag. *Acta Archaeologica Hungarica* 43, 267–9.

Chapter 9

TECHNOLOGICAL EFFICIENCY AS AN ADAPTIVE BEHAVIOR AMONG PALEOLITHIC HUNTER-GATHERERS: EVIDENCE FROM LA-CÔTE, CAMINADE EST, AND LE FLAGEOLET I, FRANCE

Stephen C. Cole

MACTEC Engineering and Consulting, Inc.

ABSTRACT

Underlying many studies of Paleolithic chipped stone technology is the idea of improvement, either gradual or abrupt, in technological efficiency. However, as there appears to be no empirical support or theoretical warrant for this idea as a principle of culture change, it is here is eschewed in favor of a different concept, which is based on Darwinian theory and adaptation. The possible adaptive significance of lithic raw material acquisition and conservation is discussed. Two main hypotheses are offered as possible expanations for observed differences in the character of local and non-local raw material in Paleolithic assemblages: blank portability and distance attrition. These are related to the manner in which raw material was transported from source to find site. After deriving the empirical implications of these hypotheses, lithic data

Lithic Materials and Paleolithic Societies, 1st edition. Edited by B. Adams and B.S. Blades, ©2009 Blackwell Publishing. ISBN 978-1-4051-6837-3.

from three French assemblages that date between *ca.* 38 000 BP and 32 000 BP are examined. The results show variability in technological efficiency in terms of acquisition and conservation of raw material. The methods explicated here could usefully be applied to better understand technological efficiency among mobile hunter gatherers in many contexts, including Paleoindian adaptations in the Eastern Woodland.

INTRODUCTION

Did hunter-gatherers become more efficient throughout the Pleistocene, leading to a zenith of efficiency in the Upper Paleolithic? Historically, some writers suggested or implied such a progression and although few cling to that notion, today its influence can still be seen in the archaeological literature. The idea was clearly expressed by Leroi-Gourhan (1964), who claimed that the rise of blade technology was the late manifestation of a long, gradual increase in the amount of cutting edge that could be obtained from a given quantity of raw material. Bordes seems to have agreed with the notion of increasing efficiency, but cited a different mechanism. For him, the appearance and spread of first flake technology and later blade technology were linked "to a change in the conception of the tool kit, in which different functions, initially fulfilled by a hand ax, were partitioned into various flake tools," and later, various blade tools (Bordes 1967: 41). A problem with this hypothesis is its reliance on an assumed linkage between form and function, which is no longer accepted as valid. Seeking to resolve this issue, Demars (1982, 1989, 1994) has argued that blade technology became widespread in certain Upper Paleolithic industries because blades offer an ideal combination of utility and portability, qualities that were valued by mobile Upper Paleolithic hunter-gatherers who often transported their raw material long distances.

Contrasting somewhat with these models, the replacement and discontinuity model of the Middle to Upper Paleolithic transition (*ca.* 40 000–30 000 BP) proposes a sudden, radical shift in efficiency. This model proposes that a set of traits representing "fully modern behavior" appeared suddenly in Europe during this period, and that this resulted from the replacement of an indigenous Neandertal population with an emigrant population of *Homo sapiens sapiens* (Binford 1989; Mellars and Stringer 1989; Klein 1995, 2000; Mellars 1996, 1989; Pfieffer 1990; Tatersall and Schwartz 1999; McBrearty and Brooks 2000; Bisson 2002). The suite of traits asserted to distinguish modern from primitive behaviors includes: systematic blade technology; a large number of standardized lithic tool types; first use of bone, antler, and ivory as a raw material for artifacts; first objects of personal adornment; elaborate burials with evidence of ritual; more efficient hunting techniques; fishing; increased use of non-local lithic raw materials; existence of widespread social networks; expansion into harsh periglacial environments; and structured living spaces (Klein 1995, 2000; Mellars 1996; Miracle 1998; McBrearty and Brooks 2000). Most of these traits represent an increase in innovation, efficiency, or effectiveness to a level qualitatively superior to that seen in the Middle Paleolithic, and so the model is clearly based on the assumption of a significant increase in efficiency, albeit a rapid jump rather than a gradual progression.

Serious flaws in this model have been revealed recently, some of which throw doubt on the notion of either gradual or punctuated increases in hunter-gatherer technological efficiency. Grayson and Cole (1998) showed that the apparent increase in toolkit diversity arises from a statistical relationship between assemblage size and assemblage diversity rather than from behavioral change. Kuhn and Bar-Yosef (1999) summarized the substantial evidence for systematic blade production in Middle Paleolithic assemblages, some pre-dating the Last Interglacial. It is now well-established that blade technology came and went throughout the Pleistocene. Moreover, few Upper Paleolithic assemblages are composed entirely of blade tools; most show a combined use of blade and flake technology (e.g., Sackett 1999; Cole 2002). As for the premise that Upper Paleolithic people were more specialized and/or more efficient hunters (Mellars 1996), strong evidence of hunting prowess has emerged in some Middle Paleolithic sites (Jaubert *et al.* 1990; Baryshnikov and Hoffecker 1994; Gardeisen 1999). These studies show how misleading it is to infer the efficiency of Paleolithic behavior based on empirical generalizations at a relatively gross level. Increasingly sophisticated analysis of lithic and faunal data are rapidly eroding the empirical basis of both the discontinuity and replacement model and the idea of an increase, either gradual or punctuated, in technological efficiency.

This writer is aware of no scientific theory that would predict an increase in the efficiency of that hunting and gathering throughout the Pleistocene without reference to any causal factor. The idea seems strikingly similar to the notion of the law of progress advocated by Mortillet (1883), and by cultural evolutionists in this century (Carneiro 1970). Without such a law, it is inconceivable that technology (or other behaviors) would have tended constantly toward increasing efficiency over such a long time span, particularly given that the significant oscillations in global climate and accompanying changes in ecological settings were generally not directional. Furthermore, there is nothing dramatically different in climate or environmental change during the Würm Interstadial, when the Middle to Upper Paleolithic transition took place, compared with other Pleistocene warm episodes.

To the extent that the idea of progressive increases in technological efficiency derives from the cultural evolution paradigm, it should be abandoned because that paradigm has been effectively demolished (Blute 1979; Dunnell 1988). In sum, the problems are that the analytic units of this paradigm (band, tribe, chiefdom, state) fail to describe adequately synchronic variation in economic and social dimensions; those units are overly bounded units that fail to describe variation in the temporal dimension, necessitating a transformational model of change; the proposed sequence of change from band to state level has not been demonstrated; and the reasoning by which Carneiro (1970), for example, concluded that a progressive sequence occurred is seriously flawed (Dunnell 1978, 1988). Moreover, the paradigm's reliance on a law of culture change is based on a flawed notion of explanation in the historical sciences, one based on the essentialist ontology and space-like view of reality characteristic of the ahistorical physical sciences rather than the materialist ontology and time-like view on which historical sciences depend (Blute 1979; Dunnell 1986).

For some (e.g., Kuhn 1994), the answer to these problems lies in evolutionary ecology and the concept of optimality. This theory is attractive for many reasons; for example, it has greater empirical relevance for archeology than cultural evolution has. However, because it assumes that humans generally act to improve their situation by pursuing the most optimal behavioral strategies (Binford 1979; Torrence 1983, 1989; Bleed 1986), this approach seems to make the same mistake that Huxley did when he expressed Darwinian evolution as the survival of the fittest. The phrase does not well describe either Darwin's conception of change or what actually occurs, which is better described as "the survival of the fit." That is, there is no evolutionary force compelling people to behave in optimal ways, or condemning to extinction those who are slightly less

efficient than their neighbors. Ethnographic evidence shows tremendous variation in human behaviors even within small societies, and while it is inescapable that much of this variation has fitness consequences it is far from clear that all patterned behaviors are adaptively optimal. That being the case, it is unclear whether a search for optimal behaviors offers the potential to generate powerful explanations of cultural change.

Darwinian theory would seem to hold the most promise in addressing questions about the adaptive nature of technology, as the notion of adaptation is most completely developed and most usefully applied in biological science. As Dunnell (1978) noted, the concept of adaptation has played no real role in archeological explanation despite being a key part of Binford's (1968) definition of culture, and despite the evolutionary language of archeology throughout the twentieth century. However, the concept adaptation has not been ignored by the Darwinian theory of socio-cultural evolution that has been developed over the last two decades (Dunnell 1978, 1980; Rindos 1985, 1989; Leonard and Jones 1987; Lyman and O'Brien 1998; Madsen *et al.* 1999). Adherents of this theory have attempted to demonstrate its superiority over other theoretical frameworks with applications to archeological cases (e.g., Feathers 1990; Maschner 1996; Lipo 2001). In the process, many theoretical and methodological issues have been raised, some of which remain unsolved, such as the relevant unit of selection in cultural phenomena, the precise mechanisms by which cultural transmission occurs, and how to generate relevant descriptions of past phenotypic variation. The current study has nothing to contribute to these fascinating issues and is not intended as a programmatic statement about this school of thought. Instead, the main goal is to develop a method for deriving and testing the empirical expectations of Darwinian theory as it concerns the question of technological efficiency among Paleolithic hunter-gatherers. It is directed toward the question of whether we can understand variation in lithic raw material use and technological efficiency in a way that is consistent with Darwinian theory in its most general expression.

By this last phrase, I mean the major principles of Darwinism – variation, heritability, and fitness differentials leading to natural selection as the major, but not sole, process shaping variation over time; and the notion of adaptation as "a characteristic of an organism [or cultural unit] advantageous to it or the conspecific group in which it lives" (Simpson 1953: 160), including behavioral characteristics. Within any population, there is variation among individuals in their characteristics, or traits; and since some of these are fitness-affecting, there must also be variation in adaptedness. Adaptive differences can also be identified at the group level, and this is useful for distinguishing between "narrow" and "wide" adaptations (Simpson 1955); that is, between specialized and generalized adaptations. A narrow or specialized adaptation is one that depends on a narrow range of possibilities, such as a hunting specialization on one kind of ungulate, or a focus on blade technology to the exclusion of other blank production techniques. A wide or generalized adaptation is one characterized by many variants, such as a lithic technology that relies on several different blank production techniques. To rephrase what was said earlier in Darwinian terms, Paleolithic industries range from technologically specialized to technologically generalized, but there is no clear evidence of any long-term trend toward or away from specialization over the Pleistocene. Nevertheless, it is possible that there were differences in the efficiency of the various blade production techniques, that such differences were significant to a group's adaptation, and that technological evolution occurred as a result of these differences.

Here, I approach the question of efficiency from the angle of raw material use. Efficiency is conceived as the ratio between useful product and unused raw material. Useful product consists of blanks struck from a core and any formal tools, whereas unused material consists of the mass left in the wasted core and unused debitage and shatter. Methods of blank production, such as recurrent Levallois or blades struck along the edge of thick flake, tend to differ in efficiency. This is clearly implied by the widespread notion that Levallois techniques tend to produce fewer useable blanks than blade techniques, although this has been overstated. Whether a given technological tactic is more efficient than an extant alternative is an empirical question that could be addressed with data such as core : blank ratios and the ratio of shatter and unused blanks to retouched tools and utilized flakes. Whether selection favored the more efficient alternate is another matter, to be addressed by observing changes over time in the relative frequencies of technological variants differing in inherent efficiency, and how these relate to environmental variables.

More specifically, the goal here is to present a method for testing certain hypotheses concerning the efficiency of lithic raw material exploitation among hunter-gatherers who rely on chipped stone tools. Three hypotheses are offered as possible explanations for observed patterns in the sizes of local and non-local tools and blanks in Paleolithic assemblages. I refer to these as the blank portability, distance attrition, and nodule size hypotheses. Testing them helps determine the extent to which the groups in question conserved energy and raw material while exploiting tool stone sources, as I attempt to show by presenting the results from four Initial Upper Paleolithic assemblages from the Perigord region of France.

ECONOMICS OF LITHIC RAW MATERIAL ACQUISITION AND TRANSPORT

There are two obvious facts about chipped stone technology that directly affect the economy of hunter-gatherers. First, the raw material is heavy. Secondly, the process of converting raw material to useable products (tools) is relatively inefficient because it is a reductive process in which roughly half the volume of the raw material is discarded. This helps explain why hunter-gatherers so frequently located their camps near to sources of useable toolstone. However, the location of toolstone sources is not the only factor affecting site locations or landscape use. Many environments have various kinds of toolstone arrayed in irregular patterns on the landscape. It is very unlikely that other resources needed by hunter-gatherers (such as game herds, plant food, shelter, and water) were always in close proximity to the most abundant sources of toolstone, or the sources with the highest quality toolstone (Marks *et al.* 1991). Therefore, people often chose whether to move toolstone to the sites where water and/or food were, or move food to the sites with toolstone. In southern France and many other regions during the Pleistocene, many habitation sites are located several kilometers from the nearest occurrence of lithic raw material. Toolstone was moved to the habitation sites, and for this reason virtually every Paleolithic assemblage contains variable amounts of non-local raw material, not only in France but throughout Eurasia (Kozlowlski 1982; Hahn 1987; Straus and Heller 1988; Zilhão 1993; Soffer 1994; Conard and Adler 1997; Féblot-Augustins 1997).

It is clear that Paleolithic hunter-gatherers transported toolstone. It is also clear that this was accomplished in various ways. Two basic ideas underlying this study are that (i) how people carried raw material measurably affected their overall energy use, and (ii) that different ways of carrying raw material have clear empirical consequences. Raw material can be carried from a source to a camp as whole blocks (nodules or cobbles), worked cores, or products (blanks and/or shaped tools). The transport of whole blocks is the least efficient method of getting raw material to the site of manufacture and/or use. If only the products (flake and blade blanks, and tools) are carried, they may be an unsorted collection of all the debitage that was produced during field processing at the acquisition site, or a sorted collection consisting only of products that will be most useful later – such as a cache of long, thin blades. This last strategy would be close to the most efficient.

There are other ways that toolstone may have been conserved. One of them is by transforming a higher percentage of the material into useable tools. Of the entire population of products struck from a core, one might choose to make useable tools out of all, some, or none. Clearly, the trouble of obtaining the raw material, preparing a core, and striking blanks has been wasted effort if no tools were made of it. Inefficiency of this sort would be particularly costly in the case of non-local raw materials that were transported a considerable distance. The ratio of retouched tools to unretouched blanks gives a rough approximation of the percentage of useable product extracted from a given mass of raw material. Differences in this ratio amount to differences in technological efficiency.

THREE HYPOTHESES AND THEIR EMPIRICAL CONSEQUENCES

Evidence from several studies in France (Demars 1982, 1994; Chadelle 1983; Geneste 1985; Meignen 1988; Turq 1996; Lucas 2000; Cole 2002) shows that Middle and Upper Paleolithic stone workers there transported non-local materials mostly in the form of prepared cores and products, and that those products become increasingly well-sorted (homogeneous with regard to kinds of products) with increasing distances from the raw material source. In other words, faced with an array of chipped stone products (flakes, blades, and tools), people transported those that were smaller and more useful. This strongly suggests that throughout the Middle and Upper Paleolithic there was selection in favor of acquisition tactics that were cost-reducing compared with other tactics. I refer to this as the "blank portability hypothesis", and it applies to mobile hunter-gatherers who obtained at least some toolstone from non-local sources. However, this remains a suggestion only, because the precise empirical consequences of the blank portability hypothesis have not previously been deduced. I derive those empirical consequences below.

Archeologically, there are two general empirical consequences of the biased transport of smaller, more useful products. One is that, as implied above, the transported products are a well-sorted, biased subset of the array of products that were struck from raw blocks. The other is that transported products are smaller than those made of local raw material. Under different conditions, such as a lack of efficiency-favoring selection, one might expect to see a variety of raw material transport tactics represented in and among assemblages. On the face of it, the blank portability hypothesis may seem simple to test: first, segregate an assemblage into local and non-local raw materials (however those categories are defined); then quantify the mean sizes of the tools and blanks of each, and see whether they differ. If the non-local products are smaller than those made of local raw materials, then the blank portability hypothesis is supported. However, when we consider the many factors that affect assemblage raw material composition, things are not so simple. Even if non-local blanks and tools are smaller, the reason may have nothing to do with an emphasis on blank portability, or with the efficiency of raw material use.

Many studies of raw material transport, especially those designed to test the idea of direct acquisition as opposed to embedded acquisition (e.g., Henry 1995; Roth and Dibble 1998), assume that non-local raw material was transported in toto from its source to the site where it was processed and used to make tools, and where the artifacts were excavated. This assumption has been applied to lithic evidence, leading to interpretations consistent with Binford's notions of direct and embedded acquisition. Binford (1979: 259) defined "embedded procurement strategy" as the acquisition of raw materials for implement manufacture "incidentally to the execution of basic subsistence tasks." This is opposed to direct procurement, in which people go "out into the environment for the express and exclusive purpose of obtaining raw material for tools" (Binford 1979: 259–60). Direct procurement was very rare among the Nunamiut hunters he studied. Their embedded procurement strategy made the costs of raw material acquisition negligible. Thus, whether a given foraging group relied on an embedded or a direct procurement strategy would have significant implications for acquisition costs, which would in turn influence the frequency with which non-local raw materials were used and the forms in which raw materials were imported to residential sites.

What Chadelle (1983), Geneste (1985, 1988, 1990), Turq (1996), and I (2002) have shown is that, in southwest France, (i) artifacts transported from sources more than a few hours away were processed prior to transport (what Metcalfe and Barlow 1992 refer to as "field processing"), (ii) that the diversity of this transported assemblage (in terms of technological composition) decreases as transport distance increases; and in some cases (iii) that tools were manufactured, used, and repaired along the way. From this last, it follows that there was reduction in the sizes of the transported tools and blanks – size attrition – between toolstone source and archeological site. If products were consumed (used, repaired) during transport, then it follows that size attrition would increase with increasing transport distances. I refer to this as the "distance attrition hypothesis". Distance attrition results in a negative correlation between tool size and raw material source distance. It implies nothing about whether raw material consumption was logistically organized or opportunistic. The existence of distance attrition would simply show that raw material was consumed during transport. Therefore, if distance attrition alone can account for the smaller size of transported products, then the smaller size of non-local products is not automatic evidence of increased efficiency.

An important implication of distance attrition is that it only affects retouched products (tools). Unretouched products (blanks) are unaffected, by definition. The blanks therefore provide a way to gauge the magnitude of size attrition. This is critical for deriving testable implications. However, the precise nature of those implications depends on the relationship between blank size and tool size in the population of products struck from non-local cores (either during field processing at the source or during transport). If blanks and tools started out with the same average sizes and tools were subsequently reduced in size, the lower mean size of tools would be the result of distance attrition. However, there is compelling evidence of systematic differences in size between the blanks that were culled and retouched, and those that remained unretouched. Several studies of Paleolithic assemblages (e.g., Chadelle 1983; Cole 2002; Blades 1999, 2001; Geneste 1985) have found that tools are usually larger and heavier than raw blanks, regardless of raw material type, and regardless of whether they are flakes or

blades. This shows that people generally selected the largest blanks for making tools, with the obvious exception of certain functional types that are very small, such as microliths and caminade endscrapers. Therefore, when distance attrition occurs, it will generally act to shrink the average tool size until they are more similar to the average sizes of blanks. No such effect is expected for products made of local raw materials. Furthermore, distance attrition implies nothing about whether the selection of products for transport was biased; the technological composition of the non-local products may be similar to or different from that of local products.

In sum, when the products (tools and blanks) made of non-local raw materials are smaller, on average, than those made of local raw materials, there are two main hypotheses that could account for the difference: blank portability and distance attrition. If only one of these behaviors was practiced, then it will be possible to discriminate between them. However, it is more realistic to consider the case in which both occurred simultaneously; people transported a biased sample from the products produced in the field, and also consumed them during transport. This would result in an assemblage in which non-local products are, on average, smaller than local ones; the difference in size between tools and blanks is smaller among non-local raw materials; and non-locals have a restricted technological composition.

Table 9.1 summarizes the empirical expectations of each of the three hypotheses for three dimensions: mean size (all products combined), magnitude of the tool: blank size difference, and technological composition (the kinds of products present and their frequencies, including cores, kinds of blanks, and shatter). The two keys to discriminating between hypothesis 1 and 2 are the tool: blank size difference and technological composition. In the case in which both blank portability and distance attrition occurred, the result would show a mixture of the characteristics of both: an assemblage showing the smaller size and more sorted composition resulting from biased selection of more portable blanks, plus the effects of distance attrition on the tool: blank size differential.

There is yet another factor that can affect these data. In southwest France, and in many other places, there is significant variation in the sizes of cobbles and nodules of cherts from one location to another. If, by chance, the non-local raw materials in an assemblage happened to occur naturally in smaller nodules than local raw materials, then we would expect the blanks and tools made of non-locals to be smaller also. Obviously,

Table 9.1 Hypotheses considered here and their empirical consequences.

Hypothesis	Overall size	Tool: blank size difference	Technological composition
1. Blank portability	Non-locals smaller	Similar for locals and non-locals	Non-locals less diverse
2. Distance attrition	No necessary difference	Less among non-locals	No necessary difference
3. (1) and (2) combined	Non-locals smaller	Less among non-locals	Non-locals less diverse

this would have nothing to do with raw material conservation or technological efficiency. There are clear empirical expectations for this hypothesis, but it is not always easy to discriminate this nodule size hypothesis from the above two hypothesis. For the sake of clarity, I reserve consideration of this hypothesis for later, when it becomes significant in the case of level III of La-Côte.

METHODS

I explored this issue with assemblages that date to the Middle to Upper Paleolithic transition (40 000–30 000 BP), and that were collected in the Perigord region of southwest France (Cole 2002). Here, I present some of the data from my study of La-Côte (level III), Caminade Est (level G), and Le Flageolet I (levels XI and IX). La-Côte is a stratified, open-air site that yielded a Châtelperronian assemblage which has not been dated but almost certainly falls within the range 33 000–38 000 BP. The other three are Aurignacian assemblages that have been dated by radiocarbon to between approximately 37 000 BP and 27 000 BP, and that were excavated from stratified rockshelter deposits.

The data generated from an intensive examination of these assemblages includes technological composition, which was generated by the use of list of 21 technological types. For present purposes, certain types (cores, chunks, burin spalls) are ignored; only the blank types are examined. For certain purposes, such as testing for differences in tool : blank sizes, the dorsal cortex variable can be ignored and the blank types can be collapsed into several broad categories: ordinary flakes, ordinary blades, crested pieces, rejuvenations, and fragments (broken pieces that could come from either flakes or blades). Raw material composition was identified using more or less standard lithological description (the vast bulk of each assemblage consists of from 8 to 12 kinds of chert). Occurrences of most of the chert types are mapped (Demars 1982, 1994;

Geneste 1985; Turq 1996), and this provides a means of estimating transport distance. For certain purposes, raw material types are collapsed into local (transported distances of <12 km) and non-local (≥12 km). In addition, the presence/absence of retouch provides a somewhat crude means of assessing the consumption of raw materials. All these data were recorded individually for each of the tens of thousands of pieces, and so it is possible to examine the relationships between blank type, size, retouch, and transport distance in the dataset.

The analysis begins by comparing each raw material type with regard to the average sizes of local and non-local raw materials. The next step is to test for a significant difference in size between the raw blanks, on the one hand, and the retouched tools, on the other hand within each raw material type. I carried out this test using a Model I two-way analysis of variance (ANOVA) with mean weight as the independent variable and local/nonlocal and presence/absence of retouch as the dependent variables. When appropriate, technological composition was analyzed using cross-tabulations and likelihood-ratio test of independence (G^2).

Finally, the retouch intensity of each raw material type was compared in order to determine whether non-local materials were exploited more fully. For locally reduced raw materials, the ratio of retouched to unretouched blanks measures retouch intensity. In contrast, the retouch intensity of non-local raw materials cannot be measured directly, because is unlikely that the entire population of blanks struck from cores of non-local raw materials was transported to the site where they were found. In level XI of Le Flageolet, for example, there was a clear selection of smaller blanks for transport. This means that the size of the initial population of blanks from the reduction of non-local raw materials is unknown, and so therefore is the ratio of retouched to unretouched objects. In all likelihood, the observed ratio of retouched to unretouched blanks

Table 9.2 Mean weights (grams), local and non-local raw materials.

	Local	Non-local	Difference	2-tailed *p*	df
Le Flageolet I, XI	7.25	5.65	1.60	**0.05 > *p* < 0.01**	2067
Caminade Est, G	8.75	12.50	−3.75	**0.05 > *p* < 0.01**	2024
La-Côte, III	7.66	5.05	2.61	**0.10 > *p* < 0.05**	537

Source: Cole (2002).

exaggerates the true retouch intensity of non-local raw materials. For this reason, the possible effects of interaction effects between retouch frequency and other variables (such as blank type, weight, and cortext amount) was examined using partitioning G^2 analyses. The results of these analyses are presented in detail elsewhere (Cole 2002) and summarized below.

RESULTS

Le Flageolet I, level XI (Aurignacian)

Table 9.2 shows that non-local products are significantly smaller than local products in this assemblage. Table 9.3 shows that among both local and non-local raw materials, tools are significantly larger than blanks. This difference is somewhat smaller among non-local products (5.5 g for locals, 4.8 g for non-locals). Although the absolute magnitude of the difference is not large, these data are consistent with the blank portability hypothesis. An analysis of variance (not shown) indicated that the effects of raw material type and retouch on weight are significant. The interaction term is not significant, indicating that the effects of each variable are independent.

Furthermore, there are differences in the technological composition of local and non-local raw materials, as shown in Table 9.4. A likelihood-ratio test of independence (G^2) produced a significant result, and this indicates there are significant differences in the technological composition of local and non-local products. Specifically, non-local blanks have lower percentages of several cortical and semi-cortical types (2a, 2b, 2c, 3a), and higher percentages of non-cortical types (4a–4c). Standardized deviates (not shown) indicate that the differences are statistically significant for cortical flakes (2a), semi-cortical fragments (3a), non-cortical fragments (4b), and non-cortical blades (4c). This pattern is consistent with the hypothesis that blanks made of non-local raw materials are a non-

Table 9.3 Mean weights of products by product type and transport distance, level XI of Le Flageolet I.

	Tools	Blanks	2-tailed *p*	df
Local	12.5	7.0	<0.000	1964
Non-local	9.4	4.6	<0.02	101

Source: Cole (2002).

random sample of some initial blank population, a sample biased in favor of blanks with less cortex. Smaller blanks were selected from the blank population available at off-site locations, perhaps in order to increase their portability. This selection resulted in smaller blanks with less cortex. Moreover, the reduction in blank size that resulted when non-local raw materials were used and repaired during transport did not greatly exceed that resulting from the in-place use and repair of blanks made from local raw materials. All these data are consistent with the blank portability hypothesis. However, the difference is not absolute: non-local products are only somewhat less diverse in their technological composition than local materials, and it is clear that significant reduction took place after transport.

Table 9.5 shows the cross-tabulation of frequencies and percentages for raw material type and retouch. Retouch frequencies are high for Bergeracois chert, jasper, Mussidan chert, Jurassic chert, and porcelainite, significantly so for Bergeracois. All these raw materials appear, by their technological compositions, to have been reduced off-site, and as explained above, the transported objects represent a biased selection of products. In this assemblage and many others, certain kinds of products tend strongly to be retouched in higher frequencies than others, probably because their morphology was more desirable (e.g., non-cortical flakes and blades). If the kinds of products that were transported also happen to be the kinds

Table 9.4   Le Flageolet I, level XI: technological composition of blanks made of local and non-local raw materials. $G^2 = 41.4$, df = 14, $p = 0.000$.

Technological type	Local		Non-local	
2a: Cortical flake	254	12.9%	4	3.9%
2b: Cortical fragment	63	3.2%	1	1.0%
2c: Cortical blade	31	1.6%		
3a: Semi-cortical flake	564	28.7%	18	17.5%
3b: Semi-cortical fragment	114	5.8%	7	6.8%
3c: Semi-cortical blade	80	4.1%	5	4.9%
4a: Non-cortical flake	493	25.1%	29	28.2%
4b: Non-cortical fragment	181	9.2%	18	17.5%
4c: Non-cortical blade	115	5.8%	16	15.5%
5a: Levallois flake	1	0.1%		
5d: éclat débordant	1	0.1%		
6: Kombewa flake	2	0.1%	1	1.0%
7a: Crested blank	4	0.2%	1	1.0%
7b: Core rejuvenation tablet	19	1.0%	2	1.9%
7c: Other core rejuvenation	44	2.2%	1	1.0%
Grand total	1966	100.0%	103	100.0%

Source: Cole (2002).

Table 9.5   Le Flageolet I, XI: cross-tabulation of observed frequencies and percentages for raw material type and retouch. $G^2 = 78.2$, df = 12, $p = 0.000$.

Raw material type	Retouched		Unretouched		Total
Local					
Senonian, black	82	(6%)	1266	(94%)	1348
Senonian, beige	17	(4%)	447	(96%)	464
Senonian, indeterm.	8	(3%)	238	(97%)	246
Chalcedony/millstone	1	(3%)	35	(97%)	36
Quartzite			58	(100%)	58
Basalt			49	(100%)	49
Non-local					
Bergeracois	19[a]	(19%)	83	(81%)	102
Gavaudun			2	(100%)	2
Mussidan	2	(100%)			2
Jurassic	2	(100%)			2
Unknown distance					
Jasper	3	(21%)	11	(79%)	14
Porcelainite	4	(33%)	8	(67%)	12
Unidentified	19	(5%)	397	(95%)	416
Total	157	(6%)	2594	(94%)	2751

[a] Expected count = 5.8, standardized deviate = 5.46.

Source: Cole (2002).

that were generally retouched, the elevated retouch frequencies in the non-local materials might be due to that factor alone, and not to an effort to extract more utility from the raw material.

One way to resolve this question is to test the following null hypothesis:

> H0: The proportion of retouched objects in the raw material in question is equal to the proportion of retouched objects in the assemblage as a whole, given the technological composition of that raw material in the assemblage and the proportions of retouched objects in each technological type.

Typically, the retouched blanks in an assemblage do not resemble a simple random sample of blanks. Data gathered from several Paleolithic assemblages has revealed that tools tend to be made on large semi- or non-cortical blanks, and in some cases, blades are more likely than flakes to be retouched. These patterns result in significant associations between Retouch Frequency and various technological variables, such as Cortex Amount, Object Type, Blank Type, or Weight Class. At the same time, as the above examination has shown, there are significant differences between raw materials in frequencies of technological types and blank types. These relationships can be exploited to test the above null hypothesis for raw materials showing significantly high retouch frequencies. If those frequencies arise solely because of the chance co-occurrence of two kinds of associations, then they represent a spurious association. This possibility can be ascertained with partitioning G^2 applied to cross-tabulated data, to see whether apparent associations hold when the data are broken down by levels of the variable in question.

A thorough analysis was carried out involving partitioning G^2, in which the individual and combined effects of technological type, cortex amount, and weight on retouch frequency were examined (see Cole 2002 for details). These analyses showed that the higher frequencies of retouched objects in five raw material types are at least partly the function of a true retouch bias in favor of those raw materials, even when the technological composition and blank composition in each raw material are accounted for. Differential transport may have biased the technological composition of each raw material, and it may account for some of the variation in retouch frequencies among raw material types and among cortex categories. However, those biases would not fully explain the significant differences in retouch frequency. In the case of non-local raw materials (Bergeracois chert, Mussidan chert, and Jurassic chert),

this suggests that technological efficiency was more important in the case of more costly raw materials.

Caminade Est, level G (Aurignacian)

This assemblage presents a different picture. Table 9.2 (above) shows that there is a significant difference when all kinds of products are pooled, but the nature of the difference is puzzling: non-local products are larger than local ones. As Table 9.6 shows, this is almost entirely due to the much larger size of tools made of non-local raw materials, compared with tools of local materials. In both locals and non-locals, tools are significantly larger than blanks, but the difference is much more pronounced among non-locals because of the large size of non-local tools. The nature of this difference is the converse of that expected under the distance attrition hypothesis, which predicts that the difference in size between blanks and tools will be smaller among non-local materials, and that non-local products will be smaller overall than local ones (a result of conserving energy by transporting smaller pieces).

Nor is this result consistent with the nodule size hypothesis, which predicts that the products made from smaller nodules will be smaller. What is clear is that there is an absence of evidence for raw material conservation, for the use and consumption of raw material during transport from distant sources, and for systematic differences in nodule size. In fact, it suggests that local raw materials were more intensively consumed than non-local ones.

The only non-local raw material in this assemblage that occurred in sufficient quantities for statistical analysis is Bergeracois chert ($n = 55$), occurrences of which are located no closer than 45 km from Caminade Est. The technological composition of Bergeracois differs in several important respects from the local raw materials, notably in: lacking cores (which are relatively

Table 9.6 Mean weights of products by product type and transport distance, level G of Caminade Est.

	Tools	Blanks	2-tailed p	df
Local	16.4	8.1	<0.001	1961
Non-local	20.0	8.8	$0.001 < p < 0.01$	63

Source: Cole (2002).

abundant in the locals); having a significantly higher frequency (signaled by the high standardized residual in a G^2 analysis) of non-cortical blades; and having a significantly lower frequency of fragments (broken flakes and blades) with >50% dorsal cortex. These and other data strongly suggest that the Bergeracois in the assemblage was imported in the form of blanks and tools exclusively, and that no primary manufacture occurred at the site, unlike local materials which were reduced at the site. All in all, these results partially support the blank portability hypothesis in that a sorted collection of products was transported. However, those products appear to have been selected for their large size, their shape, and their small amount of dorsal cortex, rather than just for small size.

Table 9.7 shows clear differences in retouch frequency between raw material types. In this table, figures in boldface indicate significant positive standardized deviates, while underlined figures indicate significant negative standardized deviates. The highest values occur in one non-local raw material (Bergeracois chert)

and two for which source distances are unknown (jasper and porcelainite). Beige Senonian chert was not retouched at all, even though locally available and relatively well-represented in the assemblage. Does the high retouch frequency in Bergeracois chert result from more intense retouching, or is it merely a spurious effect that arises from the preferential transport of technological products tending to be retouched the most? The partitioning G^2 analysis revealed that the higher retouch frequency of Bergeracois chert, jasper, and porcelainite is real, not spurious. These raw materials are retouched more than others, regardless of the technological composition of each raw material type. This is clear evidence of the importance of technological efficiency to the makers of this assemblage.

La-Côte, level III (Châtelperronian)

The lithological landscape surrounding this site is somewhat different from that at Le Flageolet I and

Table 9.7 Caminade Est, G: cross-tabulation of observed frequencies and percentages for raw material type and retouch. $G^2 = 131.5$, df = 10, $p = 0.000$.

Raw material type	Retouched		Unretouched		Total
Local					
Senonian, black	100	(9%)	1066	(91%)	1166
Senonian, beige	0[a]	(0%)	262	(100%)	262
Senonian, indeterm.	53	(8%)	578	(92%)	631
Chalcedony/millstone	1	(6%)	15	(94%)	16
Quartzite			5	(100%)	5
Basalt			6	(100%)	6
Non-local					
Bergeracois	**20[b]**	**(36%)**	35[f]	(64%)	55
Fumelois	1	(11%)	8	(89%)	9
Unknown distance					
Jasper	**26[c]**	**(23%)**	85	(77%)	111
Porcelainite	**33[d]**	**(21%)**	122	(79%)	155
Unidentified	42[e]	(6%)	626	(94%)	668
Total	276	(9%)	2808	(91%)	3084

[a] Expected count = 23.4, standardized deviate = –4.84.
[b] Expected count = 4.9, standardized deviate = 6.80.
[c] Expected count = 9.9, standardized deviate = 5.10.
[d] Expected count = 13.9, standardized deviate = 5.14.
[e] Expected count = 59.8, standardized deviate = –2.30.
[f] Expected count = 50.1, standardized deviate = –2.13.

Source: Cole (2002).

Caminade Est. At those two sites, raw materials are either available within 5 km of the site, or more than 30 km away. At La-Côte, raw materials cluster into three groups: local (<5 km), intermediate (8–9 km), and non-local (*ca.* 20 km). In Tables 9.8 and 9.9, intermediate raw materials were grouped with the non-locals (they show similar patterns of acquisition and use). Both non-local and intermediate products are smaller than locals in this assemblage, although the difference is not significant at the 0.05 level (Table 9.2, above). The data in Table 9.8 would seem to support the distance attrition hypothesis: in local raw materials tools are nearly twice as large as blanks, whereas tools of non-local materials are the same size or smaller than blanks. However, a closer examination suggests this result may be spurious. Table 9.9, which breaks these

data down by blank type, raw material type, and presence/absence of retouch, shows that among jasper, pink speckled chert, and Bergeracois chert (the white cells), the tools are larger than the raw blanks. The mean weights in Table 9.8 were calculated on the basis of all blanks within each category (local/retouched, local/unretouched, etc.). What Table 9.9 shows is that the mean weight of non-local blanks (5.2 g) includes the large values for blanks (ordinary flakes and ordinary blades) of pink speckled chert and unretouched ordinary flakes of Bergeracois chert. Since there are no tools made of ordinary flakes in these raw materials, ordinary flakes can be removed from the comparison. Restricting the comparisons to the six shaded cells in the table, it becomes clear that for any given blank type, tools are larger than raw blanks among non-local raw materials. Furthermore, the difference in size between tools and raw blanks is smaller among non-local materials than local materials. Moreover, among local raw materials the biggest tools were made on fragments and ordinary flakes of black Senonian chert, the material that makes up the bulk of the assemblage. In other cases such as blades made of black Senonian chert, beige Senonian chert, and silicified wood (all local materials), tools are the same size or smaller than blanks. This is true of both non-local and local raw materials.

Table 9.8 Mean weights of products, La-Côte level III.

	Tools	Blanks	2-tailed *p*	df
Local	13.7	6.6	**0.03**	458
Non-local	3.6	5.2	**0.63**	48

Source: Cole (2002).

Table 9.9 La-Côte level III. Mean blank weights (in grams), by blank type, retouch, and raw material type. For explanation of shading and superscripts, see text.

Blank type	Retouch	Raw material type							
		Local				Intermediate		Non-local	
		Black Senon.	Beige Senon.	Indet. Senon.	Silicified wood	Jasp.	Pink speck.	Bergerac.	Grand total
Ordinary flake	Retouched	35.4		2.6		3.2			32.4
	Unretouched	7.6	10.7	8.1	2.7	2.7	6.3	6.5	7.6
Ordinary blade	Retouched	4.4	2.4		2.9	1.8		4.8	4.2
	Unretouched	4.5	4.6		3.8	1.3	8.9	3.6	4.4
Fragment	Retouched	14.4							14.4
	Unretouched	7.3		1.8					7.0
Crested blade	Retouched				6.0				6.0
	Unretouched	10.1							10.1
Rejuvenation	Retouched								
	Unretouched	5.4				1.1		6.7	5.3

Source: Cole (2002).

Thus, while the pooled data suggest distance attrition holds for non-local raw materials in this assemblage, a more detailed analysis suggests that this is only true when non-locals are compared with fragments and ordinary flakes made of black Senonian chert. Other blank type/raw material type combinations show a pattern similar to non-local materials, in which tools are smaller than or the same size as blanks. These data provide only equivocal support for both blank portability and distance attrition. What about the technological composition of local and non-local raw materials?

Table 9.10 shows remarkable similarity in the technological composition of local and non-local products (here, the intermediate and non-local materials have again been combined, and cores and chunks are excluded). The absence among non-local materials of several types of products that occur among locals can probably be attributed to the smaller sample size of the non-local materials. This fits the pattern we would expect if one of two things happened: either non-local raw materials were transported as whole blocks, which were subsequently reduced *in situ* along with local raw materials; or there was field processing of non-locals prior to transport, and then people picked up all the products, or a representative sample, and transported them to the site as-is, without using them along the way. That is, they transported an unsorted collection of products. When cores and chunks are considered, technological composition indicates that most of the raw materials, including local and non-local ones,

were imported as nodules or cobbles and reduced *in situ*. Therefore, the larger size of black Senonian products probably resulted from a difference in the size of the chert nodules. Whatever these data may mean, they show no clear evidence of economizing behavior.

If raw material was not conserved by the inhabitants of La-Côte in this way, were they conserved through a more intensive use of the imported products? There is a significant association between Retouch Frequency and RM Type (Table 9.11). However, despite wide variation in the frequencies of retouch among the various raw material types, the only significant standardized deviate is a negative one for beige Senonian and "Retouched". The dearth of significant standardized deviates may result from the low sample sizes of all raw material types except black Senonian. Is the significant association real or spurious? Closer examination shows it to be spurious. The raw material with the lowest retouch frequency (beige Senonian, which is local) has high frequencies of one of the least retouched technological types (semi-cortical flakes), and low frequencies of the most commonly retouched type (non-cortical blades). In contrast, Bergeracois chert, which is the most highly retouched raw material (and is non-local), is significantly high in non-cortical blades. This and other analyses suggest strongly that the high retouch frequency of Bergeracois chert is due to the close association between technological composition and retouch frequency, and not to a preference for tools of this material. In sum, there is no evidence

Table 9.10 La-Côte, level III: technological composition of local and non-local blanks.

Technological type	Local		Non-local	
2a: Cortical flake	65	14.1%	11	14.1%
2b: Cortical fragment	5	1.1%		
2c: Cortical blade	5	1.1%	3	3.8%
3a: Semi-cortical flake	110	23.9%	15	19.2%
3b: Semi-cortical fragment	12	2.6%		
3c: Semi-cortical blade	51	11.1%	7	9.0%
4a: Non-cortical flake	64	13.9%	19	24.4%
4b: Non-cortical fragment	9	2.0%		
4c: Non-cortical blade	129	28.0%	20	25.6%
7a: Crested blank	2	0.4%		
7b: Core rejuvenation tablet	1	0.2%		
7c: Other core rejuvenation	8	1.7%	3	3.8%
Grand total	461	100.0%	78	100.0%

Source: Cole (2002).

Table 9.11 La Côte, III: cross-tabulation of observed frequencies and percentages for raw material type and retouch. $G^2 = 16.4$, df $= 7$, $p = 0.026$.

Raw material type	Retouched		Unretouched		Total
Local					
Senonian, black	67	(16%)	364	(84%)	431
Senonian, beige	1[a]	(2%)	48	(98%)	49
Senonian, indeterm.	1	(11%)	8	(89%)	9
Silicified wood	3	(21%)	11	(79%)	14
Intermediate					
Jasper	4	(25%)	12	(75%)	16
Pink translucent			12	(100%)	12
Non-local					
Bergeracois	5	(9%)	51	(91%)	56
Unknown distance					
Unidentified	3	(13%)	20	(87%)	23
Total	84	(14%)	526	(86%)	610

[a] Expected count = 6.8, std. dev. = −2.21.

Source: Cole (2002).

in this assemblage for economizing behavior, either in the form of blank portability or of an increase retouch frequency among non-local raw materials.

DISCUSSION AND CONCLUSIONS

I have striven to generate a method for describing technological efficiency that is relevant to the concept of adaptation, but also independent of notions about the adaptive history of Paleolithic cultures. I believe this is important, because advances in our science depend on the development of falsifiable hypotheses that are relevant to theory, especially evolutionary theory. Whether I have succeeded is for others to judge. Certainly, the present work is not a complete description of all the technological variables that affected the adaptive success of past populations; other dimensions could and should be examined. Moreover, there may be better ways to describe technological efficiency than the methods presented here.

Assuming the present work has at least partially captured technological efficiency, the three assemblages discussed show clear differences. They can be arrayed in order of increasing evidence for technological efficiency, thus: level III of La-Côte, level G of Caminade Est, and level XI of Le Flageolet I. The first shows no

evidence of economizing behavior; the second shows a lack of attention to blank portability, but some attention to retouch intensity; while the third shows evidence of elevated technological efficiency in both respects. Does this support the notion of an increase in technological efficiency through time? The answer depends on the chronological placement of these three assemblages. Unfortunately, no material from La-Côte was dated, and the site, recovered in a salvage operation, has been destroyed. Moreover, the chronological position of Châtelperronian industry is in dispute. Radiocarbon dates on other assemblages suggest a range of approximately 38 000–33 000 BP. Radiocarbon ages from this period are susceptible to systematic errors (Pettit 1999) and are probably 2000–3000 years too young. If one believes Zilhão and d'Errico (1999), the industry pre-dates 38 000 BP, but many dispute this (e.g., Mellars 1989). Level G of Caminade Est has been dated by AMS radiocarbon on bone collagen to 37 200 ± 1500 BP (F. Delpeche, pers. comm.). The same caveats about radiocarbon ages apply here. In addition, this age was obtained from collagen extracted from an assemblage of bones that were collected from different positions within the level, and this could have introduced additional error. In short, there is no way to know which of these two assemblages is the oldest. However, both are probably older than level

Table 9.12 Frequencies of non-local raw materials (RM) in the six assemblages arranged in approximate chronological order.

Site, stratum	Lithic tradition	Approximate age BP	Non-local raw materials
Le Moustier, H8	Mousterian (MAT)	42 500	1.2%
Caminade Est, M3	Mousterian (Ferrassie)	37 200–47 000	5.3%
La-Cote, III	Châtelperronian	33 000–43 000	11.9%
Caminade Est, G	Aurignacian I	37 200	2.5%
Le Flageolet I, XI	Aurignacian I	33 400	2.2%
Le Flageolet I, IX	Aurignacian I	28 500	5.1%

Source: Cole (2002).

XI of Le Flageolet I. Four radiocarbon assays were done by three labs on charcoal and bone from this level. Including the 2-σ error terms, the range indicated is between 31 190 BP and 35 600 BP (Rigaud 1982). A liberal interpretation of the findings outlined here would be, therefore, that there was an increase in technological efficiency sometime between 38 000 BP and 32 000 BP.

A more conservative conclusion would be simply that there is limited evidence of variation in technological efficiency in the Perigord during the late Interpleniglacial, related to raw material acquisition and conservation. To test this idea further, additional data are needed and they must include other dated assemblages from the Perigord. To extend this conclusion to other parts of Europe, and thus make it relevant to the debate over cultural replacement or continuity, would require that the kind of analyses I carried out here be applied to many assemblages throughout Europe. I am not prepared to extend that conclusion based on the limited data I have presented. However, I think the effort to continue this analysis would be worthwhile.

The main goal of this paper has been to develop improved methods for studying the adaptive significance of technological efficiency. Had I addressed, instead, the proposition that the use of non-local raw materials increased markedly during the Middle to Upper Paleolithic transition, and that this is evidence of the replacement of Neandertals by *H. s. sapiens* with the capacity for "modern" behavior (Mellars 1996; Klein 2000), then I might have chosen to present data on the frequencies of non-local raw materials in various assemblages. Table 9.12 shows the frequencies of non-local raw materials in six assemblages from the Perigord, arranged in approximate chronological order. Clearly, these data do not support the replacement and discontinuity model's proposed increase in reliance on non-local raw materials. However, as I hope to have demonstrated, those data are also clearly inadequate to describe adaptively significant variation in technological behaviors. The frequency of non-local raw materials is not, by itself, evidence of technological efficiency, social networks, or any particular behavior. Such frequencies are only meaningful, and can only be evidence for or against a particular hypothesis, when tied to a theoretically-informed model of lithic raw material exploitation.

ACKNOWLEDGMENTS

I gratefully acknowledge the financial assistance of the Graduate School of the University of Washington, and the Scholarly Exchange Program in the College of Liberal Arts at the University of Washington, in enabling me to perform this research. I am very grateful to Jean-Philippe Rigaud, Jean-Pierre Texier, and Denise de Sonneville-Bordes respectively for permission to examine the assemblages from Le Flageolet I, La-Côte, Caminade Est. I also thank Brooke S. Blades and Brian Adams for inviting me to contribute this manuscript.

REFERENCES

Bar-Yosef, O. and Kuhn, S.L. (1999) The big deal about Blades: Laminar technologies and human evolution. *American Anthropologist* 101, 322–38.

Baryshnikov, G. and Hoffecker, J. (1994) Mousterian hunters of the NW Caucasus: Preliminary results of recent investigations. *Journal of Field Archaeology* 21, 1–14.

Binford, L.R. (1968) Archaeological perspectives. In: Binford, L.R. and Binford, S.R. (eds) *New Perspectives in Archaeology*, Aldine, Chicago, pp. 5–32.

Binford, L.R. (1979) Organization and formation processes: looking at curated technologies. *Journal of Anthropological Research* 35(3), 255–73.

Binford, L.R. (1989) Isolating the transition to cultural adaptations: An organizational approach. In: Trinkaus, E. (ed.) *The Emergence of Modern Humans: Biocultural Adaptations in the Later Pleistocene*, Cambridge University Press, Cambridge, England, pp. 18–41.

Bisson, M.S. (2000) Nineteenth century tools for twenty-first century archaeology? Why the middle paleolithic typology of François bordes must be replaced. *Journal of Archaeological Method and Theory* 7, 1–48.

Blades, B.S. (1999) Aurignacian lithic economy and early modern human mobility: New Perspectives from classic sites in the Vézère Valley of France. *Journal of Human Evolution* 37, 91–120.

Blades, B.S. (2001) *Aurignacian Lithic Economy: Ecological Perspectives from Southwestern France*. Kluwer Academic Press/Plenum Publishers, New York.

Bleed, P. (1986) The optimal design of hunting weapons: Maintainability or reliability. *American Antiquity* 51, 737–47.

Blute, M. (1979) Sociocultural evolutionism: an untried theory. *Behavioral Sciences* 24, 46–59.

Bordes, F. (1967) Considerations sur la Typologie et les Techniques dans le Paléolithique. *Quartar* 18, 25–55.

Carneiro, R.L. (1970) A theory of the origin of the state. *Science* 169, 733–8.

Chadelle, J.-P. (1983) *Technologie et Utilisation du Silex au Périgordien Supérieur: l'Exemple de la Couche VII du Flageolet I.*, Mémoire, École des Hautes Etudes en Sciences Sociales, Toulouse.

Cole, S.C. (2002) *Lithic raw material exploitation between 40,000 BP and 30,000 BP in the Perigord, France*. PhD dissertation, Department of Anthropology, University of Washington, Seattle, WA.

Conard, N.J. and Adler, D.S. (1997) Lithic reduction and hominid behavior in the middle Paleolithic of Rhineland. *Journal of Anthropological Research* 53, 147–75.

Demars, P.-Y. (1982) *L'Utilisation du silex au Paléolithique supérieur: choix, approvisionnement, circulation*. L'example du Bassin de Brive, Cahiers du Quaternaire, Centre National de la Recherche Scientifique, Paris.

Demars, P.-Y. (1989) L'Indice Laminaire de l'Outillage dans le Paléolithique Supérieur en Périgord. *Paléo* 1, 17–30.

Demars, P.-Y. (1994) L'Economie du Silex au Paléolithique Supérieur dans le Nord de l'Aquitaine. Synthèse et Interpretations. PhD Dissertation (Docteur d'Etat ès Sciences), l'Université de Bordeaux I.

Dunnell, R.C. (1978) Style and function: A fundamental dichotomy. *American Antiquity* 43(2), 192–202.

Dunnell, R.C. (1980) Evolutionary theory and archaeology. *Advances in Archaeological Method and Theory* 3, 35–99.

Dunnell, R.C. (1986) Methodological issues in Americanist Artifact Classification. *Advances in Archaeological Method and Theory* 9, 149–207.

Dunnell, R.C. (1988) The concept of progress in cultural evolution. In: Nitecki, M.H. (eds) *Evolutionary Progress?* The University of Chicago Press, Chicago, pp. 169–94.

Féblot-Augustins, J. (1997) *La Circulation des Matières Premières au Paléolithique. Synthèse des Données, Perspectives Comportmentales*. Études et Recherches Archéologiques de l'Université de Liège, No. 75. Université de Liège, Belgium.

Feathers, J.K. (1990) *Explaining the evolution of prehistoric ceramics in southeastern Missouri*. PhD dissertation, Department of Anthropology, University of Washington, Seattle.

Gardeison, A. (1999) Middle Palaeolithic subsistence in the West Cave of "Le Portel" (Pyrénées, France). *Journal of Archaeological Research* 26, 1145–58.

Geneste, J.-M. (1985) *Analyse Lithique d'Industries Moustériennes du Périgord: Une Approche Technologique du Comportement des Groupes Humaines au Paléolithique Moyen*. Ph.D. dissertation (Thèse pour l'obtention du titre du Docteur), Bordeaux I, Université de Bordeaux I.

Geneste, J.-M. (1988) Les Industries de la Grotte Vaufrey: Technologie du Debitage, Economie et Circulation de la Matière Première Lithique. In: Rigaud, J.-Ph. (ed.) *La Grotte Vaufrey* (Dordogne), pp. 441–93.

Geneste, J.-M. (1990) *Développement des Systèmes de Production Lithique au Cours du Paléolithique Moyen en Aquitaine Septentrionale*. Paléolithique Moyen Récent et Paléolithique Supérieur Ancien en Europe: Actes du colloque international de Nemours, 910–11 mai 1988, Sous la direction de Catherine Farizy, Nemours, France, pp. 203–14.

Grayson, D. K. and Cole, S. C. (1998) Stone tool assemblage richness during the middle and early upper Palaeolithic in France. *Journal of Archaeological Science* 2, 927–38.

Hahn, J. (1987) Aurignacian and Gravettian settlement patterns in Central Europe. In: Soffer, O. (ed.) *The Pleistocene Old World, Regional Perspectives*, Plenum Press, New York, pp. 251–62.

Henry, D.O. (1995) Late Levantine Mousterian patterns of adaptation and cognition. In: Henry, D.O. (ed.) *Prehistoric Cultural Ecology and Evolution: Insights from Southern Jordan*, Plenum Press, New York, pp. 107–33.

Jaubert, J., Lorblanchet, M., Laville, H., Slott-Moller, R., Turq, A., and Brugal, J.-P. (1990) *Les Chasseurs d'Aurochs de La Borde*. La Maison des Sciences de l'Homme, Paris.

Klein, R.G. (1995) Anatomy, behavior, and modern human origins. *Journal of World Prehistory* 9, 167–98.

Klein, R.G. (2000) Archeology and the evolution of human behavior. *Evolutionary Anthropology* 9, 17–36.

Kozlowski, J.K. (1982) *Excavations in the Bacho Kiro Cave*. Panstwowe Wydawnictwo Naukow, Warsaw.

Kuhn, S.L. (1994) A formal approach to the design and assembly of mobile toolkits. *American Antiquity* 59, 426–42.

Leonard, R.D. and Jones, G.T. (1987) Elements of an inclusive evolutionary model for archaeology. *Journal of Anthropological Archaeology* 6, 199–219.

Leroi-Gourhan, A. (1964) *Le Geste et La Parole: Technique et Langage*. Albin Michel, Paris.

Lipo, C.P. (2001) Science, style and the study of community structure: An example from the Central Mississippi River

Valley. *British Archaeological Reports*, International Series 918, Oxford.

Lucas, G. (2000) *Les Industries Lithiques du Flageolet I (Dordogne): Approche Économique, Technologique, Fonctionnelle et Analyse Spatiale*. PhD Dissertation (Thèse pour obtenir le grade de Docteur), L'Université de Bordeaux I.

Lyman, R.L. and O'Brien, M.J. (1998) The goals of evolutionary archaeology. *Current Anthropology* 39(5), 615–52.

Madsen, M.E., Lipo, C.P. and Cannon, M. (1999) Reproductive tradeoffs in uncertain environments: explaining the evolution of cultural elaboration. *Journal of Anthropological Archaeology* 18(3), 251–81.

Marks, A.E., Shokler, J. and Zilhão, J. (1991) Raw material usage in the paleolithic. The effects of local availability on selection and economy. In *Raw Material Economies among Prehistoric Hunter-Gatherers*, Montet-White, A. and Holen, S. (eds). University of Kansas, Lawrence, KS, pp. 127–40.

Maschner, H.D.G. (ed.) (1996) *Darwinian Archaeologies*. Plenum Press, New York.

McBrearty, S. and Brooks, A.S. (2000) The revolution that wasn't: A new interpretation of the origin of modern human behavior. *Journal of Human Evolution* 39, 453–563.

Meignen, L. (1988) Un exemple de comportement technologique différentiel selon les matières premières: Marillac, couches 9 et 10. In: Binford, L. and Rigaud, J.-Ph. (eds) *L'Homme de Neandertal. La Technique*, Université de Liège, Liège, Belgium, pp. 71–80.

Mellars, P.A. (1989) Technological changes across the middle-upper palaeolithic transition: Economic, social and cognitive perspectives. In: Mellars, P.A. and Stringer, C. (eds) *The Human Revolution: Behavioral and Biological Perspectives in the Origins of Modern Humans*, Edinburgh University Press, Edinburgh, pp. 338–65.

Mellars, P.A. (1996) *The Neanderthal Legacy: An Archaeological Perspective from Western Europe*. Princeton University Press, Princeton, NJ.

Mellars, P.A. and Stringer, C. (eds) (1989) *The Human Revolution: Behavioral and Biological Perspectives in the Origins of Modern Humans*. Edinburgh University Press, Edinburgh.

Metcalfe, D. and Barlow, K.R. (1992) A model for exploring the optimal trade-off between field processing and transport. *American Anthropologist* 94, 340–56.

Miracle, P.T. (1998) The spread of modernity in Paleolithic Europe. In: Omoto, K. and Tobias, P.V. (eds) *The Origins and Past of Modern Humans – Towards Reconciliation*, World Scientific, Singapore, pp. 171–87.

Mortillet, G. de (1883) *Le Préhistorique: Antiquité de l'Homme*. C. Reinwald, Paris.

Pettitt, P.B. (1999) Disappearing from the world: an archaeological perspective on neanderthal extinction. *Oxford Journal of Archaeology* 18, 217–41.

Pfieffer, J. (1990) The emergence of modern humans. *Mosaic* 21(1), 15–23.

Rigaud, J.-Ph. (1982) *Le Paléolithique en Périgord: Les Données du Sud-Ouest Sarladais et Leurs Implications*. PhD Dissertation

(Thèse de Doctorat d'Etat ès Sciences), Université de Bordeaux I.

Rindos, D. (1985) Darwinian selection, symbolic variation and the evolution of culture. *Current Anthropology* 26, 65–88.

Rindos, D. (1989) Undirected variation and the Darwinian explanation of cultural change. *Archaeological Method and Theory* 1, 1–45.

Roth, B.J. and Dibble, H.L. (1998) Production and transport of blanks and tools at the French middle paleolithic site of Combe-Capelle Bas. *American Antiquity* 63, 47–62.

Sackett, J. (1999) *The Archaeology of Solvieux. An Upper Paleolithic Open Air Site in France*. Institute of Archaeology, University of California, Los Angeles, CA.

Simpson, G.G. (1953) *The Major Features of Evolution*. Columbia University Press, New York.

Simpson, G.G. (1955) *The Meaning of Evolution*. New American Library, Muller.

Soffer, O. (1994) Ancestral lifeways in Eurasia – the middle and upper Paleolithic Records. In: Nitecki, M.H. and Nitecki, D.V. (eds) *Origins of Anatomically Modern Humans*, Plenum Press, New York, pp. 101–19.

Straus, L.G. and Heller, C.W. (1988) Explorations of the twilight zone: the early Upper Paleolithic of Vasco-Cantabrian Spain. In: Hoffecker, J.F. and Wolf, C.A. (eds) *The Early Upper Paleolithic: Evidence from Europe and the Near East*, B.A.R. International Series 437. British Archaeological Reports, Oxford, pp. 97–133.

Tattersall, I. and Schwartz, J.H. (1999) Commentary: Hominids and hybrids: The place of Neanderthals in human evolution. *Proceedings of the National Academy of Science USA* 96, 7117–19.

Torrence, R. (1983) Time budgeting and hunter–gatherer technology. In: Bailey, G. (ed.) *Pleistocene Hunters and Gatherers in Europe*, Cambridge University Press, Cambridge, pp. 11–22.

Torrence, R. (1989) Tools as optimal solutions. In: Torrence, R. (ed.) *Time, Energy, and Stone Tools*, Cambridge University Press, Cambridge, pp. 1–6.

Turq, A. (1996) L'Approvisionnement en Matière Première Lithique au Moustérien et au Début du Paléolithique Supérieur Dans le Nord Est du Bassin Aquitanian: Continuité ou Discontinuité? In: Carbonell, E. and Vaquero, M. (eds) *The Last Neandertals, the First Anatomically Modern Humans: A Tale About the Modern Diversity*, Barcelona, Spain, pp. 355–62.

Zilhão, J. (1993) Aurignacien et Gravettien au Portugal. In: *Actes du XIIe Congrès International des Sciences Préhistoriques et Protohistoriques*, Bratislava, 1–7 septembre 1991, vol. 2: Aurignacien en Europe et au Proche Orient, pp. 154–162. Colloque organisé par la Commission 8 sous la direction de Ladislav Bánesz et Janusz K. Kozlowski. Institut Archéologique de l'Académie Slovaque des Sciences, Bratislava.

Zilhão, J. and d'Errico, F. (1999) The chronology and taphonomy of the earliest Aurignacian and its implications for the understanding of Neandertal Extinction. *Journal of World Prehistory* 13, 1–68.

Chapter 10

TRASH: THE STRUCTURE OF GREAT BASIN PALEOARCHAIC DEBITAGE ASSEMBLAGES IN WESTERN NORTH AMERICA

Rebecca A. Kessler[1], Charlotte Beck[2], and George T. Jones[2]

[1]Department of Anthropology, University of Washington
[2]Department of Anthropology, Hamilton College

ABSTRACT

Lithic manufacture studies and reduction stage analyses are commonly used to address issues such as prehistoric mobility and land use practices. In the Great Basin Paleoarchaic, these studies are largely restricted to sites containing an abundance of bifaces. These tools appear to result from a single reduction strategy carried out across multiple sites on an archeological landscape, with different stages of manufacture occurring at different sites. This pattern results from the knappers' decision to differentially reduce raw material as a way to mitigate the costs associated with transporting that material, so that the degree of completion of each biface is a function of the distance that biface was transported. Unfortunately, few Paleoarchaic sites are rich enough in bifaces to allow for such a study, although debitage is plentiful. Here, we use debitage from sites with rich, previously analyzed biface components to highlight the role debitage analysis can play in modern lithic studies. If biface reduction is indeed governed by transport costs and reflects those constraints, this phenomenon should also be visible in the more abundant debitage created during reduction.

Lithic Materials and Paleolithic Societies, 1st edition. Edited by B. Adams and B.S. Blades, ©2009 Blackwell Publishing. ISBN 978-1-4051-6837-3.

In recent years, analyses of Great Basin Paleoarchaic (11 100–8000 rcy BP) stone tool assemblages have contributed inferences of two sorts. At a lower level these analyses are used to describe the mix of reduction technologies used to produce characteristic toolkits at particular sites. On a broader scale, studies considering how lithic manufacture was staged across space inform on aspects of mobility and the geographic extent of land-use practices. The focus of this paper mainly concerns the latter problem.

Paleoarchaic lithic technology in the central Great Basin appears to parallel in character what Bamforth (2002) describes as a "high-tech forager" pattern. Constrained by high mobility in a landscape in which lithic sources are neither uniformly distributed nor of equivalent quality, Paleoarchaic knappers like Paleoindian knappers elsewhere devised a tool technology comprised of a narrow suite of multipurpose tools. These tools, most of which were constructed of high-quality microcrystalline silicates, were over-designed so as to be less prone to breakage and lent themselves to resharpening or recycling. Because of their long use-lives, many of these tools "traveled" great distances from their sources before they were discarded. The centerpiece of this technology, however, the long, contracting-stemmed point (Figure 10.1), was made primarily from fine-grained volcanics (e.g., dacite and andecite) and obsidian rather than microcrystalline silicates. Not only are these tools frequent components of assemblages, but they also appear to be the sole

product of reduction in many Paleoarchaic assemblages (Beck and Jones 1994; Beck and Jones 2008). In fact, the dacite and andesite components of lithic assemblages derive nearly exclusively from a biface reduction strategy directed at the manufacture of these points. And yet, even though this one lithic reduction system dominates, it does not generate identical archeological patterns across different landscapes. In large measure this is because knappers differentially staged reduction in response to the costs associated with transporting raw material (Beck *et al.* 2002). While it does not appear that especially large, roughly shaped blanks were transported very far from quarry workshops – in contrast to the patterns attributed to southern Plains Folsom groups, for instance – intermediate-stage blanks, which were capable of being used as tools in their own right, were carried to residential sites. As we detail below, the degree of completion of these bifaces appears to have been a function of transport distance (see Beck *et al.* 2002).

Studies of the geographic patterns of staging to date have been limited to a few Paleoarchaic sites that contain substantial biface assemblages. In their 2002 paper, Beck *et al.* address the issue of quarry assemblage variability as it relates to mobility. They employ a central place foraging model to explore the conditions under which it is beneficial to invest more effort in reduction and manufacture at a quarry and those under which it is better to transport raw material to a residential site for this purpose. While Beck *et al.*'s study relies on large assemblages of bifaces, most Paleoarchaic assemblages have only a few bifaces and thus are not so easily evaluated. Debitage, on the other hand, is usually abundant and reflects the tool production and use activities that occurred at a site even if the tools themselves are no longer present (Andrefsky 1998: 217; see also, Carr and Bradbury 2001). Debitage analysis can thus complement, and perhaps offer an alternative to, biface analysis. If, as we suspect, the use-life of bifaces follows a simple sequence of manufacture and maintenance, the reductive by-products of these acts should similarly align in a series. In this paper we explore the characteristics of dacite flake assemblages from the four sites used by Beck *et al.* as well as two additional assemblages from eastern Nevada to evaluate their effectiveness in tracing lithic reduction across space. If successful, this approach will provide a way to draw upon those assemblages comprised mostly of debitage to expand and evaluate our reconstructions of Paleoarchaic landscape use.

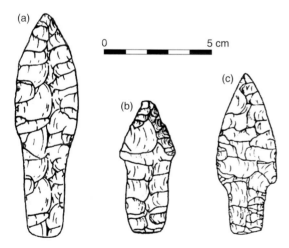

Figure 10.1 Paleoarchaic stemmed points. (a) Cougar Mountain; (b) Lake Mohave; (c) Parman.

REVIEW OF BIFACE REDUCTION AND TRANSPORT COSTS: THE BECK *ET AL.* STUDY

Central place foraging models, which have been applied primarily in the study of subsistence remains (e.g., Metcalfe and Barlow 1992; Bird and Bliege Bird 1997), suggest that as the distance between the central place and foraging location increases, field processing of resources to reduce the fraction of low utility content becomes more cost-effective. Recently these models have been extended to the study of technology (e.g., Bettinger *et al.* 1997; Beck *et al.* 2002; Bright *et al.* 2002; Ugan *et al.* 2003; Bettinger *et al.* 2006).

In general, most resources are comprised of high and low utility components: parts that contribute energy and those that do not. To use a familiar example, a nut is composed of nut meat of high utility and a shell of low utility, the latter of which adds considerably to the weight and volume of a load that is to be transported from a foraging location. Central place foraging models predict that beyond a certain distance from the residential site, the nuts will be shelled in the field, adding to the time spent foraging but substantially increasing the amount of usable nut meat that can be transported in a single load. The decision of whether or not to spend time processing a resource in the field depends on (i) the field processing time for that resource, (ii) the gain in resource utility by processing it in the field, and (iii) the distance to the residential site (Bettinger *et al.* 1997: 888).

The simplest case is one in which a resource has only two components, such as the nut and shell example above. The more complex case is one in which utility is continuous and thus the high and low utility components are not so easily separated; the question then becomes "not whether a low-utility component of a resource is expected to be transported, but rather how much of it will be transported" (Metcalfe and Barlow 1992: 348). Cortex, for example, is of low utility. Unlike the nut, from which the shell can be removed in a single act, however, the removal of cortex requires a number of sequential acts. Also unlike the shell, which always has low utility, the cortex appearing on a flake may or may not change that flake's utility. A flake with some cortex on its dorsal surface may, in fact, have the same utility as one that is completely devoid of cortex, as long as the cortex does not intersect the edge. In some instances, the presence of cortex on a flake may enhance its overall utility (see Close 2006). Cortex can

perform as a natural back to a tool, strengthening the non-use edge with the overall effect of making the tool easier to handle. Some reduction will probably always take place at a toolstone source simply as a result of cobble assay, but under what circumstances does a knapper go beyond cobble assay (and how far beyond) before that core is transported?

Beck *et al.* examine this question using the reduction of bifaces at two Paleoarchaic quarries and their associated residential sites (Figure 10.2). The Cowboy Rest Creek Quarry (CRCQ) is located in Grass Valley, central Nevada. This quarry occurs on an extensive alluvial fan containing abundant large cobbles of fine-grained dacite. Data were collected from two areas of high artifact density, the first (Locality 1) in 1999 and the second (Locality 2) in 2001. At Locality 1, all bifaces ($n = 58$) within a 50×50 m area were collected with point provenience using an electronic total station. For collection of debitage and other artifacts this 2500 m^2 area was subdivided into 100 5×5 m units and four were randomly selected. Artifacts were collected in 1×1 m grids within these four units.

At Locality 2, which lies about 1 km northwest of Locality 1, analysis was done in the field. All bifaces ($n = 613$), as well as all artifacts of chert ($n = 47$) and obsidian ($n = 4$) and two possible hammerstones, within a 150×250 m area were recorded and analyzed. No debitage was collected from Locality 2.

The Knudtsen site lies 9 km northeast and down slope of CRCQ on an east–west oriented spit reaching out into the valley floor. This site consists of two artifact scatters at the eastern end of the spit. These localities (Knudtsen 1 and Knudtsen 2) were collected in 1999. All bifaces at Knudtsen 1 ($n = 215$), an area of 200×140 m, were given point provenience. This $28\,800$ m^2 area was divided into 280 10×10 m grid units and a 10% sample of these units was randomly selected, and collection of all artifacts was done in 2×2 m grids within each selected 100 m^2 area. Knudtsen 2, approximately 250 m southeast of Knudtsen 1, covers an area of 60×100 m. All bifaces ($n = 656$) were collected with point provenience while all other artifacts were collected within 2×2 m grid units.

Given the proximity of Knudtsen to CRCQ, Beck *et al.* made the assumption that all dacite artifacts were made from CRCQ material. Since the publication of this paper, a number of the Knudtsen bifaces and flakes along with source rocks from CRCQ were compared using x-ray fluorescence (XRF). In selecting the sample to be analyzed, an effort was made to pull every artifact

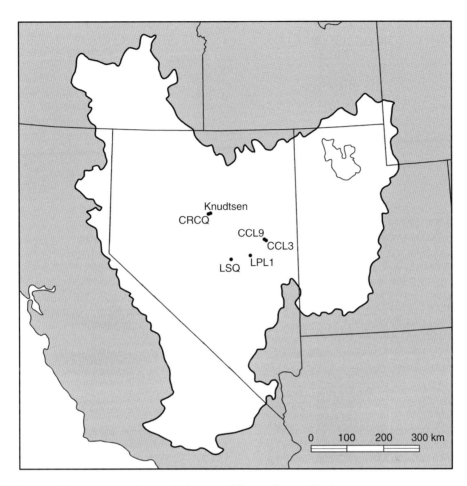

Figure 10.2   Map of the Great Basin showing the location of the sites discussed in the text.

that appeared different visually, which amounted to three specimens. Of 40 artifacts (both bifaces and flakes) that were analyzed, only these three were not of CRCQ dacite. Therefore, we are confident that 100% of those bifaces analyzed in this study are, in fact, made from CRCQ dacite.

The Little Smoky Quarry (LSQ) is located in the southeastern portion of Little Smoky Valley in eastern Nevada (Figure 10.2). Like CRCQ, LSQ coincides with an extensive alluvial fan containing dacite cobbles. Again, data were collected from two localities, the first in 1992 and the second in 2001. At Locality 1 all bifaces ($n = 130$), unifaces, and cores were collected in 2×2 m grid units from a $480\,m^2$ area (see Beck and Jones 1994). Because high flake density precluded

total collection of debitage, four 2×2 m grid units were randomly selected for collection, yielding 172 flakes. A second area, Locality 2, was examined in 2001, with analysis being done in the field. All bifaces ($n = 210$) within an $1800\,m^2$ area were analyzed; 200 are of LSQ dacite, 7 are of obsidian, and 3 are of chert. No debitage was collected from Locality 2.

The closest substantial Paleoarchaic site to LSQ, Limestone Peak Locality 1 (LPL1), lies about 60 km to the east in Jakes Valley (Figure 10.2). In 1991 a total of 6932 artifacts were collected from the site surface. Although this assemblage contains 268 fine-grained volcanic bifaces, only those made of LSQ dacite ($n = 54$) were chosen for study. LSQ dacite exhibits a distinctive pattern of large phenocrysts and weathering, and thus

Beck *et al.* relied on visual identification of artifacts made from this source. Since their analysis, source provenance studies of flakes and bifaces using both XRF and LA-ICP-MS[2] has been applied to a sample of 125 dacite artifacts from LPL1, comprising 100 flakes and 25 bifaces. Twenty six of the flakes analyzed are of LSQ while 12 of the bifaces are of that material. Using these artifacts, blind tests were conducted on bifaces (by Jones) and flakes (by Beck). These tests resulted in greater errors in identification than expected, 20% for bifaces and 25% for flakes. We return to this error factor in our discussion at the conclusion of the chapter.

In all, 1925 dacite bifaces from four sites were examined by Beck *et al.* (Table 10.1). These bifaces were

characterized using two approaches: a stage typology and a thinning index proposed by Johnson (1981: 13). The stage typology, derived from that proposed by Callahan (1979: 10–11), is based on symmetry, number, shape, and regularity of flake scars, edge sinuosity, and thickness (Figure 10.3, Table 10.2). Stage classifications can be problematic, however, because they segment what are believed to be theoretical continua (Sullivan and Rosen 1985; Teltser 1991; Whittaker 1994). The Johnson Thinning Index (JTI), a ratio between weight and plan surface area, has the advantage of measuring reduction along a continuum. Thus, these two independent approaches, while complementing one another, also act to limit bias that may be introduced by simply using one or the other.

Because the data were collected from two different localities at CRCQ, Knudtsen, and LSQ, statistical comparisons between each of the pairs (*t*-tests of JTI values, K-S tests of biface stages represented) were made in order to determine if each pair represented the same population. The two localities at both CRCQ and Knudtsen were significantly different with respect to JTI but not biface stage. Beck *et al.* treated each as a separate assemblage in their analyses, but since the two Knudtsen assemblages showed the same trends as the combined Knudtsen assemblage when compared with CRCQ, and vice versa, we present data on only the combined assemblages here.

Table 10.1 Biface samples represented at the two quarries and their associated residential sites analyzed by Beck *et al.* (2001).

Site	Sample size
Cowboy Rest Creek Quarry	671
Knudtsen	871
Little Smoky Quarry	330
Limestone Peak Locality 1	54
Total	**1925**

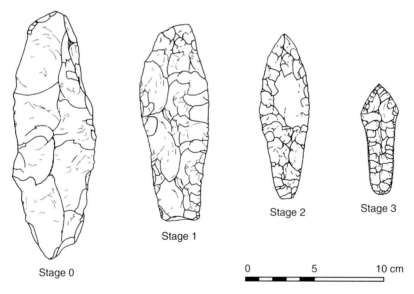

Stage 0

Stage 1

Stage 2

Stage 3

0   5   10 cm

Figure 10.3 Biface reduction sequence represented in Great Basin Paleoarchaic assemblages. Definitions for stages (Table 10.2) are based on Callahan (1979).

Table 10.2 Definitions of biface stages used by Beck *et al.* (2001).

Stage	Definition
0	Large biface with irregular shape and low symmetry; few very widely and/or spaced flake scars; very wide edge offset (very sinuous); very thick and irregular cross-section.
1	Large biface with irregular shape; widely and/or variably shaped flakes; wide offset; thick and irregular cross-section.
2	Large biface with semi-regular and symmetrical shape; closely and/or semi-regularly spaced flake scars; edge offset moderate; cross-section semi-regular.
3	Regular, symmetrical biface; closely and/or quite regularly spaced flake scars (pressure flaking often present); offset close (little edge sinuosity); cross-section thin and regular; later edge grinding evident on haft.

Table 10.3 Statistical tests between quarry and residential sites studies by Beck *et al.* (2001).

Comparison	Separate variances *t*-test of JTI		K-S test of biface stage
	t	*p*[a]	*p*[a]
CRCQ and LSQ	−6.980	<0.001	<0.001
CRCQ and Knudtsen	−20.07	<0.001	<0.001
LSQ and LPL1	8.42	<0.001	0.900

[a]Significant values in bold.

As stated above, the distance between CRCQ and Knudtsen is 9 km while that between LSQ and LPL1 is 60 km (see Figure 10.2). The central place foraging model employed by Beck *et al.* predicts that foragers will invest greater or lesser time processing toolstone at a quarry depending on how far they intend to travel upon their departure. Therefore, the model predicts that reduction at LSQ should have proceeded further than that at CRCQ, given the greater distance between LSQ and LPL1. This is exactly what the data show. A K-S test reveals a probability of <0.0001 that the biface stages represented at the two quarries are from the same statistical population (i.e., the same stage), with a predominance of later stages represented at LSQ than at CRCQ. These results are supported by a *t*-test between JTI values at the two quarries ($t = 6.980$, $p < 0.0001$) (Table 10.3).

A second expectation of the model is that the quarry assemblages should contain higher frequencies of earlier-stage bifaces than their associated residential sites. Such is the case for Knudtsen and CRCQ but the results were mixed for LPL1 and LSQ, with a significant difference in biface stages but not in JTI values (Table 10.3). Beck *et al.* conclude, however, that overall the predictions of the model were met.

We now turn to the analysis of debitage from these assemblages in order to evaluate the accuracy of the analysis of these artifacts in providing information about the staging of biface reduction as distance increases between the toolstone source and residential site.

Debitage analysis

In the case just reviewed, the geographic distribution and the large size of the biface assemblages made a statistical evaluation of the central place foraging model possible. In most Paleoarchaic assemblages, however, bifaces comprise only a small fraction of the assemblage; the largest portion is comprised of debitage. Consequently, these assemblages do not lend themselves as neatly to an examination of how transport costs influence manufacturing choices or how these choices contribute to assemblage structure and regional patterning. Some researchers have argued that the analysis of debitage can be as useful as that of formal tools as it reflects the tool production and use activities that occurred at a site (e.g., Shott 1994; Andrefsky 1998). Shott (1994) outlines several properties of debitage that make it, in fact, more useful than formal tools for certain kinds of analyses. Shott (1994: 71) states that aside from its sheer abundance, "debris is the product of, and therefore registers the kind and amount of, reduction and resharpening performed by artisans." Several studies, for example, have demonstrated the relationship between flake attributes and the type of core reduction (e.g., Sullivan and Rozen 1985; Ingbar *et al.* 1989; Patterson 1990; Teltser 1991; Carr and Bradbury 2001). Shott goes on to point out that debitage is rarely removed from the site by the knappers and is generally not the focus of artifact collectors; "the abundance, distribution, and composition, therefore, of recovered debris assemblages should not be biased by curation or collection" (Shott 1994: 71). Debitage analysis can thus complement, and perhaps offer an alternative to, biface analyses in studies such as that of Beck *et al.* For the remainder of this paper we compare flake assemblages from the four sites examined by Beck *et al.* to investigate if this indeed is the case.

Debitage samples were collected from one locality at each quarry site and from Knudtsen 1, Knudtsen 2, and LPL1. The biface analyses showed quite clearly that cobble assay and flake blank production, as well as initial thinning and shaping of these blanks, were undertaken at both quarries. Some bifaces were further refined at LSQ (but not at CRCQ), but most of the final shaping, use, and maintenance/repair were performed at the residential sites. Should it be possible to link these activities with the dacite fraction of the corresponding debitage assemblages, we expect to show that quarry assemblages contain earlier-stage debitage relative to the residential sites. Moreover, the debitage from the two quarry sites will differ, reflecting the fact that knappers working at LSQ more fully prepared bifaces as they anticipated a long travel distance to LPL1.

The present analysis also incorporates assemblages from two additional sites. These sites, CCL3 and CCL9, lie in Butte Valley, some 15 km north/northeast of LPL1 and about 90 km northeast of LSQ. These assemblages are far more typical of Paleoarchaic surface assemblages in the region (Beck and Jones 1990; Jones *et al.* 2003). That is, they contain several hundred pieces of debitage and a handful of retouched tools. As is the case for LPL1, the debitage from these sites should indicate late-stage reduction and resharpening.

The sample

Many researchers have stressed the importance of analyzing only complete flakes (e.g., Sullivan and Rozen 1985; Teltser 1991). Others (e.g., Magne 1985; Ahler 1989) have argued that this approach can greatly under-represent the total number of flakes in the assemblage and, instead, recommend analyzing all flakes in the assemblage. Carr and Bradbury (2001) argue that analyzing all flakes, broken and complete, may over-estimate the total number because fragments of the same flake may often be counted individually. One solution to this problem is to analyze only flakes with platforms (Carr and Bradbury 2001: 132). Using an experimental dataset of flakes $\geq 1/4$ inch in size, they found that approximately 66% of percussion events in bifacial core reduction produce flakes with platforms. We have not found such a high percentage in the samples we have analyzed from eastern Nevada. For example, of 5596 flakes collected at the Sunshine Locality (Figure 10.2) in 1993 (see Beck and Jones 2008), only 1847 (33.0%) had complete or nearly complete platforms. The CRCQ and Knudtsen assemblages are of sufficient size that they contain relatively large samples of flakes with platforms; this is not the case for the LSQ, LPL1, CCL3, and CCL9 assemblages. If only flakes with platforms from these assemblages were selected, the sample size would be very small (Table 10.4). Only flakes with platforms were analyzed from the CRCQ and Knudtsen assemblages while samples from the remaining assemblages, by necessity, are comprised of both flakes with and without platforms. This difference requires that analysis be conducted somewhat differently for these latter assemblages than for the CRCQ and Knudtsen assemblages (see protocol below).

Stratified random samples of flakes were selected from the CRCQ, Knudtsen 1, and Knudtsen 2 assemblages. A 30% sample of the 1×1 m units within

Table 10.4 Number of flakes analyzed from each site assemblage.

Assemblage	Flakes with platforms	Flakes without platforms	Total flakes
CRQC	183		183
Knudtsen 1	117		117
Knudtsen 2	130		130
Total CRC flakes	430		430
LSQ	64	188	252
LPL1	34	103	137
CCL3	8	8	16
CCL9	21	19	40
Total LSQ flakes	127	318	445
Total flakes analyzed	**557**	**318**	**875**

each 5 × 5 m unit that contained artifacts was selected from CRCQ. This yielded 183 flakes with platforms (Table 10.4). A 2.5% sample of collected grid squares (one 2 × 2 m^2 within each 10 × 10 m grid) from Knudtsen 1 was taken, producing 117 flakes with platforms. Knudtsen 2 was divided into 15 20 × 20 m units and two 2 × 2 m^2 within each 20 × 20 m unit were randomly selected; these units contained a total of 130 flakes with platforms. The total number of flakes with platforms selected from all three localities was 430 (Table 10.4). All of the debitage is assumed to be from CRCQ given the proximity of this quarry to Knudtsen, which, as stated above, has been confirmed through XRF analysis.

Since LPL1, CCL3, and CCL9 are quite distant from LSQ, dacite from a number of sources is present at these sites and thus only those artifacts made from LSQ material were selected. Selection was made using visual identification, which, as we noted above, can result in between 20% and 25% error, which we address in the Discussion.

All flakes from LSQ ($n = 252$) were analyzed, 64 of which have platforms (Table 10.4). A 5% sample of the LPL1 assemblage was selected, and within this sample, only flakes identified as LSQ material ($n = 137$, 34 with platforms) were analyzed. All LSQ flakes from CCL3 ($n = 16$, 8 with platforms) and CCL9 ($n = 40$, 21 with platforms) were analyzed. Together this sample is comprised of 445 flakes of LSQ material. In sum, a total of 875 flakes were analyzed in this study, 430 of CRCQ dacite and 445 of LSQ dacite (Table 10.4).

Analytical protocol

A series of attributes that together can be used to indicate reduction stage were selected. As reduction of a core proceeds, overall flake size and the amount of cortex decreases while the number of dorsal scars increases (Teltser 1991; Andrefsky 1998, 2001; Carr and Bradbury 2001). In addition, as the striking platform is increasingly prepared, it will change from one that may have cortex and/or no facets to one that is faceted and eventually ground and/or abraded (Magne 1985; Teltser 1991; Bradbury and Phillip 1995; Andrefsky 1998; Andrefsky 2001; Carr and Bradbury 2001; Sievert and Wise 2001). Thus, the variables chosen for measurement on flakes with platforms include several size measurements: weight, maximum length, maximum width, and thickness at the midpoint. Three categorical variables were chosen as well: the amount of cortex on the dorsal surface, the number of dorsal scars, and the degree of platform preparation (Table 10.5).

All of these variables were recorded for each flake in the CRCQ and Knudtsen assemblages, as well as flakes with platforms in the LSQ, LPL1, CCL3, and CCL9 assemblages. For the flakes from the latter assemblages that did not have platforms, only weight, the amount of dorsal cortex, and size category (Figure 10.4) were recorded for these artifacts. This last measurement was also recorded for those flakes with platforms in these assemblages.

If, as the results of Beck et al.'s analysis suggest, bifaces were reduced to a later stage at the residential sites than at the quarries, then the variables measured

Table 10.5 Variables measured on flakes.

Flake type	Type of variable	Variable
With platforms	Metric	Weight (g)
		Maximum length (mm)
		Maximum width (mm)
		Thickness at the midpoint (mm)
	Nominal	Dorsal cortex
		1. None
		2. >0% but <50%
		3. ≥50% but <100%
		4. 100%
		Number of dorsal scars
		1. No scars
		2. 1 scar
		3. 2 scars
		4. >2 scars
		Platform preparation
		1. Cortical
		2. Unfaceted (flat)
		3. Faceted
		4. Abraded and/or ground
Without platforms	Metric	Weight (g)
	Nominal	Dorsal cortex (as above)
	Ordinal	Size category[a] (see Figure 10.4)

[a] Also measured for flakes with platforms from LSQ, LPL1, CCL3, and CCL9.

should differ between the sites in specific ways. At the quarries, where earlier steps in the reduction sequence would have been carried out, flakes should be heavier, longer, thicker, and wider. Cortex should be more prevalent on flakes from quarry sites than on those from residential sites. Finally, the quarries should have more flakes with cortical or flat platforms and with fewer dorsal scars than the residential sites.

In addition, because the travel distance from LSQ to LPL1 is much greater than that from CRCQ to Knudtsen, the LSQ assemblage should represent later-stage reduction than the CRCQ-L1 assemblage, but earlier-stage reduction than the residential site assemblages.

ANALYSIS AND RESULTS

Before proceeding with the analysis, we need to establish whether or not the debitage assemblages from the two localities at Knudtsen differ with respect to the variables measured, as Beck *et al.* found to be the case for the biface assemblages. Comparisons of the two assemblages on both metric and categorical variables yielded only one significant difference: the number of dorsal scars. Because this was the only difference, we combine these two assemblages for our analysis.

Table 10.6 presents statistical summaries for each of the metric variables while Tables 10.7 through 10.10 show the data for the categorical variables recorded. As noted above, the LSQ, LPL1, CCL3, and CCL9 assemblages were analyzed differently depending on the presence or absence of platforms, and thus the data from Tables 10.6 through 10.9 reflect only those flakes with platforms. Table 10.10 shows the size categories for all flakes from LSQ, LPL1, CCL3, and CCL9.

We begin with a comparison of the two quarries. Restating our predictions, because of the greater distance between LSQ and LPL1, reduction should have proceeded further at LSQ than at CRCQ. With more advanced biface reduction at LSQ than at CRCQ

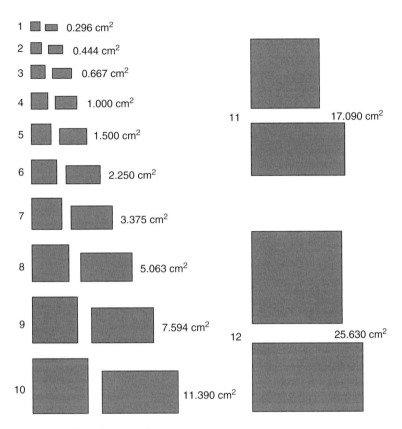

Figure 10.4 Size categories used in debitage analysis.

Table 10.6 Descriptive statistics for measurements taken on flakes with platforms for each of the six site assemblages.

Variable		CRCQ	Knudtsen	Site			
				LSQ	LPL1	CCL3	CCL9
Sample size		183	247	64	34	8	21
Weight (g)	Mean	35.156	1.175	18.196	0.983	1.301	1.057
	SD	69.764	2.158	27.958	1.078	1.278	0.567
Maximum length (mm)	Mean	38.406	13.926	35.454	13.400	16.486	14.940
	SD	21.902	7.555	17.521	5.896	6.152	4.439
Maximum width (mm)	Mean	47.672	15.952	40.656	13.459	15.986	17.495
	SD	28.657	8.644	17.493	7.257	6.495	4.096
Midpoint thickness (mm)	Mean	8.279	2.164	7.004	2.956	3.107	3.150
	SD	5.457	1.287	3.241	2.250	1.377	0.956

Table 10.7 Number of flakes from each site assemblage that fall into each dorsal cortex category.

Site	Category								Total
	0%		>0% and ≤50%		>50% and ≤100%		100%		
	N	%	N	%	N	%	N	%	
CRCQ	105	57.4	44	24.0	21	11.5	12	6.6	183
Knudtsen	246	99.6	1	0.4					247
LSQ	219	86.9	27	10.7	4	1.6	2	0.8	252
LPL1	130	94.9	5	3.6	2	1.5			137
CCL3	16	100.0							16
CCL9	39	97.5	1	2.5					40
Total	755		78		27		15		875

Table 10.8 Number of flakes from each site assemblage that fall into each dorsal scar category.

Site	Category								Total
	0 Scars		1 Scar		2 Scars		>2 Scars		
	N	%	N	%	N	%	N	%	
CRCQ	13	71.0	24	13.1	42	23.0	104	56.8	183
Knudtsen			2	0.8	23	8.3	222	89.9	247
LSQ	1	1.6	1	1.6	6	9.4	56	87.5	64
LPL1			1	3.0	5	1.5	27	81.8	33
CCL3			1	14.3			6	85.7	7
CCL9					1	4.8	20	95.2	21
Total	14		29		77		435		555

Table 10.9 Number of flakes from each site assemblage that fall into each category of platform preparation.

Site	Category								Total
	Cortex		Single facet		Multiple faceted		Abraded		
	N	%	N	%	N	%	N	%	
CRCQ	35	19.1	92	50.3	56	30.6			183
Knudtsen	1	0.4	15	6.1	112	45.3	119	48.2	247
LSQ	2	3.1	21	32.8	40	62.5	1	1.6	64
LPL1			8	24.2	25	75.8			33
CCL3			1	14.3	5	71.4	1	14.3	7
CCL9			2	9.5	16	76.2	3	14.3	21
Total	38		139		254		124		555

Table 10.10 Number of flakes from LSQ, LPL1, CCL3, and CCL9 in each size category.

Site	1		2		3		4		5		6		7		8		9		10		11		12		13		14		Total
	N	%	N	%	N	%	N	%	N	%	N	%	N	%	N	%	N	%	N	%	N	%	N	%	N	%	N	%	
LSQ					3	1.2	3	1.2	6	2.4	12	4.8	39	15.5	36	14.3	39	15.5	45	17.9	28	11.1	26	10.3	9	3.6	5	2.0	252
LPL1	5	3.6	8	5.8	20	14.6	23	16.8	17	12.4	32	23.4	14	10.2	11	8.0	5	3.6			2	1.5							137
CCL3					3	18.8	4	25.0	2	12.5	4	25.0	2	12.5	1	6.3													16
CCL9					1	2.5	2	5.0	8	20.0	19	47.5	9	22.5	1	2.5													40
Total	5		8		27		32		33		68		64		49		44		45		30		26		9		5		445

(Beck *et al.* 2002), there should be a greater range in the size of the debitage at LSQ than at CRCQ. In addition, flakes should exhibit a smaller amount of dorsal cortex, a greater number of dorsal scars, and multifaceted and abraded/crushed platforms. The results of *t*-tests (Table 10.11) and chi-square tests (Table 10.12) indicate that our predictions hold true for all variables.

Table 10.11 *P*-values for separate variance *t*-tests between CRCQ and LSQ debitage assemblages.

Variable	Probability[a]
Weight	**<0.001**
Maximum length	0.280
Maximum width	**0.022**
Midpoint thickness	**0.027**

[a] Significant values in bold.

Table 10.12 Results of chi-square tests between the two quarries.

Variable	Statistic	Value
Dorsal cortex	χ^2	54.228
	p[a]	**<0.001**
Dorsal scars	χ^2	20.187
	p	**<0.001**
Platform preparation	χ^2	26.537
	p	**<0.001**

[a] Significant values in bold.

Turning to the comparisons between the quarries and the residential sites, the data in Table 10.6 show that the debitage from CRCQ and LSQ indicates more early-stage reduction than does that from Knudtsen, LPL1, CCL3, and CCL9. *Post-hoc* pair-wise ANOVA comparisons (e.g., Zar 1999) reveal significant differences between both quarry sites and their associated residential sites with respect to all metric variables measuring flake size (Table 10.13). No significant differences, however, occur between the LPL1, CCL3, and CCL9 assemblages. Moreover, all of the differences between the quarries and the residential sites occur in the direction predicted (see Table 10.6).

Regarding the categorical variables, chi-square tests reveal significant differences in the predicted direction between the CRCQ and Knudtsen assemblages with respect to all four variables (Table 10.14). These results further support the hypothesis that later-stage reduction was carried out at Knudtsen than at CRCQ.

LPL1, CCL3, and CCL9 all differ significantly from LSQ in terms of the size categories represented (Tables 10.10 and 10.14). Significant differences in size category also exist between LPL1 and CCL9 and between CCL3 and CCL9, but not between LPL1 and CCL3. Beyond flake size, however, there are few significant differences between the LSQ and any of the residential site assemblages. In fact, the only statistically significant differences occur between CCL9 and LSQ ($p = 0.021$) and CCL9 and LPL1 ($p = 0.045$) with respect to platform preparation. The lack of statistical differences among these sites is perhaps caused by the "gearing up" activities at LSQ necessitated by the long travel distance between this quarry and any of the residential sites around it. As predicted above, biface reduction

Table 10.13 *P*-values[a] for ANOVA *post-hoc* pair-wise comparisons between quarry and residential site debitage assemblages.

Variable	CRCQ and Knudtsen	LSQ and LPL1	LSQ and CCL3	Pair LSQ and CCL9	LPL1 and CCL3	LPL1 and CCL9	CCL3 and CCL9
Weight	**<0.001**	**<0.001**	**0.024**	**<0.001**	1.000	1.000	1.000
Maximum length	**<0.001**	**<0.001**	**0.003**	**<0.001**	1.000	1.000	1.000
Maximum width	**<0.001**	**<0.001**	**<0.000**	**<0.001**	1.000	1.000	1.000
Midpoint thickness	**<0.001**	**<0.001**	**0.002**	**<0.001**	1.000	1.000	1.000

[a] Significant values in bold.

Table 10.14 Results of chi-square tests between quarry and residential site debitage assemblages.

Site pair	Statistic	Cortex	Dorsal scars	Platform	Size
CRCQ and Knudtsen	χ^2	125.021	76.811	222.892	N/A
	P^a	**<0.000**	**<0.000**	**<0.000**	N/A
LSQ and LPL1	χ^2	7.112	1.466	2.653	196.287
	p	0.068	0.690	0.448	**<0.000**
LSQ and CCL3	χ^2	2.389	4.394	4.623	78.313
	p	0.496	0.222	0.202	**<0.000**
LSQ and CCL9	χ^2	3.820	1.171	9.715	111.129
	p	0.282	0.760	**0.021**	**<0.001**
LPL1 and CCL3	χ^2	0.857	2.534	4.983	3.193
	p	0.652	0.282	0.083	0.956
LPL1 and CCL9	χ^2	0.727	2.149	6.216	25.221
	p	0.695	0.342	**0.045**	**<0.003**
CCL3 and CCL9	χ^2	0.407	3.385	0.127	11.292
	p	0.523	0.184	0.938	**<0.046**

[a] Significant values in bold.

would have been carried out further at this quarry, so that the assemblage should contain a relatively large proportion of middle- to-late-stage reduction flakes exhibiting the same characteristics as flakes from residential sites.

DISCUSSION

As expected, the patterns in the debitage mirror those of the biface assemblages in our sample. As no surprise to the reader, we are sure, these results reflect the fact that biface manufacture is highly constrained mechanically, an observation given further weight by experimental studies. Yet obvious as these predictions may seem, demonstrating them archeologically is another matter altogether. This is because the sources of variation that influence the morphology of debitage – raw material, core reduction technique, intended product – and that can be controlled in an experimental study, are more difficult to control in archeological situations. Even when they can be, the resulting samples may be so small as to limit confidence in the results of comparative analyses.

For the most part, we have been able to control these sources of variation. Yet, among the assemblages we have described, several are prone to error, including source mis-assignment (i.e., the 20–25% error discussed earlier regarding visual identification of LSQ dacite). The effects of this error, while difficult to estimate, will be most strongly felt in small assemblages like CCL3, which exhibit high source diversity (CCL9, in contrast, is represented by a single source of dacite, LSQ dacite, as judged from XRF analysis as well as visual inspection). Thus, we are reluctant to develop our interpretations much beyond the simple prediction of the transport model; that is, these assemblages reflect the sequential staging of biface reduction and "gearing up" activities prior to their formation.

As we noted above, among Paleoarchaic lithic assemblages, those from CCL3 and CCL9 are more common than the far larger ones from Knudtsen and LPL1. In agreement with the transport model, both contain biface reduction debitage of smaller size (and with later-stage reduction morphologies) than corresponding quarry assemblages. Yet, neither site would be judged a habitation site or a long-term resource extraction site like LPL1. They do not possess the diversity of tool types found in such sites. Instead, they might better be described as short-term extraction sites where the most archeologically visible activity was tool manufacture and refurbishment. We believe these sites represent the distal segments of logistical systems whose hub might include a far larger site like the nearby Sunshine Locality (Figure 10.2) (Jones *et al.* 1996; Huckleberry *et al.* 2001; Beck and Jones in press). Depending on the way groups at the hub gear up for moves to smaller extraction sites, we could expect realistically that source representation would vary a good deal among the latter sites. Thus, one simple prediction drawn from the transport model, that there will be diminished frequencies of a material type with increasing distance from its source, will probably not hold true.

Perhaps we should qualify this last statement. The transport model cannot predict variation expressed at a local level. It should be better able to organize variation at a broader geographic scale encompassing a series of residential moves and visits to quarries to replenish supplies of raw material. This seems to us especially important when the question surrounds establishing the geographic boundaries of a settlement system. Reliance on exotic raw materials principally represented among highly curated formal tools may not be the most definitive evidence. A far stronger case is made when such evidence is aligned with debitage patterns arising in a predictable manner from reduction stages under constraints of transport costs.

CONCLUSION

The acquisition of stone at the source and its subsequent transformation into tools often occurs as a series of steps, each one increasing the utility of the product. Sometimes these steps are performed in one place and sometimes they take place at a different location. Central place foraging models predict that the likelihood of completing this task at one location increases as the anticipated travel distance from that location increases. The study by Beck *et al.* (2002) of Paleoarchaic biface assemblages from the central Great Basin confirms model predictions. It is far from common, however, to find biface assemblages of sufficient size to test this model. Thus, this study examines if the predictions of the model hold equally well for flake debitage, products of this manufacturing sequence. Results from analyses of size characteristics and morphology confirm predictions of the model: first, there are differences between quarry and residential extraction site assemblages and second, reduction will be carried further at the quarry as the distance to the residential site, and hence the cost of carrying low utility components, increases.

The results of our analysis, then, reinforce the general agreement that has been observed between the results of tool analyses and debitage analyses (Whittaker and Kaldahl 2001: 59), and support the proposition that an analysis of debitage should yield the same results about the degree to which bifaces were reduced at a site as will an analysis of the bifaces themselves. This suggests, in turn, that debitage analysis can be used in place of biface analysis in situations where large bifacial assemblages do not exist. Whether in conjunction with, or in place of, biface analysis, however, it is clear that debitage analysis can contribute to a greater understanding of lithic technologies and their roles in the lifestyles of prehistoric peoples (Carr and Bradbury 2001: 127).

ACKNOWLEDGEMENTS

We would like to thank Mike Cannon, Leslie Cecil, Sloan Craven, Cynthia Fadem, Roberta McGonagle, and Amanda Taylor for their help and encouragement in this project.

ENDNOTE

1   LA-ICP-MS: laser ablation, inductively coupled plasma, mass spectrometry.

REFERENCES

Ahler, S.A. (1989) Mass analysis of flaking debris: Studying the forest rather than the trees. In: Henry, D.O. and Odell, G.H. (eds) *Alternative Approaches to Lithic Analysis*, Archaeological Papers of the American Anthropological Association No. 1, Washington, DC, pp. 85–118.

Andrefsky, Jr., W. (1998) *Lithics: Macroscopic Approaches to Analysis*. Cambridge University Press, Cambridge.

Andrefsky, Jr., W. (2001) Emerging directions in debitage analysis. In: Andrefsky, Jr., W. (ed.) *Lithic Debitage: Context, Form, Meaning*, The University of Utah Press, Salt Lake City, pp. 2–14.

Bamforth, D.B. (2002) High-Tech foragers? Folsom and Later Paleoindian Technology on the Great Plains. *Journal of World Prehistory* 16, 55–98.

Beck, C. and Jones, G.T. (1990) Late Pleistocene/early Holocene occupation in Butte Valley, eastern Nevada: Three seasons work. *Journal of California and Great Basin Anthropology* 12, 231–61.

Beck, C. and Jones, G.T. (1994) On-site analysis as an alternative to collection. *American Antiquity* 59, 304–15.

Beck, C. and Jones, G.T. (2008) *The Archaeology of the Eastern Nevada Paleoarchaic, Part 1: The Sunshine Locality.* The University of Utah Anthropological Papers, Salt Lake City. (With and contributions by Gary Huckleberry, Michael Cannon, Amy Holmes, Stephanie Livingston, Jack Broughton, Donald Tuohy, and Amy Dansie.)

Beck, C., Taylor, A., Jones, G.T., Fadem, C.M., Cook, C.R., and Millward, S.A. (2002) Rocks are heavy: Transport costs and paleoarchaic quarry behavior in the Great Basin. *Journal of Anthropological Archaeology* 21, 481–507.

Bettinger, R.L., Malhi, R., and McCarthy, H. (1997) Central place models of acorn and mussel processing. *Journal of Archaeological Science* 24, 888–99.

Bettinger, R.L., Winterhalder, B., and McElreath, R. (2006) A simple model of technological intensification. *Journal of Archaeological Science* 33, 538–45.

Bird, D.W. and Bliege Bird, R.L. (1997) Contemporary shell-fish gathering strategies among the Meriam of the Torres Strait Islands, Australia: Testing Predictions of a Central Place foraging Model. *Journal of Archaeological Science* 24, 39–63.

Bradbury, A.P. and Phillip, J. (1995) Flake typologies and alternative approaches: An experimental assessment. *Lithic Technology* 20, 100–15.

Bright, J.A., Ugan, A., and Hunsaker, L. (2002) The effects of handling time on subsistence technology. *World Archaeology* 34, 164–81.

Callahan, E. (1979) The basics of biface knapping in the eastern fluted point tradition: A manual for flint knappers and lithic analysts. *Archaeology of Eastern North America* 7, 1–180.

Carr, P.J. and Bradbury, A.P. (2001) Flake debris analysis, levels of production and the organization of technology. In: Andrefsky, Jr., W. (ed.) *Lithic Debitage: Context, Form, Meaning,* The University of Utah Press, Salt Lake City, pp. 126–46.

Close, A.E. (2006) *Finding the People who Flaked the Stone at English Camp (San Juan Island).* The University of Utah Press, Salt Lake City.

Huckleberry, G.A., Beck, C., Jones, G.T., *et al.* (2001) Terminal Pleistocene/early Holocene environmental change at the sunshine locality, North-central Nevada, U. S. A. *Quaternary Research* 55, 303–12.

Ingbar, E.E., Larson, M.L., and Bradley, B.A. (1989) A non-typological approach to debitage analysis. In: Amick, D.S. and Mauldin, R.P. (eds) *Experiments in Lithic Technology,* BAR International Series 528, Oxford, pp. 117–36.

Johnson, J.K. (1981) *Yellow Creek Archaeological Project v.2, Tennessee Valley Authority Publications in Anthropology No. 2.,* University of Mississippi, Tennessee Valley Authority, Norris, TN.

Jones, G.T., Beck, C., Jones, E.E., and Hughes, R.E. (2003) Lithic source use and Paleoarchaic foraging territories in the Great Basin. *American Antiquity* 68, 5–38.

Jones, G.T., Beck, C., Nials, F.L., Neudorfer, J.J., Brownholtz, B., and Gilbert, H. (1996) Recent archaeological and geological investigations at the Sunshine Locality, Long Valley, Nevada. *Journal of California and Great Anthropology* 18, 48–63.

Magne, M.P.R. (1985) *Lithics and Livelihood: Stone Tool Technologies of Central and Southern Interior B. C.,* Archaeological Survey of Canada, Mercury Series No. 133., Ottawa.

Metcalfe, D. and Barlow, K.R. (1992) A model for exploring the optimal trade-off between field processing and transport. *American Anthropologist* 94, 340–56.

Patterson, L.W. (1990) Characteristics of bifacial-reduction flake-size distribution. *American Antiquity* 53, 550–8.

Shott, M.J. (1994) Size and form in the analysis of flake debris: Review and recent approaches. *Journal of Archaeological Method and Theory* 1, 69–110.

Sievert, A.K. and Wise, K. (2001) A generalized technology for a specialized economy: Archaic period chipped stone at kilometer 4, Peru. In: Andrefsky, Jr., W. (ed.) *Lithic Debitage: Context, Form, Meaning,* The University of Utah Press, Salt Lake City, pp. 80–105.

Sullivan, III, A.P. and Rozen, K.C. (1985) Debitage analysis and archaeological interpretation. *American Antiquity* 50, 755–79.

Teltser, P.A. (1991) Generalized core technology and tool use: A Mississippian example. *Journal of Field Archaeology* 18, 363–75.

Ugan, A., Bright, J., and Rogers, A. (2003) When is technology worth the trouble? *Journal of Archaeological Science* 30, 1315–29.

Whittaker, J.C. (1994) *Flintknapping: Making and Understanding Stone Tools.* University of Texas Press, Austin.

Whittaker, J.C. and Kaldahl, E.J. (2001) Where the waste went: A knapper's dump at Grasshopper Pueblo. In: Andrefsky, Jr., W. (ed.) *Lithic Debitage: Context, Form, Meaning,* The University of Utah Press, Salt Lake City, pp. 32–60.

Zar, J.H. (1999) *Statistical Analysis.* Prentice Hall, Upper Saddle River.

Part III

MICRO-LANDSCAPE PERSPECTIVES

RECONSTRUCTING LANDSCAPE USE AND MOBILITY IN THE NAMIBIAN EARLY STONE AGE USING *CHAÎNE OPÉRATOIRE*

Grant S. McCall

Department of Anthropology, Tulane University

ABSTRACT

This chapter examines the organization of stone tool technology at the Lower Paleolithic site of Tsoana, Northeastern Namibia. This study employs a *chaîne opératoire* approach in order to address the relationship between Lower Paleolithic stone tool technology, mobility, and site use. The chapter argues that the observed archeological patterns of relatively short knapping sequences are not consistent with "home base" site use typical of modern forager residences. Instead, based on patterns of flake cortex and refitting, this study suggests that Lower Paleolithic archeological sites like Tsoana were combinations of geological palimpsests and non-domestic activity areas. The chapter explores several existing models of non-residential site use and stone tool accumulation in light of this evidence, and discusses concomitant theoretical implications.

INTRODUCTION

It has become almost axiomatic to observe that stone tool technology is vitally important to the reconstruction of early hominid behavior because of the relative persistence of lithics compared to other archeological

Lithic Materials and Paleolithic Societies, 1st edition. Edited by B. Adams and B.S. Blades, ©2009 Blackwell Publishing. ISBN 978-1-4051-6837-3.

remains. Despite the central role of stone tools in the analysis of the Paleolithic, this data source remains problematic. The intellectual landscape of Lower Paleolithic archeology is one divided by considerable schisms, many stemming from the important critique of existing big-game hunting-models of humans presented by Binford (1981). As for stone tools, Binford considers them difficult to deal with because of the problems with inferring functional behavior. In other words, Binford's work on faunal remains is based on the premise that it is hard to link stone tools with the patterns in which they were used. This perspective led to the large-scale popularity of faunal analysis within Lower Paleolithic archeology and a revamping of typological approaches to lithic technology.

This chapter begins with this Binfordian premise that it is difficult to specify the functional behavior associated with stone tools beyond the need for a sharp or durable edge. Instead, this chapter focuses on the relationship between raw material economy – the need to maintain a supply of stone suitable for making a sharp, durable edge – and patterns of core reduction. This chapter also argues that various patterns of mobility constrain access to raw material supplies, control raw material economies, and strongly influence strategies of core reduction. Therefore, while functional behavior may be problematic to infer from stone tools, the organization of lithic technological systems is a highly productive source of information concerning behavioral variables controlling raw material economy.

This chapter presents a case study from the Lower Paleolithic (or Early Stone Age [ESA], in the African terminology) site of Tsoana in the Nhomadom valley in the Kaudom National Park of Northeastern Namibia. I use the concept of *chaîne opératoire* to make inferences about the relationship between lithics manufacture, mobility, and site use. I also place Tsoana in the context of other Lower Paleolithic sites where comparable data have been collected, and discuss hypotheses to explain the reasons for the observed variability. I conclude that early hominids in the Nhomadom were moving about the landscape and using sites in ways unlike modern forager home bases; I find frequent movement between sites, extremely short-term site use, and limited recognizable movement beyond a limited territory defined by proximity to lithic raw material availability. Stone tool accumulations represent the long-term conflations of diverse activities involving manufacture and use. My findings suggest considerable variability in the behaviors that produced Lower Paleolithic lithic assemblages, but mainly contradict expectations for home base sites.

FRAMING THE PROBLEM

In attempting to link stone tools with behavior, it is easier to make inferences about the technology of manufacture than patterns of use. Stone tools are simply not specifically designed for discrete tasks and,

like all technology, are flexible in their patterns of use (Binford 1976, McCall 2005). There is, of course, a substantial and burgeoning literature concerning use-wear analysis (more, in fact, than I could usefully cite here), ranging from the early optimism of Keeley (1980), to the revision and ambiguity of Vaughan (1985). However, as the experimentation of Vaughan strikingly demonstrates, it is difficult, if not impossible, to link damage and wear on stone tools with patterns of use in all but the most exceptional cases. Vaughan shows the diverse and ubiquitous set of processes that can produce pseudo-wear patterns, rendering many assumptions problematic. This is especially the case in Africa, where suitable fine-grained raw material is rare. Keeley and Toth (1981) provide a lonely example of successful use-wear analysis by examining obsidian lithics from Koobi Fora. For the most part, the basalts, quartzites, and other coarse-grained cryptocrystaline silicates preclude any such analytical procedures.

The behavior involved in stone tool manufacture is much easier to infer. This has been an idea crucial to lithic analysis since the boom in experimental studies dealing with manufacture (Crabtree 1966; Bordes 1969; Dibble 1985; Whittaker 1994). This fact rests upon two related concepts in French and American archaeology: the *chaîne opératoire*, and its American counterpart, the *sequence of reduction*. The French *chaîne opératoire* refers specifically to the sequence of operations carried out in the manufacture of a specific stone tool. Leroi-Gourhan (1964) is most frequently credited as the progenitor of the concept, whose structuralist concern was building a grammar-like syntax of technological operations. As Sellet (1993) details, the term expanded to include all of the systematics of behavior that occurred at a site and between sites, and became a mainstay of French Paleolithic archeology. Perhaps because of Leroi-Gourhan's focus on the reconstruction of sequences of mental processes, the French *chaîne opératoire* has remained driven by a focus on the final product. The notion of syntax implies a structured set of rules involved in producing a referent that cannot be altered or contradicted during the process: a precise ordering of operations in order to produce a predetermined end-product. Therefore, such mental templates play a significant role in the French *chaîne opératoire*.

The American *sequence of reduction* has a parallel genealogy to its French counterpart. Dibble (1988) and Bleed (1991, 2001) are closely associated with this idea, while it has its roots in the breakthrough experiments of Crabtree (1966). Most of this concept overlaps with the French version, however, since the work of Dibble, the American sequence of reduction has lacked the same focus on the final product. The sequence of reduction is not driven by a mental template or focus on the final product. Instead, this concept sees the sequence of operations as an ongoing flexible process producing a diversity of stone tool morphologies under a diverse set of technological situations. Hence, process is more important than the final product, which is often discarded as mere refuse. As Sellet (1993) observes, this has some of its roots in the transformational systemics of Schiffer (1972). The French and American versions are only minimally different, and can be profitably combined. In this chapter, I use the term *chaîne opératoire* to contain both concepts.

The concept of *chaîne opératoire* has a great deal of potential for approaching Lower Paleolithic stone tool technology because it provides an analytical framework for understanding core reduction strategies and their relationships to external variables, such as raw material economy. This chapter uses stone tool assemblages to investigate how early hominids moved around the landscape and how they used sites in the Nhomadom valley. This chapter tests a few specific models of Lower Paleolithic hominid mobility and site use: central place site use models, such as Isaac's (1971, 1978) "home base" model, Schick's (1987) revision of this model, and Rose and Marshall's (1996) "focal site" model – and special activity location models, such as Binford's (1981, 1984, 1987) "scavenging station" model and Pott's (1988, 1992) "stone cache" model.

Isaac (1971, 1978) is among the first and strongest proponent of the central place or "home base" idea, and represents continuity with the initial behavioral interpretations made of Lower Paleolithic archeological assemblages. His ideas are closely tied to mate provisioning, sex-based division of labor and food sharing, which rely upon increased hunting and central place site use (Washburn and Devore 1961; Lovejoy 1981). In a recent review, O'Connell *et al.* (2002) refer to this as the "Washburn/Isaac synthesis." Isaac states that hominids repeatedly occupied the same location for long periods of time, with males acquiring food (through predatory activity) and returning to the home base, sharing with the less mobile child-rearing females. This considers the sites as all-purpose activity areas, as all activities (except resource extraction) would be carried out at home base locations.

Schick (1987) revamps the "home base" hypothesis by discarding the food provisioning and sex-based division of labor, and recognizing some of the taphonomic difficulties with certain sites. She states that artifacts accumulated over long periods of time at central place foraging locations, as hominids carried with them the raw material to construct the necessary tools for various foraging situations, and slowly discarded refuse pieces. She suggests that some tools may have been deposited in specific places and reused at later times. In addition, faunal remains accumulated at specific locations as hominids returned with various animal parts. These locations were placed as strategic centers on the landscape, equidistant to sites of frequent resource availability. Likewise, Rose and Marshall (1996) redefine central place site use by presenting a "resource defense" model. This idea sees Lower Paleolithic sites as focal points on the landscape where most economic and social activity took place. Again, this model incorporates some of the formational complexity noticed by critical researchers; it sees focal sites as places where archeological remains were pooled, while not every activity took place there. While more complex, it retains most features of earlier central place concepts.

Isaac's (1978) version of the "home base" model is archeologically testable but this has been done largely on the basis of faunal remains (Binford 1981, 1984). In thinking about my own data, if this model were true, I would expect to find large and relatively complete segments of the *chaîne opératoire*, since hominids were using the site without interruption over long periods of time. It is more difficult to build testable conditions for Schick's (1987) and Rose and Marshall's (1996) versions of the central place model, because of their recognition of problems of the taphonomic complexity of these site.

Binford (1981), of course, presented the first major attack on the "Washburn/Isaac synthesis" (O'Connell *et al.* 2002). However, Binford's attack was more aimed at questions concerning hunting. In considering site use more directly, Binford (1981, 1983, 1984, 1987) suggests that Lower Paleolithic sites were the result of activities surrounding waterholes and other attractive features, and that marginal scavenging was the primary way in which hominid were interacting with carcasses. Hominids deposited artifacts used in a very limited capacity for processing carcasses at locations on the landscape where scavengable carcasses frequently accumulate through nocturnal carnivore predation.

Furthermore, both hominid artifacts and bones, as well as bones accumulated by other predator-scavengers, accumulated at the same strategic locations on the landscape, such as near waterholes and under shade trees.

More recently, O'Connell *et al.* (2002) offer their "near-kill accumulation" version of this idea. They use ethnographic observations on the Hadza to show that the characteristics of early hominid faunal assemblages share a great deal with locations of bone deposition from initial butchery near kills or scavenged carcasses, and have very little in common with "home base" sites. This conception of site use overlaps a great deal with Binford's scavenging station idea, while attributing to early hominids higher frequencies of hunting and earlier access to carcasses through scavenging.

Potts' (1988, 1992) model of hominid mobility shares some aspects with that of Schick (1987). Potts suggests that accumulations of lithics were caches of raw material. He sees raw material sources and caches as strong attractors of hominid activity. Potts suggests hominids formed caches of raw material at strategic locations on the landscape and moved carcasses to these locations for processing. Moving carcasses away from the original kill locations to strategic stone caches solved problems of returning predator/scavengers, as well as transportation of butchery implements. Potts suggests that these locations are especially important in regions where suitable raw material is not ubiquitous. He also makes the important observation that these locations were likely not sleeping places, as was assumed in Isaac's (1971, 1978) and Schick's (1987) model, because of the related attraction of dangerous predators and scavengers. By moving lithics across the landscape and accumulating them in caches, strategic locations for carcass processing were produced.

These special activity location models have some clear conditions in terms testing with the Tsoana lithics analysis. Unlike the central place models, which assume long-term and continuous use of sites over significant durations of time, these models instead conceive of site use as immediate, short-term, and discontinuous. Such uses of sites should result in very little deposition of archeological remains per visit, and only the aggregate of innumerable visits could account for large sites. Therefore, I would not expect to find long and complete *chaînes opératoire* at Lower Paleolithic sites. Because of this complexity, the total reduction of a single core may involve several sites, dozens of actors in many different situations, and may span thousands of years.

DATA ANALYSIS AND INTRA-SITE COMPARISONS

During the 2000–2 field seasons at Tsoana, I collected 1359 lithics from 6 1×1 m excavation units. For the sake of brevity, I will forgo any discussion of the chronology or stratigraphy of the site, except to say that the assemblage is currently being dated through optical stimulated luminescence (OSL), it is typologically ESA.

The lithics from Tsoana are manufactured from two main types of raw material: silcrete and quartzite. Quartz is also found occasionally, mainly in the form of hammerstones and hammerstone spall. This raw material is derived from nodules eroding out of the basal Kalahari bed – an ancient conglomerate containing silcrete, quartzite, and quartz pebbles of varying sizes. The Kalahari bed is exposed in several locations near Tsoana and nodules of raw material are also spread across the landscape as the result of erosion.

In considering the patterns of cortex on the dorsal surfaces of flakes, I adapted Villa's (1978, 1983) and Toth's (1982, 1987) six-category system of classification (see Figure 11.1). On the basis of my own experimentation and the relationship of cortex patterns with place in the sequence of core reduction, I have re-numbered the categories slightly. In seeking to understand the patterning in the Tsoana assemblage, I experimentally reduced six chert cobbles from Eastern Missouri. These cobbles were of similar size and had similar cortex coverage to the Tsoana material, and I reduced them in fashion seeking to maximize the size and cutting edge length of each subsequent flake. What I determined is that Tsoana has relatively fewer of the

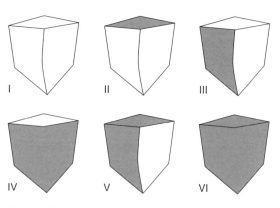

Figure 11.1 Classifications for patterns of cortex on dorsal surfaces of flakes.

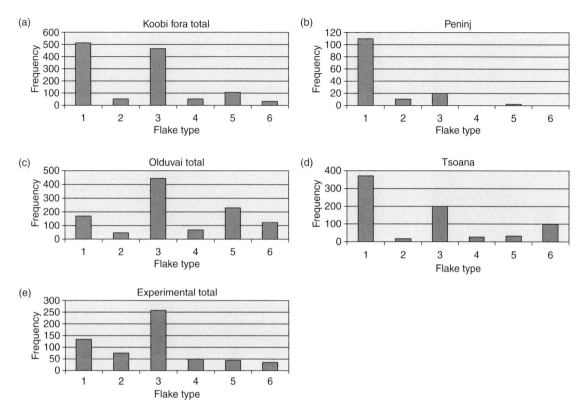

Figure 11.2 Comparison of cortex frequencies for (a) Koobi Fora sites (Toth 1987), (b) Peninj (De la Torre *et al.* 2003), (c) Olduvai sites (Kimura 1999), (d) Tsoana, and (e) experimental core reduction (Toth 1987 and McCall, this chapter).

early stages of flake removals (cortex types 3–6), but is otherwise extremely similar to the experimentally-produced flake assemblage containing all stages of reduction.

I also looked at Toth's (1987) experimental and archeological data from Koobi Fora, Kimura's (1999) data from Olduvai, and De la Torre *et al.*'s (2003) data from Peninj. These comparisons are also quite striking. Tsoana is extremely similar to Peninj and the Koobi Fora sties, and it is not terribly different from the Olduvai sites. Figure 11.2 shows a graphical comparison of cortex frequencies for the various archeological and experimental assemblages. Figure 11.3 shows a dendrogram produced from a hierarchical cluster analysis based on cortex frequencies. Peninj, Tsoana, and Koobi Fora have fewer of the highly cortical flakes produced in the first stages of core reduction. Olduvai is very similar to the summed experimental frequencies. It is important to add that none of these sites are wildly different from the experimentally derived expected frequencies.

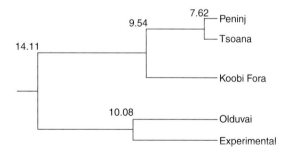

Figure 11.3 Dendrogram of hierarchical cluster analysis based on flake cortex frequencies.

The implication of this is that there is a range of variability for the cortex patterns on flakes, ranging from slightly earlier in the reduction sequences to slightly later in the reduction sequences. Toth (1987; Schick and Toth 1993) suggests that some sites at Koobi Fora are missing the early stage of flaking because of the testing of cores at raw material sources. In other words, one would expect the removal of several flakes from the core in order to assure its quality; this might account for the missing flakes early in the sequence. At Olduvai, in contrast, the flakes from the later stages of reduction are missing. This may indicate that some of the Olduvai assemblages were locations of raw material sources and that transport was occurring.

I also examined the ratio of flakes to cores for these archeological assemblages. Tsoana has 13.53 flakes per core when the entire assemblage is considered and 7.67 when flakes larger than 2 cm are considered. As with the cortex frequencies, Peninj is similar to Tsoana with 10.52 flakes per core (n.b. this number represents my calculation, however the authors report 7.3 flakes per core – see De la Torre *et al.* 2003). For Koobi Fora, there are 4.56 flakes per core (my calculation), however this number climbs to 6.55 when the core-heavy FxJj 33 (called the "megacore" site by Toth), is excluded. For the Olduvai assemblages discussed by Kimura (1999), there are 4.22 flakes per core (my calculation).

These numbers are interesting, but I caution that their usefulness is somewhat limited to regional or site-specific conclusions. My experiments confirm that the number of flakes per core is directly related to the size of the original raw material object. In other words, as is common sense, a bigger core yields more flakes. In addition, low ratios of flakes to cores may also be indicative of taphonomic problems, such as differential size sorting. Of the sites considered, Olduvai seems particularly suspect in this regard. Nonetheless, this comparison in interesting, and these values complement the cortex analysis. Once again, they suggest that flakes from later stages of core reduction were transported (in low frequency) to Peninj and Tsoana. In contrast, the low value suggests that the later stages of flaking were removed from the Olduvai assemblages discussed by Kimura (1999). Koobi Fora remains somewhat intermediate and ambiguous. This again points to variability in the formation of these lithics assemblages.

These findings have interesting implications for the refitting analysis I did with Tsoana lithics. I went about refitting a sample of the Tsoana material: 269 flakes, 2 cores, and 3 core fragments from a 1 × 1 m unit within the main excavation. I chose to examine a sample of the total excavated material because of the lengthy amount of time and effort involved in refitting. In addition, while the number of refits may increase with a larger sample, the refitting **rate** should remain relatively stable (assuming truly random sampling), and refitting rate is what I am truly interested in for testing the continuity of site use. I was not able to find any true sequential refits in the assemblage. I was able to find only one refitting flake, which was broken apparently as the result of endshock.

In addition to the physical refitting of flakes, I used a technique borrowed from Wyckoff (1992) and Larson and Ingbar (1992), who used the variability between raw material nodules in terms of color and grain size to reconstruct "analytical" cores. This offers a way of seeing relationships between flakes that might not be visible with physical refitting. Table 11.1 presents the findings of this analytical refitting. Using this technique, I was able to find several long sequences of related flakes. These sequences include one of 15 flakes and another of 10 flakes, with several others of significant length. At the other end of the spectrum, there were 143 unique flakes and 23 pairs of related flakes. This study shows that 90.4% of the analytical cores had either one or two flakes in their sequence. This is largely congruent with the finding of no sequential refits. There is apparently little continuity in the flaking sequences at Tsoana.

It is interesting to compare the findings of the analytical refitting from Tsoana with the results of this same technique from Laddie Creek, Wyoming, and Lowrance, Oklahoma, reported by Larson and Ingbar (1992) and Wyckoff (1992) respectively (see Table 11.2). Laddie Creek is an early archaic site deposited near a perennial spring, probably the result of numerous unrelated

Table 11.1 Analytical refitting of Tsoana refitting sample.

Number of flakes/ analytical core	Number of cases	Percentage total analytical cores (%)
15	1	0.5
10	1	0.5
6	2	1.1
5	1	0.5
4	5	2.7
3	6	3.3
2	23	12.7
1	143	77.7

Table 11.2 Comparison of selected refitting analyses of lithics assemblages.

Site	Refitting rate	Total *N*	Description	References
Tsoana	0.4%	269	See this chapter...	
Boxgrove	26.0%	735	Middle Pleistocene site dominated by bifacial thinning and shaping activity. Handaxe manufacture???	Brees 2002
Gran Dolina	1.1%	268	Spanish Lower Paleolithic site, at least 780 kyr. Small lithics and faunal assemblage.	Carbonell *et al.* 1999
Maastricht-Belvedere	21.5%	???		Roebroeks and Hennekens 1990
Kalambo Falls	0.2%	???	Late African ESA site. Lots of geological disturbance.	Schick 1992
8-B-11, Sai Island, Sudan	About 4%	About 200	African, late ESA/early MSA. Stable taphonomy, located next to ancient river.	Van Peer *et al.* 2003
Isoyama	<1%	More than 13 000	Japanese UP	Bleed 2002
Iwato	<1%	More than 9000	Japanese UP	Bleed 2002
Mosanru	3.0%	7350	Japanese UP	Bleed 2002
Solvieux	10.0%	Around 10 000	French UP rock shelter site.	Grimm and Koetje 1992
Abri Dufaure	Many	Many	French UP rock shelter site.	Petraglia 1992
Allen Site	2.0%	More than 11 228	Paleoindian site. Lots of reoccupation of the same location over long periods of time	Bamforth and Becker 2000
Laddie Creek, WY	1.5%	Over 5600	Early Archaic site, around 8600 BP. 172 analytical nodules, with only 30 that contained refits. 1.5 refits per refitting analytical nodule.	Larson and Ingbar 1992
Cave Spring, TN	5.0%	At least 8100	Archaic site with some vertical mixing	Hofman 1992
Stewart's Cattle Guard, CO	<1% (most hammers/anvils)	11 644	Archaic site with some vertical mixing	Jodry 1992
Lowrance, OK	30.6%	157	Protohistoric knapping with stockpiled raw material and expedient flake removal. 21 analytical cores, 12 with refits. Average 4 refits per analytical core	Wyckoff 1992

uses of the site. The flaking is defined by conservative bifacial reduction. In contrast, Lowrance is a late prehistoric site, marked by longer-term sedentary occupation, stockpiling of raw material, and expedient flaking. Laddie Creek is very similar to Tsoana, showing very low physical and analytical refitting rates. Larson and Ingbar identify 172 analytical nodules, with only 30 of these contain a refitting flake or core. In these cases, only there was on average only 1.5 refits per analytical core. In contrast, the Lowrance site is very unlike Tsoana, which has a rate of 0.006 refits per analytical core. From a sample of 157 flakes, Wyckoff was able to find only 21 analytical cores. In addition, he was able to find 48 physical refits within this sample, including one sequence of ten refitting flakes. This is an average of 4 physical refits per analytical core. This is obviously very different from both Laddie Creek, and more importantly for this chapter, Tsoana.

The extremely low refitting rate at Tsoana should best be viewed within the range of variability for lithics assemblages. Table 11.2 summarizes a number of refitting studies that offer good points of comparison. There is a considerable amount of variability in these cases, with refitting rates ranging from more than 30% to almost nothing at all. Schick (1986) reports essentially no refits from Kalambo Falls, Zambia, and blames mainly taphonomic factors for this finding. Likewise, Carbonell *et al.* (1999) report a refitting rate of 1.1% at Gran Dolina, Spain. Here, the authors seem to point to the transport of flakes onto the site and the short core reduction sequences as the cause of the low rate. It is also interesting to compare these low refitting rates to those from the Allen Site and Cave Spring, from the North American Paleoindian and Early Archaic (Hofman 1992; Bamforth and Becker 2000). These time periods are defined by very high mobility and it makes intuitive sense that frequent mobility would correspond with low refitting rates.

In contrast to these sites with low refitting rates, there are several with very high rates. For example, Boxgrove has a refitting rate of 26% (Brees 2000). Brees describes the flake assemblage as dominated by thin, spreading bifacial thinning flakes, and links the high refitting rate to *in situ* handaxe manufacture at a source of raw material availability. Another interesting comparison is Lowrance, mentioned earlier, which boasts a refitting rate of greater than 30% (Wyckoff 1992). Here, I suspect that the sedentary mobility pattern, stockpiling raw material economic strategy, and expedient flaking cause such a high refitting rate. This variability

in refitting rate points to some possibilities concerning the formation processes of Tsoana.

As Schick (1986) suggests, taphonomy may often explain low refitting rates. However, there are no other signs of taphonomic sorting at Tsoana. In addition, Tsoana appears mostly undisturbed according to the other criteria that Schick lays out. The patterns of microdebitage, flake size frequency, and artifact orientation all suggest little disturbance. I will spare the details of this analysis for the sake of brevity. While I hold out the possibility that taphonomic disturbance may explain the low refitting rate, I think a better explanation involves the complexity of hominid use of the site. Sites with high refitting rates like Boxgrove and Lowrance are defined by either high rates of flaking or continuous use of the site.

The use of Tsoana by early hominids is marked by the complex movements of flakes and cores both onto and away from the site. By comparison, Tsoana is like the number of North American Paleoindian and Archaic sites with low refitting rates. In Figure 11.4, I show my conception of the relationship of frequency of movement and rate of flake removal with refitting rate. I infer from this set of findings that the lithics of Tsoana were the result of high rate of mobility, frequency of movement, and infrequent flaking. I suggest that Tsoana is the result of a huge number of unrelated uses of the site, marked by infrequent but expedient flaking.

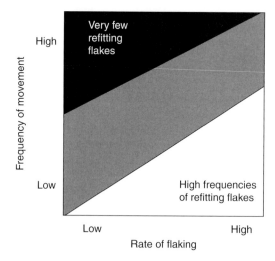

Figure 11.4 Refitting rate as related to mobility and rate of flaking.

In summary, I found cortex patterns tilted toward the later stages of core reduction and a high ratio of flakes to cores. I also found a very low physical refitting rate and a very low number of refitting flakes for every analytical core during the analytical refitting exercise. In the context of a wide diversity of other sites, I interpret these findings as indicating a complex pattern of movement of flakes and cores onto and away from Tsoana, and as a result of a large number of very short-duration uses of the site with low rates of flake production. I will now use these findings to examine some of the previously discussed models of early hominid site use and mobility patterns.

DISCUSSION

These findings from Tsoana clearly challenge the central place models of early hominid site formation. Even at Lower Paleolithic sites with high refitting rates, such as Boxgrove, it is only certain parts of the *chaîne opératoire* that are present. It is not the complete *chaînes opératoire* that I would expect from longer-term central place sites. What is also interesting in comparing the Lower Paleolithic sites presented in this chapter is that there seems to be a substantial range of variability to the activities carried out at these sites. Certainly, it seems unwise to treat every lithics assemblage or every conflation of stones and bones as a single site type – central place.

The findings of this chapter have been more consistent with special activity area models of site use, as the uses of Tsoana seem to have been brief and discontinuous. In this way, it is more similar to the models put forward by Binford (1981, 1984) and Potts (1988, 1992). Similarly, O'Connell *et al.* (2002) propose the so-called "near-kill" accumulation model. They also suggest that these sites accumulated over long periods of time through numerous short-term (less than one day) site uses. O'Connell *et al.* present a detailed description of the archeological and geographical characteristics of near-kill assemblages and conclude most Plio/Pleistocene archeological data are consistent with them. They state (O'Connell *et al.* 2002: 853):

> On the basis of site location and species representation alone, most [well-studied Plio/Pleistocene sites] appear to be near-kill accumulation points, *not* home bases. Body part representation patterns are consistent with hominin transport of variable, sometimes broad arrays of elements

from nearby kills and subsequent density-dependent attrition by secondary consumers and other non hominin agents. If early hominins did indeed move parts of large carcasses to more distant sites to share with others, archaeological evidence of the practice has yet to be discovered.

This assessment of the faunal evidence seems consistent with the complexity that I perceive at Tsoana, and other major early hominid stone tool assemblages. The repeated site use, perhaps with varying characteristics, for short durations seems a good explanation for Tsoana and other such sites. In this respect, Dominguez-Rodrigo *et al.* (2002) have recently argued for the idea that the Peninj sites show patterns consistent with O'Connell *et al.*'s (2002) "near-kill accumulations." This is interesting given many of the similarities seen between the Peninj sites and Tsoana.

CONCLUSION

This chapter has found that the lithic assemblages of Tsoana are not consistent with central place models of hominid site use. This chapter has argued that other special activity areas, such as the "near-kill" accumulations described by O'Connell *et al.* (2002), are a more parsimonious explanation of the characteristics of the lithics assemblage from Tsoana. This chapter has also tried to show the value of refitting studies and the *chaîne opératoire* concept more generally for examining the relationship between core reduction strategies and constraints of raw material economy. More extensive refitting studies and the application of other analytical techniques concerned with sequence to Lower Paleolithic assemblages has a great deal of potential to improve our knowledge of how early hominids moved around the landscape and how archeological sites formed.

The other challenge for Lower Paleolithic archeological research is to develop a novel evolutionary theory concerning the origins of the anatomical and behavioral characteristics of the genus *Homo* that are not based on "home base" use, hunting of large game, and social sharing of food beyond closest kin. Many lines of evidence, including the one presented in this chapter, suggest that archeological evidence for the "Washburn/Isaac synthesis" is lacking. However, the value of this archeological finding is limited without the creative generation of new evolutionary models.

REFERENCES

Bamforth, D.B. and Becker, M.S. (2000) Core/biface ratios, mobility, refitting, and artifact use-lives: A Paleolindian example. *Plains Anthropologist* 45, 273–290.

Binford, L.R. (1976) Forty-seven trips: A case study in the character of some formation processes of the archaeological record. In: Hall, E.S. (ed.) *Contributions to anthropology: The interior of peoples of northern Alaska*. National Museum of Man, Mercury Series Paper No. 49. National Museum of Canada, Ottawa, pp. 299–351.

Binford, L.R. (1978) *Nunamiut Ethnoarchaeology*. Academic Press, New York.

Binford, L.R. (1981) *Bones: Ancient Men and Modern Myths*. Academic Press, New York.

Binford, L.R. (1983) *In Pursuit of the Past*. Thames and Hudson, New York.

Binford, L.R. (1984) *Faunal Remains from Klasies River Mouth*. Academic Press, Orlando.

Binford, L.R. (1987) Searching for camps and missing evidence? Another look at the Lower Paleolithic. In: Soffer, O. (ed.) *The Pleistocene Old World: Regional Perspectives*, Plenum Press, New York, pp. 17–32.

Bleed, P. (1991) Operations research and archaeology. *American Antiquity* 56,19–35.

Bleed, P. (2001) Trees or chains, links or branches: Conceptual alternatives for consideration of sequential activities. *Journal of Archaeological Method and Theory* 8, 101–27.

Bleed, P. (2002) Obviously sequential, but continuous or staged? Refits and cognition in three late paleolithic assemblages from Japan. *Journal of Anthropological Archaeology* 21, 329–43.

Bordes, F. (1969) The Corbiac blade technique and other experiments. *Tebiwa* 12, 1–21.

Brees, D.A. (2000) The refitting of lithics from Unit 4D, Area Q2/D excavations at Boxgrove, West Sussex, England. *Lithic Technology* 25, 120–34.

Carbonell, E., Garcia-Anton, M.D., Mallol, C., Mosquera, M., Olle, A., Rodriguez, X.P., Sahnouni, M., Sala, R., and Verges, J.M. (1999) The TD6 level lithic industry from Gran Dolina, Atapuerca (Burgos, Spain): Production and use. *Journal of Human Evolution* 37, 653–93.

Crabtree, D.E. (1966) The stoneworkers approach to analyzing and replicating the Lindenmeier Folsom. *Tebiwa* 9, 3–39.

De la Torre, I., Mora, R., Dominguez-Rodrigo, M., De Luque, L., Alcala, L. (2003) The Oldowan industry of Peninj and its bearing on the reconstruction of the technological skills of Lower Pleistocene hominids. *Journal of Human Evolution* 44, 203–24.

Dibble, H.L. (1985) Technological aspects of flake variation. *American Archaeology* 5, 236–40.

Dibble, H.L. (1988) Typological aspects of reduction and intensity of utilization of lithic resources in the French Mousterian. In: Dibble, H.L. and Monet-White, A. (eds) *Upper Pleistocene Prehistory of Western Eurasia*, University of Pennsylvania Museum, Philadelphia, pp. 181–97.

Dominguez-Rodrigo, M., de la Torre, I., de Luque, L., Alcala, L., Mora, R., Serrallonga and J., Medina, V. (2002) The ST Site Complex at Peninj, West Lake Natron, Tanzania: Implications for early hominid behavioural models. *Journal of Archaeological Science* 29, 639–65.

Grimm, L.T. and Koetje, T.A. (1992) Spatial patterns in the Upper Perigordian at Solvieux: Implications for activity reconstruction. In: Hofman, J.L. and Enloe, J.G. (eds) *Piecing Together the Past: Applications of Refitting Studies in Archaeology*, BAR International Series 578, Oxford, pp. 264–86.

Hofman, J.L. (1992) Defining buried occupation surfaces in terrace sediments. In: Hofman, J.L. and Enloe, J.G. (eds) *Piecing Together the Past: Applications of Refitting Studies in Archaeology*, BAR International Series 578, Oxford, pp. 128–50.

Isaac, G.L. (1971) The diet of early man: Aspects of archaeological evidence from Lower and Middle Pleistocene sites in Africa. *World Archaeology* 2, 278–89.

Isaac, G.L. (1978) The food sharing behavior of proto-human hominids. *Scientific American* 238, 90–108.

Isaac, G.L. and Harris, J.W.K. (1978) Archaeology. In: Leakey, M.G. and Leakey, R.E.F. (eds) *Koobi Fora Research Project, Vol. 1*. Clarendon Press, Oxford, pp. 64–85.

Jodry, M.A. (1992) Fitting together Folsom: Refitting lithics and site formation processes at Stewart's Cattle Guard Site. In: Hofman, J.L. and Enloe, J.G. (eds) *Piecing Together the Past: Applications of Refitting Studies in Archaeology*, BAR International Series 578, Oxford, pp. 179–209.

Keeley, L.H. (1980) *Experimental Determination of Stone Tool Use: A Microwear Analysis*. University of Chicago Press, Chicago.

Keeley, L.H. and Toth, N. (1981) Microwear polishes on early stone tools from Koobi Fora, Kenya. *Nature* 293, 464–5.

Kimura, Y. (1999) Tool-using strategies by early hominids at bed II, Olduvai Gorge, Tanzania. *Journal of Human Evolution* 37, 807–31.

Larson, M.L. and Ingbar, E.E. (1992) Perspectives on refitting: Critique and a complementary approach. In: Hofman, J.L. and Enloe, J.G. (eds) *Piecing Together the Past: Applications of Refitting Studies in Archaeology*, BAR International Series 578, Oxford, pp. 151–62.

Leroi-Gourhan, A. (1964) *Le Geste et la Parole: Technique et Language*. Albin Michal, Paris.

Lovejoy, O. (1981) The origin of man. *Science* 211, 341–50.

O'Connell, J.F., Hawkes, K., Lupo, K.D. and Blurton-Jones, N.G. (2002) Male strategies and Plio-Pleistocene archaeology. *Journal of Human Evolution* 41, 833–72.

Petraglia, M.D. (1992) Stone artifact refitting and formation process at the Abri Dufaure, an Upper Paleolithic site in Southwest France. In: Hofman, J.L. and Enloe, J.G. (eds) *Piecing Together the Past: Applications of Refitting Studies in Archaeology*, BAR International Series 578, Oxford, pp. 163–78.

Potts, R. (1988) *Early Hominid Activities at Olduvai*. Aldine, New York.

Potts, R. (1992) Why the Oldowan? Plio-Pleistocene tool-making and the transport of resources. *Journal of Anthropological Research* 47, 153–76.

Roebroeks, W. and Hennekens, P. (1990) Transport of lithics in the Middle Paleolithic: Conjoining evidence from Maastricht-Belvedere (NL). In: Cziela, E., Eickhoff, E., Art, N. and Winter, D. (eds) *The Big Puzzle: International Symposium on Refitting Stone Artifacts. Studies in Modern Archaeology*, Vol. 1, Holos Verlag, Bonn, pp. 283–95.

Rose, L., Marshall, F. (1996) Meat eating, hominid sociality, and home bases revisited. *Current Anthropology* 37, 307–38.

Schick, K.D. (1987) Modeling the formation of Early Stone Age artifact concentrations. *Journal of Human Evolution* 16, 789–807.

Schick, K.D. (1992) Geoarchaeological analysis of an Acheulean site at Kalambo Falls, Zambia. *Geoarchaeology* 7, 1–26.

Schick, K.D. and Toth, N. (1993) *Making Silent Stones Speak: Human Evolution and the Dawn of Technology*. Simon and Schuster, New York.

Schiffer, M.B. (1972) Archaeological context and systemic context. *American Antiquity* 37, 157–65.

Sellet, F. (1993) *Chaîne Opératoire*: It concept and application. *Lithic Technology* 18, 106–12.

Toth, N. (1985) The Oldowan reassessed: A close look at early stone artifacts. *Journal of Archaeological Science* 12, 101–20.

Toth, N. (1987) Behavioral inference from Early Stone Age artifact assemblages: An experimental model. *Journal of Human Evolution* 16, 763–87.

Van Peer, P., Fullagar, R., Stokes, S., Bailey, R.M., Moeyersons, J., Steenhoudt, F., Geerts, A., Vanderbeken, T., De Dapper, M. and Geus, F. (2003) The Early to Middle Stone Age transition and the emergence of modern human behaviour at site 8-B-11, Sai Island, Sudan. *Journal of Human Evolution* 45, 187–93.

Vaughan, P.C. (1985) *Use-wear Analysis of Flaked Stone Tools*. University of Arizona Press, Tucson.

Villa, P. (1978) *The Stone Artifact Assemblage from Terra Amata: A Contribution to the Comparative Study of Acheulian Industries in Southwestern Europe*. PhD Dissertation, Department of Anthropology, University of California, Berkeley.

Villa, P. (1983) *Terra Amata and the Middle Pleistocene Archaeological Record of Southern France*. University of California Press, Berkeley.

Washburn, S.L. and Devore, I. (1961) Social behavior of baboons and early man. In: Washburn, S.L. (ed.) *Social Life of Early Man*. Aldine, Chicago, pp. 91–105.

Whittaker, J.C. (1994) *Flintknapping: Making and Understanding Stone Tools*. University of Texas Press, Austin.

Wyckoff, D.G. (1992) Refitting and protohistoric knapping behavior: The Lowrance example. In: Hofman, J.L. and Enloe, J.G. (eds) *Piecing Together the Past: Applications of Refitting Studies in Archaeology*, BAR International Series 578, Oxford, pp. 83–127.

Chapter 12

CHANGING THE FACE OF THE EARTH: HUMAN BEHAVIOR AT SEDE ILAN, AN EXTENSIVE LOWER–MIDDLE PALEOLITHIC QUARRY SITE IN ISRAEL

Ran Barkai and Avi Gopher

Institute of Archaeology, Tel Aviv University

ABSTRACT

The Middle Pleistocene flint extraction and reduction complex of Sede Ilan is presented as a model for human behavior related to both raw material economy and landscape perception. The organized, large scale quarrying and flintknapping is dated to the Lower Paleolithic (represented by handaxe manufacture) and the Middle Paleolithic (dominated by the use of the Levallois technique). No evidence of Later Paleolithic or Neolithic-Chalclithic extraction and reduction was found at the Sede Ilan complex. The chapter discusses the following issues: (i) the familiarity of Middle Pleistocene Hominins with the landscape and natural resources; (ii) long-term use of specific flint outcrops and recurrent, large-scale use of designated industrial areas; (iii) the significant alteration of the pristine landscape; (iv) quarry landscape maintenance; (v) the possibility of traditional ecological knowledge, land-use legacies, and resource management and conservation practiced at the Sede Ilan quarrying complex. Our major conclusion is that the extensive Sede Ilan quarrying landscape reflects the talent of prehistoric man to procure large quantities of flint using a rare combination of knowledge, sophistication, and care.

Lithic Materials and Paleolithic Societies, 1st edition. Edited by B. Adams and B.S. Blades, ©2009 Blackwell Publishing. ISBN 978-1-4051-6837-3.

INTRODUCTION

Prehistoric man was constantly looking for stone. As soon as ancient hominins started to use stone tools on a regular basis, some 2.6 million years ago and most probably even earlier than that, the search for stone became second nature. Recently it has been argued that the earliest African tool makers were highly familiar with the different outcrops available along the rift valley and made selective use of specific raw materials (e.g., Stout *et al.* 2005). This is actually the earliest example of a conceptual change in human–nature relationship, when people started looking at nature as a resource, as a supplier of their technological needs and demands. This is most probably the moment in human history when the conceptions of "raw material" and "resources", so familiar to our modern ears and minds, took shape. All prehistoric groups living in the old world from the Oldowan cultural complex onwards made extensive use of large quantities of stone and must have made a significant impact on the natural environment (e.g., Stiles *et al.* 1974; Biagi and Cremaschi 1998; Stiles 1998; Petraglia *et al.* 1999; Paddayya *et al.* 2000; Barkai *et al.* 2002; Vermeersch 2002; Barkai *et al.* 2006; Paddayya *et al.* 2006; Sampson 2006; Biagi 2007). The systematic production of designed, well planed, stone tools required complex behavior concerned with the location and exploitation of carefully selected stone sources. It must be stressed that alongside the procurement from selected sources, people during all times used random surface raw material as well. In other words, it appears that since the dawn of human technology two separate stone procurement strategies were applied: a simple strategy that exploited stone available on the surface of the earth (slopes or wadi beds) and a complex strategy using selected stone sources from primary geological contexts. Evidence for stone quarrying during the Early Pleistocene is still not widespread (e.g., Petraglia *et al.* 1999; Paddayya *et al.* 2000, 2006), while Middle Pleistocene quarry sites are already part of the archeological landscape at many places (e.g., Barkai *et al.* 2002, 2006, Sampson 2006). A new study of the cosmogenic isotope ^{10}Be in flint artifacts from Middle Pleistocene caves in Israel has confirmed the use of quarried stone as early as 400 kyr (Verri *et al.* 2004, 2005) and recent field work has uncovered several Middle Pleistocene extensive quarry complexes in Northern Israel (Barkai *et al.* 2002, 2006).

It seems to us that the fact that Middle Pleistocene stone-tool makers were indeed using stone from primary geological contexts is well established now, and the discussion should be focused on better understanding complex human behavior supporting such impressive operations and human capabilities reflected in these earliest industrial areas in the history of humanity.

Recent discoveries in Northen Israel revealed extensive Lower–Middle Paleolithic flint quarry complexes where immense quantities of raw material were extracted and reduced. These huge Paleolithic "industrial areas" are a testimony to large scale works of Paleolithic man and his impact on the pristine environment. Sede Ilan is one of these complexes.

We would like to use the Sede Ilan data in order to discuss the following issues:
- The familiarity of Middle Pleistocene hominins with the landscape and natural resources.
- Long-term use of specific flint outcrops and recurrent, large-scale use of designated industrial areas
- The significant alteration of the pristine landscape
- Quarry landscape maintenance
- The possibility of traditional ecological knowledge, land-use legacies, and resource management and conservation practiced at the Sede Ilan quarrying complex.

The extensive Middle Pleistocene flint extraction complex of Sede Ilan is used as a model for the extent of energy investment, knowledge, and what we see as some sort of ecological wisdom revealed in the establishment, long-term exploitation, and maintenance of this industrial area.

We will show that the specific location of the Sede Ilan extraction complex was carefully selected, based on an intimate acquaintance with the natural environment and an understanding of the long-term potential of this locality in supplying rock for the never-ending demands of Acheulian and Mousterian flint knappers. Another reason for the selection of this specific locality might stem from its physiographic setting and proximity to the most prominent topographic landscape marker in the region – Mt. Tabor. We will try to follow the logic behind the organization of quarrying operations and maintenance activities, possibly aimed at the preservation of the stone resources for long-term use. The numerous extraction and reduction localities will help us demonstrate the recurrent use of this specific industrial area throughout the generations, possibly reflecting a linkage between distinct human groups and this quarrying complex kept throughout time.

The scale of extraction and reduction activities at Sede Ilan will be used to demonstrate the extensive alteration of the natural landscape by Middle Pleistocene hominins, the creation of culturally induced megascars in the pristine environment, and the creation of a quarried landscape.

Selected archeological and anthropological studies will be used to bridge the gap between the mute data from Sede Ilan and our ambition to understand how people were actually operating at the site and the more general human behavior related to such Paleolithic industrial areas. We are fully aware of the possible methodological flaws embedded in the use of modern ethnographic records for reconstructing Paleolithic life-ways (e.g., Binford 2001) and therefore only examples directly related to cultural attitudes and conceptions towards natural resources will be used.

The search for stone played an important role in the daily activities of Middle Pleistocene hominins in the Levant, and the long-lasting bond between the prehistoric groups of the Lower Galille and the Sede Ilan industrial area must have had special significance in functional, social, ecological, and symbolic terms. Our study of the Sede Ilan Middle Pleistocene industrial complex aims at unfolding the complex relationships between man and stone and its bearing on human technologoical, social, and cognitive abilities.

THE SEDE ILAN LOWER–MIDDLE PALEOLITHIC INDUSTRIAL AREA

The Sede Ilan extraction and reduction complex is located in the Lower Galilee, Israel, some 10 km northwest of the Sea of Galilee (Barkai *et al.* 2006). This complex is comprised of hundreds of tailings piles and is similar in scale, density, and cultural characteristics to the Mt. Pua complex (Barkai *et al.* 2002). Modern activities have damaged this expansive Paleolithic industrial area. However, large parts are still available for archeological and geological investigations.

An air photograph taken before modern construction reveals the huge extent of the Sede Ilan complex (Figure 12.1). It lies on the slopes of two plunging folds, and the quarry activity is mainly focused on one flint-bearing horizon. Noteworthy is the close proximity of the most prominent topographic feature at the Lower Gallile, Mt. Tabor, to the Sede Ilan complex (Figure 12.2). Although Eocene flint is ubiquitous throughout the landscape, flint diagenesis reaches its

Figure 12.1   Air photo of the extraction landscape at Sede Ilan (white spots are tailing piles).

acme in development at Sede Ilan. Owing to geological structure and stratigraphic relationships in the area, prehistoric quarry activities are tightly focused along the limbs and hinges of these two folds. The flint-bearing horizon has been mined to the limits of Paleolithic technology, particularly where closely spaced joints have been accentuated by karstic activity.

Where the quarry fronts are developed along the limbs and hinges of the two folds, the style of mining varies depending upon elevation and bedrock structure. Mining activity focused along the hinge of the fold may be oriented towards gathering flint between joint surfaces where karstic activity has loosened them from the bedrock matrix. However, flint occurring along the limbs of the folds is mined through the purposeful development of highly organized quarry fronts. The zones of extraction appear to have been maintenanced, with mine tailings intentionally backfilled towards the opposing edges of each quarry front and on top of exhausted extraction fronts. This was performed in order to stabilize the back walls of the remaining declivity, clear the waste material from the actual extraction front, and keep the unexploited areas available for future use.

Complementing this sophisticated quarry development is a chain of operation (*chaîne opératoire*) of mining tools that are designed, manufactured, and implemented from basalt and limestone materials, attesting to the ingenuity of Lower–Middle Paleolithic mining endeavors. Specifically, basalt was gathered from the high plateau above the site and brought into

Figure 12.2 The quarried landscape of Sede Ilan with Mt. Tabor in the background.

Figure 12.3 The test excavation at tailing pile No. 3 at Sede Ilan.

the quarry fronts where at least five diagnostic tool types were manufactured. The basalt mining toolkit is complemented by an assemblage of curated wedges fashioned from siliceous limestone (some of which were recovered from open joints).

Field work at Sede Ilan was conducted by the authors in January 2005 and July 2006, and included preliminary geological mapping and a test excavation at one of the large tailings piles at the eastern edge of the complex (Figure 12.3). The excavation was carried out in the elongated pile No. 3 (henceforth SE3) in 2×2 m². The pile consisted of broken limestone blocks mixed with lithic artifacts and raw material blocks, as well as a prominent component of basalt items (Figure 12.4). SE3 is 15 m long and 8.6 m wide at its center. The test excavation is located at the southern edge of this pile, at a place where the deposits are relatively thin. At the northern edge of SE3 the tailings pile reaches an elevation of 130 cm above ground surface and slopes

Figure 12.4 A close look at knapped lithic artifacts, broken limestone blocks, and basalt items from tailing pile No. 3 at Sede Ilan.

towards the south to only 30 cm above surface level. The excavated material, which is 50 cm thick, rested on accentuated and levered limestone blocks. Flaked lithic artifacts are abundant throughout the depth of excavated smashed limestone blocks. It appears that the waste material was piled on top of either an exhausted extraction front or a specific area with low extraction potential. The pile is clearly aligned with other similar piles, located along a selected flint horizon at the eastern edge of the Sede Ilan complex (Figure 12.5). The waste piles along this line are similar in scale and are placed 20–30 m from each other. It seems to us that the extraction activity took place along a flint horizon spread to a length of dozens of meters at the slope of the Sede Ilan small spur. Several extraction localities were placed at similar distances along this line and the waste material resulting from extraction was intentionally piled at specific places, most probably in order to concentrate the waste and clear potential areas for future extraction (note a similar strategy at the Mt. Pua and site 164 complexes, Barkai et al. 2002, 2006).

The lithic industry found within the numerous workshops is rich in primary reduction products and blanks, highlighted by the predominance of Levallois cores and debitage, and large flake production with a minor component of bifacial tool preforms.

Figure 12.5 Tailing pile No. 3 at Sede Ilan (SE3) within the quarrying landscape and surrounding piles.

The Sede Ilan SE3 lithic assemblage (see Barkai *et al.* 2006 for a detailed report with figures): Noteworthy components observed on the surface, prior to the test excavation, were Levallois cores and handaxes. The SE3 assemblage from the first excavation area (2×2 m) excavated in 2005 ($n = 480$) is rich in cores ($n = 99$) and shaped items ($n = 73$), and we suggest this area was used for relatively advanced stages of blank reduction and tool shaping. Shaped items include *ad hoc* tools only, with no early-stage bifaces, handaxes, or other shaped items apart from retouched blanks. Some of the retouched flakes were shaped on Levallois blanks. Only few tested nodules were found at SE3, in contrast to the high frequency of Levallois cores, reinforcing the impression that the advanced flaking stage was the focus at SE3. The general character of the large tailing piles at Mt. Pua (Barkai *et al.* 2002, 2006) and Sede Ilan are similar and this observation is interpreted as reflecting continuous flint reduction processes, including advanced flaking stages.

In addition to the flaked flint assemblage, flint blocks ($n = 69$) with no evidence of flaking were also recovered from the SE3 excavation unit. Flint nodules still embedded within the limestone karrens ($n = 16$) were found as well. A rich basalt assemblage was also recovered, including long and narrow items ($n = 6$,); large and round basalt cobbles ($n = 7$); flat, tapering rectangular wedge-like items, some with thinned edges ($n = 28$); and basalt fragments ($n = 65$). Limestone flaked artifacts were also recovered during excavation, including thin and small limestone flakes ($n = 42$), thick and large limestone flakes ($n = 29$), and limestone cores ($n = 7$). While a few of the limestone flakes may have been the result of smashing the limestone blocks in order to extract the flint nodules, the bulk of the limestone cores, and the thin and small limestone flakes, may be indicative of a limestone production trajectory at the site. It is proposed that the limestone flakes were hammered into tightly sealed joints, thereby permitting diurnal temperature changes and seasonal weather patterns to further accentuate the joint surfaces. The limestone cores index the end of a production trajectory focused towards the protracted exploitation of the land through the use of a plug-and-feather method.

Although field investigations at Sede Ilan are only preliminary and based on initial geological mapping and a small scale test excavation, the Lower–Middle Paleolithic quarrying and workshop complex is interpreted as both expansive and sophisticated. Hominin utilization of these specific Eocene flint outcrops seems to be recurrent and a quarry toolkit of local basalt and limestone was employed. Quarry debris were piled in waste piles, and flint reduction, focusing on the Levallois technique and the production of large flakes, took place on top of these tailings piles.

Acheulian and Mousterian lithic traditions in the Levant are characterized by intensive use of flint resources for the production of handaxes during the Acheulian, and Levallois products in the Mousterian, both of which necessitate the use of high quality homogeneous flint that can be usually found in primary geological sources.

The presence of Levallois cores and debitage, as well as early-stage handaxes, suggests that the quarrying activity is related to the late phase of the Acheulian complex of the Lower Paleolithic period (e.g., Goren 1979; Goren-Inbar 1985). Large flake production is a well known Acheulian cultural marker and it is clear that the use of these quarry complexes began during Lower Paleolithic times. The use of the Levallois technique in the Levant began during the Middle Paleolithic or Late Lower Paleolithic (e.g., Goren-Inbar 1985) and a possible connection between handaxe reduction and the Levallois technology was suggested (De-Bono and Goren-Inbar 2003). Thus, the co-occurrence of handaxes and Levallois technology is attributed to the Late Lower Paleolithic period. It is possible, however, that the use of these flint sources continued during the Middle Paleolithic by Mousterian flint knappers, as indicated by the relative dominance of Levallois technology. No post Middle Paleolithic artefacts were found on the tailing piles and quarrying localities. The rocky landscape seems to have remained very much unchanged since the Middle Pliestocene apart from some possible karstic activity. The Sede Ilan complex is surrounded by agricultural activity and worked field, but the rocky hill where the Paleolithic extraction and reduction took place was never exploited in modern times. It seems to us that this complex has been surprisingly left untouched since Middle Pliestocene times.

HUMAN BEHAVIOR AT THE SEDE ILAN INDUSTRIAL AREA

The vast use of flint during the Lower and Middle Paleolithic periods in the Levant has been self evident since the early days of prehistoric research, but no real progress has been made in studying Paleolithic flint procurement strategies until the last decade. The location of several Middle Pleistocene industrial areas and the study of cosmogenic isotopes in flints from the Tabun and Qesem caves indicated the existence of systematic, large scale, and long lasting human endeavors towards obtaining the stone they needed (e.g., Barkai *et al.* 2002; Verri *et al.* 2004, 2005; Barkai *et al.* 2006).

Flint extraction complexes are a special type of archeological site, very different from habitation sites or other task specific sites like hunting or butchering localities. These early industrial areas demonstrate a narrow range of activities concentrated around locating, extracting, and reducing favorable stone sources. Notwithstanding this, the special nature of such sites holds the potential of touching issues hitherto unexplored by the main stream of Paleolithic research, thus promoting our understanding of mundane activities as well as the world views of Middle Pleistocene hominins. In the following section we would like to use the Sede Ilan site as a case study for such insights.

THE FAMILIARITY OF MIDDLE PLEISTOCENE HOMININS WITH THE LANDSCAPE AND NATURAL RESOURCES

The Sede Ilan quarrying complex is located within a specific Eocene formation in the Lower Galilee. This is actually an island of an Eocene limestone formation especially rich in flint nodules in a big "sea" of other geological formations devoid of the desired flint. The quarrying complex perfectly matches the outlines of this Eocene formation and our survey indicated evidence for Paleolithic extraction activities covering the whole extent of this specific geological formation. Other exposures of similar Eocene formations are known in the Lower Galilee, some at a very close distance from Sede Ilan. We have visited several such localities a few kilometers only from Sede Ilan and observed potential flint sources at each and every locality. At least two of these Eocene formations bear witness to small scale prehistoric flint extraction and reduction activities, but the sporadic nature of these activities did not allow for an assignment to any familiar Levantine prehistoric cultural entity. It is, however, clear that other potential flint sources exist in the vicinity of the Sede Ilan complex and we cannot rule out the possibility that some of these were exploited in Paleolithic or Neolithic–Chalcolithic times. None of these, however, is similar in scale or intensity to Sede Ilan. The Sede Ilan Eocene formation is not only rich in flint sources, but is also located adjacent to a small basalt outcrop at its northern edge. Other basaltic outcrops are rather abundant in the Lower Galilee, but at Sede Ilan a combination of two desired raw materials – flint and basalt – was extensively exploited by Middle Pleistocene hominins. One of the special characteristics of the Sede Ilan complex is the ubiquitous use of extraction tools made of basalt, and therefore the location of this quarrying complex perfectly uses these two geological formations. The familiarity of contemporary hunting and gathering societies with their surroundings is very well documented in the anthropological literature and this intimate environmental knowledge is gained by constant movements within the landscape and shared "environmental gossip" between group members (e.g., Bird-David 1992; Lye 2002). Middle Pleistocene hunter-gatherers roamed around the Levantine landscape for hundreds of thousands of years and most probably had excellent environmental knowledge and a perfect acquaintance with the natural resources. Their decision to place the major flint industrial area at Sede Ilan resulted from such environmental knowledge and the extent of quarrying operation at the site stands as a testimony to the wise decision they made. It is of note that inspite of continued later presence of people in the lower Galilee, evidence of post Lower–Middle Paleolithic exploitation was not recognized at Sede Ilan. This is rather strange, due to the fact that flint-using groups were still roaming that landscape at least during the Neolithic period and potential high quality raw material is still found in abundance at Sede Ilan. We have no explanation for this state of affairs since it is clear to us that Neolithic man also had highly intimate familiarity with the landscape, and the possibility of not recognizing the potential of Sede Ilan as a raw material source is simply out of the question. We find it hard to believe that the site escaped the experienced eye of Neolithic flint miners and the question of why Paleolithic Sede Ilan was left untouched during later periods must remain open for the time being.

LONG TERM USE OF SPECIFIC FLINT OUTCROPS AND RECURRENT, LARGE-SCALE USE OF DESIGNATED INDUSTRIAL AREAS

The Sede Ilan quarrying complex is comprised of hundreds of extraction and reduction localities (Figure 12.1), representing episodes of focused exploitation of selected flint horizons. Some of these quarrying localities, such as the Sede Ilan pile No. 3 (SE 3), are very large (over 120 m² surface area and *ca.* 1.3 m in depth) while other localities are smaller. Evidence for flint extraction and reduction were documented from both large and small tailing piles but an in depth comparative study of the whole complex had not yet been conducted. It should be mentioned that during our survey a few tailing piles with no clear evidence of quarrying activity were observed and this phenomenon awaits further investigations as well. Going back to the hundreds of quarrying localities, one remains astonished at the immensity of human activity and the duration of exploitation of this specific industrial area. It is, of course, impossible to use the number of extraction localities and their volume in a simple equation and come out with the scale and duration of human activity at the site. We have no idea regarding the time it takes to locate a desired flint outcrop, prepare suitable extraction tools, extract flint, reduce the selected nodules, and create a tailing pile such as SE 3. We do not know how many individuals were operating at each extraction and reduction locality, and at the moment we have no clue regarding the question of whether each locality was exploited independently or perhaps several localities were used simultaneously. Following our understanding of Middle Pleistocene societies (e.g., Gamble 1999; Gamble and Poor 2005) we find it hard to believe that regiments of hominins took over Sede Ilan and exploited the quarry landscape like locusts. Taking into account the available ethnographic studies of stone quarrying conducted among simple societies in Australia, Papua New Guinea, and Polynesia (e.g., Binford and O'Connell 1984; Burton 1984; Jones 1984; Jones and White 1988; Pétrequin and Pétrequin 1993a,b; Hampton 1999) it becomes clear that all these groups were repeatedly exploiting specific stone sources for generations; the mode of exploitation was rather slow according to the group's needs; quarrying expeditions were small; and the stay at the quarry site was always rather short. In addition, it appears that in all cases stone quarries were conceived as highly important, dangerous, and sacred places; in many cases ceremonies are conducted before and after quarrying; and the quarrying activity itself is performed solely by males. Although it seems that all modern ethnographic examples of stone quarrying share similar, cross-cultural, characteristics, it is hard to apply these shared practices and conceptions to Middle Pleistocene Sede Ilan. However, we believe that the ethnographic evidence reinforces our contention regarding the scale and duration of human

activity at Sede Ilan. We argue that Middle Pleistocene hominins recognized the never-ending potential of the Sede Ilan area in supplying raw material for their lithic industries and kept coming back to the area for generations and generations. Sede Ilan most probably served as a major raw material source for Lower–Middle Paleolithic communities occupying the Lower Galilee and was visited by small quarry expeditions on a regular basis for thousands of years. This is not of much help in understanding the intra-site variability in the size of extraction localities; and it would be simplistic to suggest that a large tailing pile represents a relatively long quarry operation by a relatively large quarrying expedition, while the small piles represent short visits by smaller groups. However, understanding the pace and duration of human activity at the site might help reconstruct this long and complex human endeavor. Why was Sede Ilan, of all other raw material sources, the focus of quarrying activity for so long? Why was this specific source visited and exploited throughout the generations, becoming a large scale industrial area and a quarried landscape? We believe that the simple and straightforward answer, the presence of rich flint sources and adjacent basalt outcrops, is by all means valid, but not satisfactory. The answer might lie in the topographic and geographic setting of Sede Ilan, serving not only as a raw material source but as a landmark, a landscape beacon embedded in the collective memory and tradition of Lower-Middle Paleolithic societies of the area. Sede Ilan is located in very close proximity to Mt. Tabor, the most prominent topographic landscape marker in the region. It is well known that modern hunter-gatherers orient themselves in the landscape using prominent features such as rivers, significant rock features, and mountain peaks (e.g., Lye 2002) and special landscape features might have had special significance to prehistoric communities (e.g., Bradley 2000). We therefore would like to suggest that these two extraordinary features, Mt. Tabor as a natural peak and Sede Ilan as man-made "monumental landscape" were significant landmarks in the "maps" of Middle Pleistocene hominins and were most probably used to orient individuals and groups towards traditional routes. The significance of the Sede Ilan industrial area developed with time, as one group after the other passed near Mt. Tabor and extracted flint at the same place over and over again, until the place became embedded within the collective memory not only as a supply point on the way but as a place of special significance, connecting past, present, and future activities.

THE SIGNIFICANT ALTERATION OF THE PRISTINE LANDSCAPE

The impact of Paleolithic man on the natural environment is usually considered negligible, mostly due to the rather limited technological means employed by early hominins and modern hunter-gatherers and low population densities (e.g., Simmons 1989; Goudie 1990; Redman 1999). Others have claimed, based on ethnographic studies, that hunting and gathering societies had a non-materialistic view towards nature, viewed nature as a "giving environment", or had a set of norms and conceptions that controlled the alteration of the pristine landscape (e.g., Reichel-Dolmatoff 1976; Bird-David 1990; Lye 2002, 2005). It is agreed upon by most scholars that the impact of man on the environment became significantly more pronounced during Neolithic times with the advent of agriculture, village-life, and a demographic expansion (e.g., Simmons 1989; Redman 1999). Others have emphasized the impact of Neolithic stone mining activities on the landscape, claiming this was not only the most practical and efficient method of obtaining large quantities of stone, but also that the large extent of Neolithic mining operation destroyed natural landscapes, thus demonstrating human control over nature (Field 1997). Our discoveries at Sede Ilan and other similar Middle Pleistocene industrial areas (Barkai *et al.* 2006) represent, for the first time, large scale landscape alteration by Paleolithic hominins. Although we are dealing with relatively simple technologies and low population densities, it is clear that flint extraction by Lower-Middle Paleolithic communities, over long periods at the same location, resulted in an alteration of the pristine landscape and the creation of "Mega scars" in the environment. Most of the activities of Middle Pleistocene hominins are documented at cave and open-air sites where no evidence of massive landscape alteration is evident. Quarry sites reveal a different aspect of human–nature relationship in the Paleolithic. The extensive use of flint in the Levant required the procurement of large quantities of raw material, and this transformed the landscape. Lower–Middle Paleolithic communities left a pronounced mark in the landscape and, surprisingly, some of these quarry sites escaped modern disturbances and survived to this very day. Middle Pleistocene hominins in the Levant changed the face of the earth and made a pronounced impact on the natural environments; and therefore the dichotomy between the Paleolithic and the Neolithic periods regarding the transformation of the landscape must not be regarded to be as sharp as commonly accepted. We fully agree that the monumental quarrying landscape presented by the Sede Ilan site and other similar complexes reveal an unfamiliar aspect of Middle Pleistocene hominins that stands in sharp contrast to the common view of these communities as "harmless" in terms of landscape alteration. This is why more research should be invested in Paleolithic stone procurement strategies for expanding our understanding of the wider range of activities performed by pre-modern hominins and their impact on the environment.

QUARRY LANDSCAPE MAINTENANCE

The reconstruction of quarry operation at Sede Ilan indicated a rational sequence of activities aimed, in our opinion, at using to exhaustion the selected raw material outcrops, backfilling the exhausted extraction fronts by quarry debris, and reducing the selected quarried nodules on top of the tailing pile, thus leaving the unexploited area around the exhausted extraction front undamaged and available for future extraction. This process repeats itself at the Sede Ilan quarrying complex, although investigated by excavation at one tailing pile only. One might point out the fact that a series of five tailing piles of similar size at the eastern edge of Sede Ilan are aligned along one specific flint horizon (Figure 12.5) and located at even distances from one another (20–30 m). This scenario repeats itself in other parts of the Sede Ilan complex and was observed at other Middle Pleistocene quarry complexes as well (Barkai *et al.* 2006). This seems to us a systematic quarrying procedure following strict rules aimed at maintaining the quarry landscape and preserving the unexploited outcrops for future use. The quarrying operation is highly organized and the location of the individual extraction localities is set according to a plan taking into account the geological structure of the complex and the characteristics of the available flint horizon. Looking at the air photograph of Sede Ilan (Figure 12.1), it seems that the extraction localities were strictly placed according to the bedding and the location of the outcrops. In our view, this followed a master plan and was not randomly affected by decision taken at the individual level. The fact that flint-knapping took place on top of the quarrying debris piles seemed odd to us in the first place. But when we realized that this pattern repeats itself at least in three Middle Pleistocene quarry complexes, we realized this may be a clue to understanding quarry organization and maintenance. We see no apparent reason for reducing the nodules and shaping the flint cores and tools on top of the tailing piles, apart from the wish to leave the area between the piles clean and available for future exploitation. The huge amount of flint extracted at these quarry sites resulted in large quantities of reduction debitage and debris that would easily cover the surface between the piles. Although some flints were found between the tailing piles, it is clear that most flint working took place on top of the piles, and we see this pattern as directed towards constantly maintaining the quarry landscape and following the master plan of the Sede Ilan future exploitation.

TRADITIONAL ECOLOGICAL KNOWLEDGE, LAND-USE LEGACIES, AND RESOURCE MANAGEMENT AND CONSERVATION

We believe that the data and interpretation presented thus far allow us to touch upon issues related to the conceptual environmental worldviews and practical activities of Middle Pleistocene hominins. The insights from the study of the extensive Lower–Middle Paleolithic quarry complex of Sede Ilan enables us to join the lively discussion regarding the environmental knowledge of pre-industrial societies, the complex histories of land-use, and the debate concerning the ecological awareness of small-scale societies.

It is beyond the scope of this chapter to try and cover the vast anthropological, historical, and ecological literature on the subjects mentioned above, but it should be stressed that this is the first attempt to present such evidence from the early stages of human prehistory and enlarge the perspective of modern, relevant, environmental issues, to the Middle Pleistocene. It is our contention here that profound human ecological knowledge and environmental awareness appeared at least as early as the Middle Pleistocene and accompanied humanity until recently, as long as humans kept the traditional hunting and gathering way of life. The origins of the so-called "contemporary" debate regarding ecological and nature conservation issues are rooted deep into our prehistoric past and the study presented here enlarges our perspective regarding the environmental perception of prehistoric man.

It is well accepted now that indigenous, aboriginal societies worldwide demonstrate a comprehensive and intimate acquaintance with the ecological and environmental systems, usually termed "Traditional Ecological Knowledge" (e.g., Berkes *et al.* 2000; Ellen *et al.* 2000; Turner *et al.* 2000; Davis and Wanger 2003). This knowledge is shared by contemporary "simple" hunter-gatherers ("Immediate-Return societies," e.g., Bird-David 1990, 1992; Lye 2002, 2005) and we have every reason to believe that our data from Sede Ilan reflects a well-developed traditional ecological knowledge, at least regarding the level of familiarity with the natural resources and the continuous exploitation of the chosen industrial area for generations.

While in the discussions of contemporary issues in ecology and conservation the importance of understanding land-use legacies is obvious (e.g., Foster *et al.* 2003), Paleolithic evidence for prolonged exploitation of the Levantine environments is meager despite the clear human presence throughout time. It is well established at many cave and open-air sites that Lower and Middle Paleolithic communities occupied the Mediterranean Levant through generations and generations, and recently it has been suggested that both Early Modern humans as well as Neanderthals used specific territories in a repeated manner during the Middle Paleolithic (Hovers 2001). The industrial complex of Sede Ilan is presented here as a case of land-use legacies during the Middle Pleistocene Levant, as throughout the ages this specific location was visited and exploited over and over again. The possibility that Sede Ilan was used as a traditional territorial marker is not ruled out and the question of some sort of restricted access to this rich source is left for future thought.

But does the evidence presented for the degree of quarry maintenance and the repeated use of specific, well defined, outcrops at Sede Ilan truly reflect conscious and planed behavior aimed at a degree of control over the natural resources and its conservation? Were Middle Pleistocene hominins indeed aware of the degree of exploitation and the damage they were inflicting upon Nature? We believe the answer to these questions is positive and see this awareness as a plausible explanation for the repeated and strict exploitation system at Sede Ilan and other contemporary quarry complexes, amongst the more practical explanations. We are, however, aware of the hot debate regarding the scale of Nature conservation among small-scale societies (e.g., Alvard 1993; Berkes 1999; Smith and Wishnie 2000; Ichikawa 2001; Dods 2002) and we argue that our study supports the so-called "Ecologically noble savage" hypotesis.

CONCLUSIONS

Prehistoric archeology has a large potential for making a contribution towards the understanding of contemporary environmental issues and taking sides in environmental debates. This chapter is an example of a step towards this aim. The management of modern ecosystems and landscapes could benefit from the imprints left by prehistoric societies on the landscape and the study of their impact on the natural environment (e.g., White 1967; Van der Leeuw and Redman 2002; Barnosky *et al.* 2004; Butzer 2005; Hayashida 2005).

The long-term perspective provided by the archeological record on the exploitation of the natural environment by prehistoric man shows that Middle Pleistocene hominins in the Levant used huge quantities of flint extracted from well-chosen sources. These extraction complexes or industrial areas were exploited repeatedly in a well-planed manner through constant maintenance of the quarried landscape. The quarry landscape had become a landscape beacon and must have had a central place within the conceptual and practical realms of prehistoric communities occupying the Lower Galilee.

REFERENCES

Alvard, M.S. (1993) Testing the "ecological noble savage" hypotesis: Interspecific prey choice by Piro hunters of Amazonian Peru. *Human Ecology* 21/4, 355–87.

Barkai, R., Gopher, A., and LaPorta, P.C. (2002) Paleolithic landscape of extraction: Flint surface quarries and workshops at Mt. Pua, Israel. *Antiquity* 76, 672–80.

Barkai, R., Gopher, A., and LaPorta, P.C. (2006) Middle Pleistocene landscape of extraction: Quarry and workshop complexes in northern Israel. In: Goren-Inbar, N. and Sharon, G. (eds) *Axe Age. Acheulian Too-Making from Quarry to Discard*, Equinox Publishing, London, pp. 7–44.

Barnosky, A.D., Koch, P.L., Feranec R.S., Wing, S.L., and Shabel, A.B. (2004) Assessing the causes of Pleistocene extinctions on continents. *Science* 306, 68–75.

Berkes, F. (1999) *Sacred Ecology: Traditional Ecological Knowledge and Resource Management*. Tailor and Francis, Philadelphia.

Berkes, F., Colding, J., and Folke, C. (2000) Rediscovery of traditional ecological knowledge as adaptive management. *Ecological Applications* 10/5, 1251–62.

Biagi, P. (2007) Modeling the past: The Paleoanthropological evidence. In: Henke, W. and Tattersal, I. (eds) *Handbook of Paleoanthropology*, Springer, Berlin, pp. 723–45.

Biagi, P. and Cremaschi, M. (1998) The early Palaeolithic sites of the Rohri Hills (Sind, Pakistan) and their environmental significance. *World Archaeology* 19(3), 421–33.

Binford, L.R. (2001) *Constructing Frames of Reference: An Analytical Method for Archaeological Theory Building Using Hunter-Gatherer and Environmental Data Sets*. University of California Press, Berkeley.

Binford, L.R. and O'Conell, J.F. (1984) An Alyawara day: The stone quarry. *Journal of Anthropological Research* 40, 406–32.

Bird-David, N. (1990) The giving environment: Another perspective on the economic system of Gatherer-Hunters. *Current Anthropology* 31, 189–96.

Bird-David, N. (1992) Beyond "The Original Affluent Society": A culturalistic reformulation. *Current Anthropology* 33, 25–47.

Bradley, R. (2000) *The Archaeology of Natural Places*. Routledge, London.

Burton, J. (1984) Quarrying in tribal societies. *World Archaeology* 16/2, 234–47.

Butzer, K.W. (2005) Environmental history in the Mediterranean world: Cross-disciplinary investigation of cause-and effect for degradation and soil erosion. *Journal of Archaeological Science* 32, 1773–800.

Davis, A. and Wanger, J.R. (2003) Who knows? On the importance of identifying "experts" when researching local ecological knowledge. *Human Ecology* 31/3, 463–90.

De-Bono, H. and Goren-Inbar, N. (2001) Note on a link between Acheulian handaxes and the Levallois method. *Journal of the Israel Prehistoric Society* 31, 9–23.

Dods, R.R. (2002) The death of Smokey Bear: The ecodisaster Myth and forest management practices of Prehistoric North America. *World Archaeology* 33/3, 475–87.

Ellen, R., Parkes, P., and Bicker, A. (2000) *Indigenous Environmental Knowledge and Its Transformations*. Hardwood Academic Publishers, Amsterdam.

Field, D. (1997) The landscape of extraction: Aspects of the procurement of raw material in the Neolithic. In: Topping, P. (ed.) *Neolithic Landscapes*. Oxbow Monographs 86. pp. 55–67.

Foster, D., Swanson, F., Aber, J., Burke, I., Brokaw, N., Tilma, D., and Knapp, A. (2003) The importance of land-use legacies to ecology and conservation. *BioScience* 35/1, 77–88.

Gamble, C. (1999) *The Palaeolithic Societies of Europe*. Cambridge University Press, Cambridge.

Gamble, C. and Poor, M. (2005) *The Hominid Individual in Context: Archaeological Investigations of Lower and Middle Palaeolithic Landscapes, Locales and Artefacts*. Routledge, London.

Goudie, A. (1990) *The Human Impact on the Natural Environment*. The MIT Press, Cambridge, MA.

Goren, N. (1979) An Upper Acheulian industry from the Golan Heights. *Quartär* 29–30, 105–21.

Goren-Inbar, N. (1985) The lithic assemblage of the Berekhat Ram Acheulian site. *Paléorient* 11, 7–28.

Hampton, O.W. (1999) *Culture of Stone. Sacred and Profane Uses of Stone among the Dani*. Texas A & M University Press, Texas.

Hayashida, F.M. (2005) Archaeology, ecological history and conservation. *Anual Review of Anthropology* 34, 43–65.

Hovers, E. (2001) Territorial behavior in the Middle Palaeolithic of the Southern Levant. In: Conard, N.J. (ed.) *Settlement Dynamics of the Middle Palaeolithic and Middle Stone Age*, Kerns Verlag, Tübingen. pp. 123–54.

Ichikawa, M. (2001) The forest world as circulation system: The impacts of Mbuti habitation and subsistence activities on the forest environment. *African Study Monographs Suppl.* 26, 157–68.

Jones, K.L. (1984) Polynesian quarrying and flaking practices at the Samson Bay and Falls Creek argillite quarries, Tasman Bay, New Zealand. *World Archaeology* 16/2, 248–66.

Jones, R. and White, N. (1988) Point blank: Stone tool manufacture at the Ngilipitji quarry, Arnhem land, 1981. In: Meehan, B. and Jones, R. (eds) *Archaeology with Ethnography: An Australian Perspective*, Australian National University, Canberra, pp. 51–87.

Lye, T.-P. (2002) The significance of forest to the emergence of Batek knowledge in Pahang, Malaysia. *Southeast Asian Studies* 40/1, 3–22.

Lye, T.-P. (2005) The meaning of trees: Forest and identity for the Batek of Pahang, Malaysia. *The Asia Pacific Journal of Anthropology* 6/3, 249–61.

Paddayya, K., Jhaldiyal, R., and Petraglia, M. (2000) Excavation of an Acheulian workshop at Isampur, Karnataka (India). *Antiquity* 74, 751–2.

Paddayya, K., Jhaldiyal, R., and Petraglia, M.D. (2006) The Acheulian quarry at Isampur, Lower Deccan, India. In: Goren-Inbar, N. and Sharon, G. (eds) *Axe Age. Acheulian Tool-Making from Quarry to Discard*, Equinox Publishing, London. pp. 45–74.

Petraglia, M., LaPorta, P., and Paddayya, K. (1999) The first Acheulian quarry in India: Stone tool manufacture, biface morphology and behaviors. *Journal of Anthropological Research* 55, 39–70.

Pétrequin, P. and Pétrequin, A.-M. (1993a) *Écologie d'un Outil: La Hache de Pierre en Irian Jaya (Indonésie)*. CNRS Éditions, Paris.

Pétrequin, P. and Pétrequin, A.-M. (1993b) From polished stone tool to sacred axe: The axes of the Danis of Irian Jaya, Indonesia. In: Berthelet, A. and Chavaillon, J. (eds) *The Use of tools by Human and Non-human Primates*, Clarendon Press, Oxford. pp. 359–77.

Redman, C.L. (1999) *Human Impact on Ancient Environments*. The University of Arizona Press, Tucson.

Reichel-Dolmatoff, G. (1976) Cosmology as ecological analysis: A view from the rain forest. *Man (n.s.)* 11/3, 307–18.

Sampson, G.C. (2006) Acheulian quarries at hornfels outcrops in the Upper Karoo region of South Africa. In: Goren-Inbar, N. and Sharon, G. (eds) *Axe Age. Acheulian Too-Making from Quarry to Discard*, Equinox Publishing, London. pp. 75–108.

Simmons, I.G. (1989) *Changing the Face of the Earth. Culture, Environment, History*. Blackwell Publishers, Oxford.

Smith, E.A. and Wishnie, M. (2000) Conservation and subsistence in small scale societies. *Annual Review of Anthropology* 29, 493–524.

Stiles, D. 1998. Raw material as evidence for human behaviour in the Lower Pleistocene: The Olduvai case. In: Petraglia, M.D. and Korisettar, R. (eds) *Early Human Behaviour in Global Context: The Rise and Diversity of the Lower Paleolithic Period*, Routledge, London. pp. 133–50.

Stiles, D., Hay, R. and O'Neil, J. (1974) The MNK chert factory site, Olduvai Gorge, Tanzania. *World Archaeology* 5, 285–308.

Stout, D., Quade, J., Semaw S., Rogers, M.J., and N.E. Levin. (2005) Raw material selectivity of the earliest stone toolmakers at Gona, Afar, Ethiopia. *Journal of Human Evolution* 48, 365–80.

Turner, N.J., Ignace, M.B., and Ignace, R. (2000) Traditional ecological knowledge and wisdom of aboriginal peoples in Bitish Colombia. *Ecological Applications* 10/5, 1275–87.

Van der Leeuw, S. and Redman, C. (2002) Placing archaeology at the center of socio-natural studies. *American Antiquity* 67, 597–605.

Vermeersch, P.M. (ed.) (2002) *Palaeolithic Quarrying Sites in Upper and Middle Egypt. Egyptian Prehistory Monographs*. Vol. 4. Leuven University Press, Leuven.

Verri, G., Barkai, R., Bordeanu, C., Gopher, A., Hass, M., Kaufman, A., Kubik, P., Montanari, E., Paul, M., Ronen, A., Weiner, S. and Boaretto, E. (2004) Flint mining in prehistory recorded by *in situ* produced cosmogenic ^{10}Be. *Proceedings of the National Academy of Sciences U.S.A.* 101(21), 7880–4.

Verri, G., Barkai, R., Gopher, A., Hass, M., Kubik, P., Paul, M., Ronen, A., Weiner, S., and Boaretto, E. (2005) Flint mining in the late Lower Palaeolithic recorded by in situ produced cosmogenic ^{10}Be in Tabun and Qesem Caves (Israel). *Journal of Archaeological Science* 32, 207–13.

White, L. (1976) The historical roots of our ecological crisis. *Science* 155, 1203–7.

Chapter 13

AURIGNACIAN CORE REDUCTION AND LANDSCAPE UTILIZATION AT LA FERRASSIE, FRANCE

Brooke S. Blades

Archeologist/Principal Investigator, A.D. Marble and Company

ABSTRACT

This chapter examines the Aurignacian sequence at one of the most well-known Paleolithic sites in France: la Ferrassie. Technological reduction as reflected in cores and core remnants is considered initially. Since the vast majority of cores were made on Senonian cherts that evidently were procured locally, it is postulated that there may have been a relationship between procurement of lithics and subsistence resources as reflected in faunal remains. This relationship is examined within a framework of geographical scales of analyses. A framework for partitioning exploitation of the landscape surrounding the rock shelter is introduced. Core reduction and faunal species presence are evaluated with regards to overall environmental conditions to suggest generalized strategies of movement in the vicinity of la Ferrassie.

INTRODUCTION

La Ferrassie in the Périgord of southwestern France has yielded fundamental information on the nature of Middle Paleolithic and Upper Paleolithic behaviors in Western Europe. Furthermore, excavations at the site have played an important role in the historical development of the manner in which data elsewhere have been integrated into the broader realms of Paleolithic

Lithic Materials and Paleolithic Societies, 1st edition. Edited by B. Adams and B.S. Blades. ©2009 Blackwell Publishing. ISBN 978-1-4051-6837-3.

and prehistoric research. Following initial investigations by others, Denis Peyrony and Louis Capitan undertook investigations over three decades in the early twentieth century that revealed extensive evidence of Mousterian, Aurignacian, and Periogordian occupations. Peyrony (1934) argued for a major revision of early Upper Paleolithic systematics based on these investigations.

Henri Delporte of the Musée National de Préhistoire undertook renewed excavations from 1969 to 1973 along the remaining frontal and sagittal sections that formed a right angle in the rock shelter (Delporte 1984). The Delporte project cut each section back 50 cm within 1 m *carées* or excavation squares. The project revealed the extent of stratigraphic complexity that was not recognized earlier in the century. The Peyrony sagittal section noted "Aurignacian I" occupation in level F and "Aurignacian II" occupation in level H. By contrast, the Delporte project recorded eight separate "Aurignacian II" levels. The well-provenienced artifact and faunal assemblages from the Delporte project provide an opportunity to examine various aspects of occupation in this large shelter at the base of limestone cliff in an interior valley, a location in marked contrast to the more numerous sites in the Vézère River valley approximately 4 km to the south.

AURIGNACIAN CORES AT LA FERRASSIE

The cores under consideration were derived from seven assemblages, one representing a combination of two levels. The general paleoenvironmental framework has been derived primarily from the sedimentological research of Henri Laville (Laville *et al.* 1980; Laville and Tuffreau 1984) but also from pollen analyses (Paquereau 1984). The intersite correlations for Périgord rock shelters promoted by Laville have been criticized (Texier *et al.* 1999). It is argued herein that the basic climatic interpretations are relevant for la Ferrassie, irrespective of any wider correlations, and may be summarized as follows:

• Cold and dry during the early Aurignacian (K6 and K5) with some amelioration during initial later Aurignacian K4 occupations (*ca.* 35–33 Ka);

• Warmer and moister climates during the later Aurignacian (K2 and J) that are often correlated with the Arcy interstadial (*ca.* 32–30 Ka);

• Conditions became cooler during subsequent occupations in levels I3 and I2 and colder and drier during I1 occupations (*ca.* 29–28 Ka).

A cyclical variation in core length was suggested, one that also correlated to an extent with climatic conditions (Table 13.1 and Figure 13.1). Cores were on average shorter during the colder conditions in the early (K6) and initial later Aurignacian (K4), although longer ones in level K5 varied from this trend. Those cores associated with the warmer interstadial conditions were longer (K2 and J). The larger core sizes were maintained during the cooler climatic conditions associated with the I3 and I2 occupations, but as conditions became colder as reflected in the I1 deposits, mean core length again became shorter. While level K5 varies from this pattern, the overall trend is an interesting one that will be explored later in the chapter.

Aspects of lithic reduction at la Ferrassie have been explored previously (Brooks 1979, 1995; Delporte 1984; Blades 2001), and bladelet production has been the subject of specific attention (Chazan 2001; Blades 2005). While elements of bladelet core reduction will be summarized, the discussion herein will focus on the reduction sequences oriented to the production of blades and flakes from larger cores.

The distinction between these core forms is to a certain extent an arbitrary one, particularly in terms of bladelet production. As has been indicated previously (Bon 2000, 2002, 2005; Blades 2005), bladelets were

Table 13.1 Core metrics (mm).

mm (weight g)	K6	K5	K4	K2	J	I3–I2	I1
N	42	26	68	36	39	18	39
Length mean	54.7	59.3	50.2	67.3	60.9	67.1	53.6
Length S	21.7	15.6	16.2	24.3	18.7	20.3	19.0
Width mean	50.4	54.9	50.8	57.9	57.9	61.5	56.6
Width S	16.3	16.7	15.4	18.9	14.8	14.8	14.4
Thickness mean	39.7	49.6	42.0	50.1	47.5	53.3	48.9
Thickness S	15.8	20.4	13.7	18.5	16.3	15.1	12.6
Weight mean	139.3	199.0	127.0	257.3	194.9	238.8	183.1
Weight S	135.4	143.3	98.3	208.5	122.8	152.3	160.3

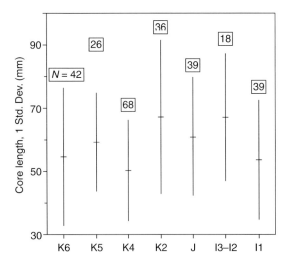

Figure 13.1 A comparison of core lengths for the Aurignacian levels at la Ferrassie under consideration in this chapter. The mean values and one standard deviation ranges are indicated. A general temporal relationship between colder climatic conditions and shorter mean core lengths is indicated, although exceptions are indicated as noted in the text.

produced from various supports following multiple reduction sequences:

• Cores on small nodules dedicated to production of bladelets and narrow blades;

• Bladelet blanks derived from larger flake or blade cores as late or final removals;

• Bladelet and/or blade blanks derived expediently from the edges of larger flakes such as *tablettes* or core platform rejuvenation flakes;

• Most significantly, particularly for the later Aurignacian, were the numerous "nontraditional" forms previously considered to be "carinated" scrapers or burins. These forms have, since the 1980s research by Jacques Tixier, Jean-Pierre Chadelle, Jacques Pelegrin, and others, been increasingly accepted as cores for the production of narrow and twisted bladelets (Le Brun-Ricalens and Brou 1997; Lucas 1997; Bon 2000, Brou and Le Brun-Ricalens 2005, to name but a few sources.)

The nature of these reduction sequences varies geographically in southern France, with greater emphasis during the earliest Aurignacian on small bladelet cores at Isturitz in the Pyrénées compared with both small cores and "carinated scrapers" at Brassempouy north

of the Pyrénées and further north still in the Dordogne (Bon 2000, 2001).

The production of Aurignacian blade blanks from cores at la Ferrassie manifested an interesting degree of continuity through time, judging from overall core morphology, platform orientations, and directions of blank removals. Qualitative and quantitative differences are also indicated. One theme that is maintained through these aspects of continuity and change, however, was the intensity of utilization that confounds attempts at determining clear boundaries between "types" of cores. Cores may, for example, pass from single platform to two or more platforms and then reveal creation of a single platform emerging from a late reorientation for final removal of a limited number of blanks.

One of the earliest Upper Paleolithic cores (F.69.4.K7a.39) recovered at la Ferrassie was found in the Aurignacian *ancien* level K7a, defined by Delporte for those artifacts lying on the uppermost surface of the underlying Aurignacian level L that yielded very few artifacts. The core exhibited a strong blade focus and a single platform orientation that emphasized lateral (width) reduction. The resulting appearance suggests a narrow two or three sided shape that may or may not reflect an original tabular or narrow nodule. A similar lateral reduction focus was observed on cores excavated by Anta Montet-White from the basal Aurignacian level N at Termo-Pialat in the Dordogne. From a temporal perspective, therefore, it is interesting to note that a core with similar morphology was recovered in the later Aurignacian level J at la Ferrassie (Figure 13.2).

Single platform orientations account for between 35 and 44% of all cores in levels K6 through J, with the exception of level K5 at 12%. The numbers of single orientations reduced sharply in levels I3–I2 and I1 in favor of opposed and angular orientations. Opposed orientations reflect blank removal in opposite directions on the same face of the core, while alternate orientations reflect reversed directions on different faces. Angular orientations result in blank production from platforms with varying orientations, up to and including right angles (Table 13.2).

Alternate and opposed orientations are illustrated in two cores with similar morphologies from the early Aurignacian level K5 and the later Aurignacian level K2. The core from level K5 combines aspects of opposed and alternating removals (Figure 13.3) while reflecting evidence of four platform phases, two at each end of the core. The level K2 prismatic core (Figure 13.4) has a stronger blade orientation on one face with probably later angular removals from the lower platform.

Certain distinct core morphologies are, however, indicated in specific periods, such as a curved pyramidal form (Figure 13.5) from level I1, the most recent occupation examined in this study. While this form reflects the fundamentals of single orientation blank production demonstrated throughout the Aurignacian sequence, it also illustrates the technological flexibility of adapting to a possibly broken base with a shift to an opposed orientation for additional blank removals. Apart from the sloping sides, the overall morphology

bears some slight resemblance to prismatic forms associated with subsequent Perigordian occupations (Inizan *et al.* 1992: 61, 62). The presence of a bladelet negative also emphasizes the extent to which bladelet (or blade) production may have been expediently derived, although greater numbers of blanks emerged from more organized *chaînes opératoires*.

Temporal variability is apparent in specific core morphologies, in platform orientations, and certainly

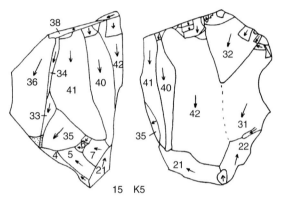

Figure 13.3 A blade core with alternate platform orientations made on brown Senonian chert with a coarse interior (F.69.0.K5.15) from level K5. The blank numbers reveal the approximate sequence of removal: blanks 4–7 from the first platform, 21–22 from the second platform, 31–38 from the third platform, and 40–42 from the fourth platform. Blade blank 3 is not visible in these views and removal of blank 33 may have preceded that of blank 31. Flakes 21 and 22 were removed to form the second platform, from which a large flake blank (41.5 mm by 41.9 mm) had been removed on another face. The core measures 84.0 mm in length and weighed 291.3 g.

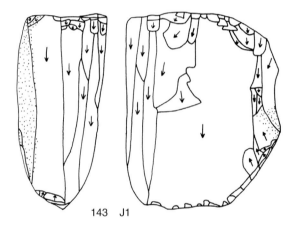

Figure 13.2 A single orientation blade core from level J1 (F.72.53.J1.143, which denotes the site of la Ferrassie, year, *carée* or square, level, and catalog number). The core was made on a tabular piece of patinated gray Senonian chert; blade blanks were removed along the front edge and side of the core, which measures 89.6 mm in length and weighs 364.8 g.

Table 13.2 Core platform orientations (percentage).

Orientations	K6	K5	K4	K2	J	I3–I2	I1
N	39	26	65	34	37	19	37
Single	43.6	11.5	41.5	35.3	40.5	10.5	18.9
Alternate	15.4	23.1	20.0	20.6	8.1	21.1	10.8
Opposed	12.8	11.5	13.8	11.8	8.1	52.6	18.9
Angular/right angle	17.9	42.3	16.9	29.4	35.1	10.5	40.5
Multiple	10.3	11.5	7.7	2.9	8.1	5.3	10.8

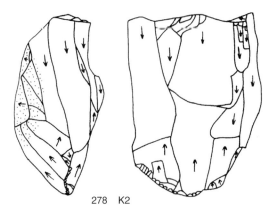

278   K2

Figure 13.4 An opposed orientations prismatic blade core on gray Senonian chert (F.70.5.K2.278) from level K2. Removals from opposite directions are indicated on the front face. The blade blanks from the top platform truncate those from the lower one, although more narrow angular removals along the side of the bottom platform may have alternated with blank removals from the top. The core measures 78.1 mm in length and weighs 189.8 g.

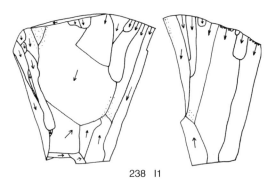

238   I1

Figure 13.5 An opposed orientations pyramidal blade core on a very fine grained piece of chalcedony-like brown Senonian chert from level I1 (F.69.6.I1.238). The core reflects blade removals on three curved faces with a flat rear surface. The core, which measures 80.1 mm in length and weighs 302.2 g, probably had a single orientation initially, with some later removals from the base that were subsequently truncated by still later removals from the top. In addition to the blade scars, a late bladelet negative measuring 11.5 mm by 6.7 mm is indicated.

in the importance of the "carinated" cores for bladelet production. The latter shift in emphasis is indicative of broader technological changes between early and later Aurignacian occupations. Bidirectional or multidirectional blank production may have been a raw material economizing strategy, as was suggested in level K6 by the tendancy for cores with single removals to be longer compared with the shorter multiple platform cores (Blades 2001: 147–9). The other Aurignacian levels did not reflect such distinctions, due in part to the creation of "new" single orientation cores on rejuvenated core fragments.

It is important to recognize, however, that single and multiple orientation cores were found in all Aurigancian levels and emphasize the importance of a reduction continuum. The random association of length with platform quantity for most levels suggests the existence of parallel strategies of single platform and multiple platform reduction and an often fluid boundary between these forms.

RAW MATERIALS AND GEOGRAPHICAL SCALES OF ANALYSES

The lithic raw materials that were exploited for core preparation and reduction were almost without exception gray and brown Senonian cherts that were presumably obtained near the shelter of la Ferrassie. Only two cores – one each from levels K6 and K4 – were derived from Bergerac chert sources 30–35 km west of the shelter, and both of these were heavily reduced. A few other cores may have been made on chalcedony or were patinated. However, the materials that were carried to the site for further reduction were available within close proximity to the shelter.

Such a pattern of local procurement calls attention to the issue of scales of analyses. The overall prehistoric landscape may be partitioned conceptually into territories defined by extents of movement as determined by resource distribution, mobility patterns, and social structure (Blades 2006). The sizes of these territories are highly variable, being conditioned by geographical context and cultural structures, and their definition requires differing scales of analyses.

The territory of material/social exchange may not be relevant in all geographical or social contexts, or may overlap with the territories to be discussed below. The geographical scale of analysis is regional, extra-regional, or even continental. This territory may be one in which materials and/or people moved over long distances. The appearance of rare lithics, marine shells, amber, or other materials from distant sources may symbolize risk-minimizing social networks

(Wiessner 1982) that would facilitate movement and be reinforced by spousal exchanges. "Directional" distributions in which materials from a specific source do not necessarily decrease relative to distance from that source (Renfrew 1977) are considered indications of such material transports (see Féblot-Augustins in this volume). The movement of marine shells from the Atlantic and Mediterranean coasts into Aurignacian sites in the Périgord including la Ferrassie (Taborin 1993) may reflect either social exchange or seasonal group movement.

The territory of seasonal/annual mobility relates ultimately to subsistence needs and generally encompasses a roughly predictable area within which social groups or portions of such groups moved. The scale of analysis for these territorial movements would be regional. When lithics or other materials are attributed to geographical locations at a distance, the presence of such materials at an archeological locus may reflect and assist in the definition of the seasonal or annual territory. A "down-the-line" or distance decay distribution (Renfrew 1977) is consistent with the movement of materials within the territory of mobility.

Raw material analyses of Aurignacian assemblages in the Périgord indicate small relative proportions of materials from more distant sources that have been interpreted as reflecting the extent and trajectory of mobility, with consequent implications for the overall settlement system. Other explanations may account for these distributions, such as stochastic or random movements for small amounts of materials and perhaps direct procurement for larger proportional amounts.

The territory of subsistence – the one of greatest relevance for this chapter – is the one within which hunter-gatherer groups obtain the daily caloric fulfillment from available faunal and floral resources. It is also the area from which the vast majority of lithic materials in most archeological occupations were obtained. The term "foraging radius" has been used to describe this exploitation area. However, the concept is one of surprising complexity as such radii vary at a given time for differing resources and often increase through time for a specific resource.

The sizes of these territories may often overlap and were certainly varied in space and time. The concept was encapsulated in the analysis by Geneste (1985, 1989) as discussed by Meignen *et al.* elsewhere in this volume (Chapter 2).

ENVIRONMENTAL COMPLEXITY AND THE LOCAL FORAGING AREA

The study of most of the lithics recovered at la Ferrassie and specifically of virtually all of the cores is therefore a consideration of local procurement that ultimately overlaps with subsistence procurement strategies and particulars of forager or collector settlement patterns (Binford 1980). La Ferrassie is located near but not at a raw material source. Lithic exploitation may generally be categorized as secondary reduction in comparison to a locus such as Termo-Pialat at a Senonian chert source. (Compare the core weights for the various la Ferrassie levels with that from Termo-Pialat level N: 369.0 g mean, 303.9 g standard deviation, $n = 17$.)

As mentioned above, core length data suggest a temporal relationship between colder assemblages and shorter cores compared with longer cores during warmer periods. The shorter core lengths would generally be interpreted as indicating greater reduction intensity, which is supported by other lithic measures (Blades 2003). This greater intensity may reflect various behaviors, including changes in the frequency or distance of group movement.

Assuming cores were initially prepared at primary quarry locations, the longer cores during the later Aurignacian occupations may mean that groups traveled shorter distances from quarries, had more regular access to quarry locations, and/or moved more frequently from the shelter to other locations without transporting the cores. Since cores were procured in the local areas, there is at least a possibility that their presence may provide insights into broader exploitation of the landscape. Further, since fauna were also likely hunted in the vicinity of the shelter, it is possible that procurement of lithic and faunal resources were in some fashion related activities.

Binford (1979: 259) has argued that lithic procurement was often "embedded" in more fundamental subsistence pursuits, but that need not necessarily have been the case for procurements to have been mutually reinforcing. Studies of skeletal elements in faunal assemblages from Aurignacian levels at Abri Pataud (Bouchud 1975: 72–3) and slightly later Perigordian occupation at le Flageolet (Enloe 1993: 107–109) – both in the Périgord – reveal a predominance of meat and marrow rich lower front and hind limbs. The general absence of lower utility elements has been interpreted by Enloe as indicating single hunters transporting only useful body parts over a distance, rather

Table 13.3   Large mammal (ungulate NISP) percentages.

Species	Levels L1, K6–K4	Levels K3–J	Levels I1–H1
Rangifer tarandus (reindeer)	77.4	6.5	16.9
Bison priscus, Bos primigenius (bison, aurochs)	16.7	51.6	34.9
Equus caballus (horse)	3.2	3.2	18.0
Equus hydruntinus (wild ass)	–	1.9	1.7
Megaloceros giganteus? (giant elk?)	–	0.7	–
Sus scrofa (wild boar)	0.5	13.6	1.2
Cervus elaphus (red deer)	1.4	18.7	18.6
Capreolus capreolus (roe deer)	–	2.6	2.3
Capra ibex (ibex)	0.9	–	3.5
Rupicapra rupicapra (chamois)	–	1.3	2.9
Ungulate N	221	155	172
Reciprocal of D (Simpson's Index)	1.55	3.07	4.64

than engaging in cooperative hunting. Therefore, faunal remains were often reduced in size to facilitate energetically efficient transport of the most useful portions, just as raw material blocks were often shaped into pre-cores at quarries for further reduction elsewhere.

The extent to which archeological megafauna are reliable indicators of the climates in which they lived or the habitats they preferred has been the subject of debate in the Périgrod. Such fauna clearly do provide a perspective on group diet, although subject to variation in meat yield for different species and biased by poor preservation or lack of recovery of floral remains. The percentages of numbers of individual specimens (NISP) in three stratigraphic groups are presented in Table 13.3. (The fauna from levels L3 and L2 were very limited in number and are not listed.)

The early Aurignacian and level K4 assemblages reflect the reindeer numerical dominance and consequent low evenness – presented as the reciprocal (1/D) of Simpon's Index of Diversity (Magurran 1988: 152–3) – that is often associated with the early Aurignacian in the Périgord. The later Aurignacian level K3-J and the assemblages I1 to H1 had greatly diminished reindeer presence, a numerical dominance of bovids, and overall much greater diversity. It is interesting to note, therefore, that these later assemblages have very different climatic associations: warm and humid (K3-J) and cold and dry (I1 to H1).

These data may reflect the range of paleoenvironmental habitats around the interior location of la Ferrassie and the extent to which these habitats were sustained regardless of changing climatic conditions. Various scholars have argued that the association of aurochs with wild boar, roe deer, and red deer suggests wooded or at least mosaic settings (White 1985: 43; Gamble 1986: 103–5; Gordon 1988: 41; Boyle 1990: 176–96). Other faunal data, specifically the presence of horses and bison, would indicate more open grasslands (Spiess 1979: 258–61; Gamble 1986: 108).

Topographical variation in the region may have contributed to habitat diversity. White (1985: 44) has emphasized that grassy steppe vegetation – perhaps with some trees – would have been present in uplands and valley bottoms, while sheltered valleys and south-facing slopes would likely have favored the growth of deciduous trees (Wilson 1975: 183). One of the reasons that groups returned to la Ferrassie for millennia may have been the access to these various topographic settings. Another may have been seasonal potential; Pike-Tay (1991) observed that red deer were exploited in the spring during Perigordian occupation of la Ferrassie, but no comparable seasonal data exist for the Aurignacian occupations in the shelter. Seasonal data from Aurignacian occupations at the nearby Vézère Valley locus of Abri Pataud indicate late fall and winter reindeer hunting (Spiess 1979). The faunal record at la

Ferrassie is sufficiently coarse-grained that the varying species may have been hunted at different times of the year in distinct or overlapping habitats.

Nelson (1986: 274–6) contended that mobility in the boreal forest was conditioned by resources that were scattered, localized, and variable. If one accepts the evolutionary ecology perspective that resources are generally found in stationary or mobile "patches" rather than in uniform distributions, the duration of occupation at a particular location – at least from a subsistence perspective – would depend upon the quantity and yield of such patches. Krebs and Davies (1978) argued that efficiency in food selection would suggest three predictable decisions:

1 Foragers should favor profitable (high energy yield) resources.

2 Groups may afford to be more selective if profitable resources are more abundant.

3 Certain resources, however, would not be selected regardless of abundance.

Kelly (1995: 135) observed that lower return resources would need to be gathered closer to camp and in quantity, resulting in a decreased effective foraging radius. Groups would thus move more frequently and over shorter distances compared with those exploiting higher return rate resources. Féblot-Augustins (1993: 243) expressed a similar argument when she suggested that a diverse mixture of plants and animals would deplete the resource base quickly and require frequent short movements.

The fauna at la Ferrassie would be comparatively high yield resources although they varied in potential live weight and meat yield per individual animal compared with reindeer:

- higher (bovids, red deer, horse)
- comparable (wild boar)
- lower (roe deer, small game, salmon)

In the absence of more specific localized paleoenvironmental data, the landscape surrounding la Ferrassie was partitioned into 1 km blocks extending from 3–5 km

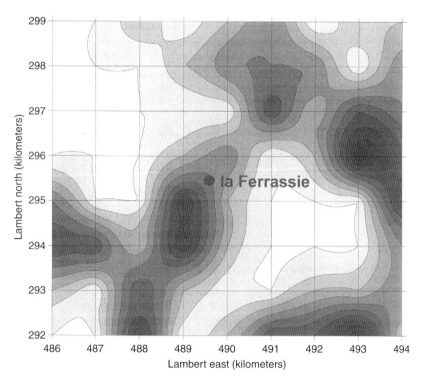

Figure 13.6 Geographical zones around la Ferrassie defined in 1 km blocks. The contour emphases reflect proximity to tributary valleys, south-facing exposures above confluences, and locations that overlooked tributary valleys and *vallons* or dry valleys. Depending on the nature of resources being sought, however, the more interior areas may have been desirable or preferred loci for exploitation.

away. Each block was categorized by the cumulative value of one or more landforms (Figure 13.6):

- stream/*vallon* confluence (value of 2)
- stream confluence (2)
- south-facing slope above confluence (2)
- stream or river valley (1)
- *vallon* (dry valley) (1)
- south-facing slope/ridge (1)
- south-facing location along stream (1)
- high ridge (1)
- interfluvial ridge (1)
- known raw material source (1)

The cumulative values would favor movement/ exploitation to the northeast and southwest, with both directions leading to the Vézère River. It must be recognized, however, that these higher values reflect geographical complexity and possibly ecotonal overlap. If a group was seeking grazing horses or bison, a ridge or valley setting lying to the northwest may have been a preferred area of exploitation.

SUMMARY

The pursuit of reindeer during the early Aurignacian may have led groups southward along valleys to the Vézère River. Little indication of forest-adapted species is apparent, which reflects to a degree the cold and dry environmental conditions. More closed environmental conditions during the later Aurignacian is suggested by the presence of aurochs. The increased diversity is tempered by the presence of high yield bovids. As a consequence, these later Aurignacian groups may have pursued a complex combination of movements utilizing resources in the closed forests and somewhat more open parklands. The longer cores would seem to reflect lower reduction intensity of raw materials blocks.

The more diverse faunal assemblage with which level I1 was one of the associated levels would seem to present the best correspondence between a diverse faunal base and greater core reduction intensity possibly related to frequency of movement. The I1 fauna reveal a presence of grazing animals – horse and possibly bison – and probably less forest, reflecting the more open environments often associated with cold conditions.

ACKNOWLEDGMENTS

I wish to express my appreciation to the National Science Foundation, the American Philosophical Society, and the Masion Suger in Paris for generous support of this research. My perspectives have been shaped through numerous discussions and communications with my friend and colleague Jehanne Féblot-Augustins, for which I am very thankful. Finally, I would like to express my love and appreciation for the support provided by my wife Meg Bleecker Blades and our daughter Emma.

REFERENCES

Binford, L. (1979) Organization and formation processes: Looking at curated technologies. *Journal of Anthropological Research* 25, 255–73.

Binford, L. (1980) Willow smoke and dogs' tails: Hunter-gatherer settlement systems and archaeological site formation. *American Antiquity* 45, 4–20.

Blades, B. (2001) *Aurignacian Lithic Economy: Ecological Perspectives from Southwestern France*. Kluwer Academic/ Plenum Publishers, New York.

Blades, B. (2003) End scraper reduction and hunter-gatherer mobility. *American Antiquity* 68, 141–56.

Blades, B. (2005) Small bladelet cores from Aurignacian levels at la Ferrassie (Dordogne, France). In: Le Brun-Ricalens, F., Bordes, J.-G., and Bon, F. (eds) *Production lamellaires attribuées à l'Aurignacien: chaînes opératoires et perspectives technoculturelles*, Archéologiques 1, Luxembourg, pp. 245–54.

Blades, B. (2006) Common concerns in the analysis of lithic raw material exploitation in the Old and New Worlds. In: Bressy, C., Burke, A., Chalard, P., and Martin, H. (eds) *Notions de territoire et de mobilité. Examples de l'Europe et des premières nations en Amérique du Nord avant le contact européen*. ERAUL 116, Liège, pp. 155–62.

Bon, F. (2000) *La question de l'unité technique et économique de l'Aurignacien: reflections sur la variabilité des industries lithiques à partir de l'étude comparee de trois sites des Pyrénées françaises (La Tuto de Camalhot, Régismont-le-Haut et Brassempouy)*. Thèse de Doctorat, Université Paris I Panthéon-Sorbonne.

Bon, F. (2001) Personal communication to the author, July 2001.

Bon, F. (2002) *L'Aurignacien entre Mer et Océan. Reflection sur l'unité des phases anciennes de l'Aurignacien dans le sud de la France*. Société Préhistorique Française Mémoire XXIV, Paris.

Bon, F. (2005) Little Big Tool. Enquête autour du success de la lamelle. In: Le Brun-Ricalens, F., Bordes, J.-G., and Bon, F. (eds) *Production lamellaires attribuées à l'Aurignacien: chaînes opératoires et perspectives technoculturelles*, Archéologiques 1, Luxembourg, pp. 479–84.

Bouchud, J. (1975) Étude de la faune de l'Abri Pataud. In: Movius, H.L., Jr. (ed.) *Excavation of the Abri Pataud, Les Eyzies (Dordogne)*, Peabody Museum, Harvard University, Cambridge, pp. 69–153.

Boyle, K. (1990) *Upper Paleolithic Faunas from South-West France – A Zoogeographic Perspective.* British Archaeological Reports International Series 557, Oxford.

Brooks, A. (1979) *The Significance of Variability in Paleolithic Assemblages: An Aurignacian Example from Southwestern France.* PhD dissertation, Harvard University, Cambridge.

Brooks, A. (1995) L'Aurignacien de l'abri Pataud niveaux 6 à 14. In: Bricker, H. M. (ed.) *Le Paléolithique supérieur de l'abri Pataud (Dordogne): les fouilles de H. L. Movius Jr.*, Éditions de la Sciences de l'Homme, Paris, pp. 167–222.

Brou, L. and Le Brun-Ricalens, F. (2005) Productions lamellaires et technocomplexes paléolithiques. Incidences: le Paléolithique supérieur revisité. In: Le Brun-Ricalens, F., Bordes, J.-G., and Bon, F. (eds) *Production lamellaires attribuées à l'Aurignacien: chaînes opératoires et perspectives technoculturelles*, Archéologiques 1, Luxembourg, pp. 489–94.

Chazan, M. (2001) Bladelet production in the Aurignacian of la Ferrassie (Dordogne, France). *Lithic Technology* 26, 16–28.

Delporte, H. (ed.) (1984) *Le grand abri de la Ferrassie 1968–1973. Études Quaternaires 7*, Université de Provence. Institut de Paléontologie Humaine, Paris.

Enloe, J. (1993) Subsistence organization in the early Upper Paleolithic: Reindeer Hunters of the Abri du Flageolet, Couche V. In: Knecht, H. Pike-Tay, A., and White, R. (eds) *Before Lascaux: The Complex Record of the Early Upper Paleolithic*, Boca Raton, CRC Press, pp. 101–15.

Féblot-Augustins, J. (1993) Mobility strategies in the late Middle Paleolithic of central Europe and western Europe: Elements of stability and variability. *Journal of Anthropological Archaeology* 12, 211–65.

Gamble, C. (1986) *The Paleolithic Settlement of Europe.* Cambridge University Press, Cambridge.

Geneste, J.-M. (1985) *Analyse lithique d'industries moustériennes du Périgord: une approche technologique du comportement des groupes humains au Paléolithique moyen.* Thèse Sc., Université de Bordeaux I.

Geneste, J.-M. (1989) Economie des resources lithiques dans le Moustérien du Sud-Ouest de la France. In: Freeman, L. and Patou, M. (eds.) *L'homme de Néandertal, tome 6: la subsistence*, ERAUL 33, Liège, pp. 75–97.

Gordon, B. (1988) *Of Men and Reindeer Herds in French Magdalenian Prehistory.* British Archaeological Reports International Series 390, Oxford.

Inizan, M.-L., Roche, H., and Tixier, J. (1992) *Technology of Knapped Stone.* Préhistoire Pierre Taillée 3. CREP in association with CNRS, Meudon.

Kelly, R. (1995) *The Foraging Spectrum: Diversity in Hunter-Gatherer Lifeways.* Smithsonian Institution Press, Washington.

Krebs, J. and Davies, N., (1978) *Behavioral Ecology: An Evolutionary Approach.* Blackwell, Oxford.

Laville, H. and Tuffreau, A. (1984) Les depots du grand abri de la Ferrassie: stratigraphie, signification climatique et chronologie.

In: Delporte, H. (ed.) *Le grand abri de la Ferrassie: fouilles 1968–1973*, Études Quaternaires 7, Université de Provence, Institut de Paléontologie Humaine, Paris, pp. 25–50.

Laville, H., Rigaud, J.-Ph., and Sackett, J. (1980) *Rock Shelters of the Périgord – Geological Stratigraphy and Archaeological Succession.* Academic Press, New York.

Le Brun-Ricalens, F. and Brou, L. (1997) Burins carénés-nucléus à lamelles: identification d'une chaîne opératoire particulière à Thémes (Yonne) et implications. In: Fosse, G. and Thévenin, A. (eds) *Le Paléolithique supérieur et le Mésolithique dans le nord-est de la France et les pays limitrophes*, Pré-actes de la Table Ronde de Valenciennes, 18–19 octobre 1997.

Lucas, G. (1997) Les lamelles Dufour du Flageolet I (Bézenac, Dordogne) dans le contexte aurignacien. *Paléo* 9, 191–220.

Magurran, A. (1988) *Ecological Diversity and its Measure.* Princeton University Press, Princeton.

Nelson, R. (1986) *Hunters of the Northern Forest*, 2nd Edition. University of Chicago Press, Chicago.

Paquereau, M.-M. (1984) Étude palynologique du gisement de la Ferrassie (Dordogne). In: Delporte, H. (ed.) *Le grand abri de la Ferrassie: fouilles 1968–1973*, Études Quaternaires 7, Université de Provence. Institut de Paléontologie Humaine, Paris, pp. 51–9.

Peyrony, D. (1934) La Ferrassie – Moustérien, Périgordien, Aurignacien. *Préhistoire* 3, 1–92.

Pike-Tay, A. (1991) *Red Deer Hunting in the Upper Paleolithic of Southwestern France: A Study in Seasonality.* British Archaeological Reports International Series 569, Oxford.

Renfrew, C. (1977) Alternative models for exchange and spatial distribution. In: Earle, T. and Ericson, J. (eds) *Exchange Systems in Prehistory*, Academic Press, New York, pp. 71–90.

Spiess, A. (1979) *Reindeer and Caribou Hunters: An Archaeological Study.* Academic Press, New York.

Taborin, Y. (1993) Shells of the French Aurignacian and Périgordian. In: Knecht, H. Pike-Tay, A., and White, R. (eds) *Before Lascaux: The Complex Record of the Early Upper Paleolithic*, CRC Press, Boca Raton, FL, pp. 211–27.

Texier, J.-P., Delpech, F., and Rigaud, J.-Ph. (1999) Programme collectif de Recherche "Litho et bio-stratigraphie de quelques sites paléolithiques de référence du Périgord." Rapport final, SRA Aquitaine.

White, R. (1985) *Upper Paleolithic Land Use in the Périgord: A Topographic Approach to Subsistence and Settlement.* British Archaeological Reports International Series 253, Oxford.

Wiessner, P. (1982) Risk, reciprocity and social influences on! Kung San economics. In: Leacock, E. and Lee, R. (eds) *Politics and History in Band Societies*, Cambridge University Press, Cambridge, pp. 61–84.

Wilson, J. (1975) The last glacial environment at the Abri Pataud: A possible comparison. In: Movius, H.L., Jr. (ed.) *Excavation of the Abri Pataud, Les Eyzies (Dordogne)*. Peabody Museum, Harvard University, Cambridge, pp. 175–86.

Chapter 14

OBŁAZOWA AND HŁOMCZA: TWO PALEOLITHIC SITES IN THE NORTH CARPARTHIANS PROVINCE OF SOUTHERN POLAND

Paweł Valde-Nowak

Institute of Archaeology, Jagiellonian University, Kraków

ABSTRACT

Recent archeological research in the Carpathian Mountains of southern Poland has revealed the presence of localized raw material sources. This chapter examines the utilization of these materials at two archeological sites: Hłomcza associated with a Late Paleolithic occupation and Obłazowa Cave with a succession of Middle and Early Late Paleolithic occupation levels. The Hłomcza site is evaluated as a hunting camp, as are several levels at Obłazowa. However, the level VIII Pavlovian occupation at Obłazowa contained a unique association of artifacts and is interpreted as a reflection of symbolic/ritualized behavior during the Late Paleolithic.

INTRODUCTION

The North Carpathians are commonly considered to contain few siliceous rocks with good cleavage characteristics. Only the radiolarite from the Pieniny rock belt

Lithic Materials and Paleolithic Societies, 1st edition. Edited by B. Adams and B.S. Blades, ©2009 Blackwell Publishing. ISBN 978-1-4051-6837-3.

has previously been noted (comp. Valde-Nowak 1995). The first serious studies of the potential for the exploitation of the local siliceous rocks by prehistoric peoples in the Polish Carpathians were undertaken a quarter of a century ago, when several workshops were discovered here (Valde-Nowak 1991). The workshops involved in this particular project reflect locations where radiolarite was worked. The skepticism which had persisted until that time among researchers as to the value of the Carpathians region for prehistoric settlement, especially during the Stone and Early Bronze Ages, prevented archeological excavation in the area. It was considered impossible for settlements to have existed in the Polish Carpathians in remote times and thus there was no question of any utilization of the local siliceous rock.

A decade after the identification of the first Polish radiolarite workshops in the Pieniny Mountains, another workshop was discovered where black menilite chert was worked. By that time numerous artifacts made of Dynów siliceous marl had been found, and increasing evidence was accumulating of Stone Age exploitation of Mikuszowice hornstone layers and of non-siliceous stone such as flysch sandstone or andesite. Finally, data suggested that other raw materials, such as Cergowa hornstones and Geza sandstones, might have also been used. Recently, when the Hłomcza Magdalenian culture assemblage was discovered, we were able to identify another new siliceous material used in the Carpathians, which was designated Bircza flint. The beds of this new flint are located in the vicinity of the village of Bircza in the Carpathian Foothills.

At present the exploitation of the following main local siliceous rocks during the Stone and Bronze Ages is documented: radiolarite, menilite and Mikuszowice hornstone, Dynów marl, and Bircza flint. As a result of new archeological research in the Polish West Carpathians some indirect evidence of mining of siliceous rocks (e.g., typical mining tools) has been published. Because the rock outcrops here have different characteristics than in the classic area of prehistoric flint mining (Cracow-Częstochowa Upland Świętokrzyskie [Holy Cross] Mountains), where flint is found in the form of nodules only and not as siliceous layers, prehistoric extraction in the Carpathians also probably depended upon different methods.

In the West Carpathians two Paleolithic sites were extensively excavated in recent years: the Magdalenian camp at Hłomcza and multicultural sequences in Obłazowa. Since the first is an open air site and the second is a cave, they reflect a representation of Pleistocene settlement in this area and provide an overview of the raw material sources used in the Stone Age.

THE OPEN AIR SITE AT HŁOMCZA

The locality 1 at Hłomcza is situated in the eastern part of the Polish Carpathians, close to the border with Ukraine, about 15 m above the present bottom of the River San at an altitude of 283.57 m above sea level. Fieldwork was conducted in 1997 and 2000–1. Investigations of topsoil and subsoil revealed 158 flint artifacts in an irregular oval-shaped pit, which also contained fragments of sandstone arranged as a pavement. The deposit in the pit fill and sediment associated with the structure were subjected to thermoluminescence dating. The old topographical surface, which predated the occupation, was estimated at 14.6 ± 2.3 years Ka BP. The fill of the settlement structure was determined to be 13.5 ± 2 Ka BP. It is worth noting that the age of the topographical surface is concurrent with the phase of the late glacial warming – identified in central Poland as the Kamionna phase – that corresponds to the Epe phase in the Netherlands (Lanezont et al. 2000: 43). It is clear that the age of the settlement activity can be placed between the two dates and thus predates the end of the Oldest Dryas.

The artifact assemblage can be divided into the following groups: cores, flakes, blades, tools, and characteristic waste (burin spalls). There are four cores in the inventory from the site. They represent a double platform, blade-like variety with steeply inclined platforms and traces of full preparation that are characteristic of the Magdalenian technocomplex (Vencl 1995: 38–52). The blades correspond to the features of the cores. Among them are items with two marked platforms. Overall 34 blades have been identified, including solid ones over 7 cm long and 3 cm wide. Flakes are relatively numerous (78 pieces). The production waste group consists of nine burin spalls. Tools ($n = 31$) compose 19.6% of the entire inventory; the dominant form is burins (22 items). The most characteristic forms are Lacan burins. Among the remaining implements there are three end scrapers, three backed points, two perforators, and a blade with marginal retouch.

What should be emphasized is the fact that Lacan burins were present at the Hłomcza site, as this item is rarely found in the eastern extent of the Magdalenian culture distribution. Even though this type of burin is a characteristic element of Magdalenian assemblages, such artifacts are rather rare (Demars and Laurent 1989: 74–5, Figure 23). If the sites where such artifacts have been discovered were indicated on a map, it would readily be apparent that such objects are more often found in western European assemblages and are almost non-existent in Moravia. Data from Belgium and Germany reveal that such tools belong to assemblages dated to no later than the Bölling interstadial. Given these facts and data referenced earlier, we can suggest that the examined remains of Magdalenian settlement at Hłomcza are relatively old. Taking into account the stratigraphy of the site and the results of TL dating, we can assume that the Hłomcza settlement existed before the end of the Oldest Dryas (Dryas I).

The topography of this site, located at a key point along the river just off its bank, as well as the fairly limited scope of the assemblage of tools, suggest that Hłomcza may have been a hunting settlement, very probably a seasonal one. Hłomcza was a camp with a stone pavement, of the type best known from the classical West European Magdalenian. The results of the analysis carried out on the Hłmocza assemblage of flint artifacts reveal several direct, far-reaching similarities with western, for example, Belgian (Orp, Canne) sites.

Recent examination of the raw material used for the stone artifacts from the Hłmocza site and numerous other archeological sites in the San valley have led to the identification of a new raw material for flint working named Bircza flint. Virtually all of the artifacts in the assemblage from Hłmocza were made of this material. One end scraper and one burin were made of Świeciechów flint and one burin spall of Volhynian flint was also recovered. The series of thin sections made from the geological outcrops as well as a selection of artifacts from the Hłomcza assemblage and from other archeological sites in the Uplands of Dynów and Przemyśl assisted a program of mineralogical, petrographical, and micropaleontological research. On this basis the examined material was identified as a silicified-foraminifera-bryozooa-lithotamnium biocalcarenite rock. This rock was attributed to the early but not earliest Palaeocene.

The comparative data collected during the above-mentioned program show that the significance of this material is regional, extending over an area around the middle courses of the San and Wistoka rivers. Individual artifacts made of this material have also been observed in the Spisz Magura area in North Slovakia. The Bircza flint sites have brought to light a variety of taxonomic units to the Stone Age and Early Bronze Age cultures. This material was used to make both small tools of the blade or flake types, as well as macrolithic core-type tools.

The discovery in Hłomcza reflects the relatively early middle European Magdalenian colonization. The well-known Maszycka Cave, near Cracow, yielded remains of a Magdalenian settlement dated from the middle of the Dryas I phase, no later that the Pre-Bölling interstadial (Kozłowski and Sachse-Kozłowska 1993). A little later is Hłomcza and last is the Upper Silesia Dzierżysław site. Both are connected with the Magdalenian penetration dated to the Dryas I, but to a younger stage of this phase. The Hłomcza settlement represents both a very early occupation and the easternmost site in the European range of the Magdalenian. The duration of Magdalenian occupation at Hłomcza must be understood as a one season occupation and exploitation of available natural resources, including local stone raw materials.

OBŁAZOWA CAVE

Obłazowa Cave was inhabited by humans several times. The settlement levels seen in a number of deposit strata indicate this interpretation. A number of Mousterian artifacts were found in the river series and at the bottom of the transitional series. Artifacts dating from the Upper Paleolithic were discovered in the upper part of the transitional series (layer XI) and within the clayey and rubble-like sequence (layer VIII and above). Another problem concerns the artifacts found outside the intact sequence series, that is, in the pit ("layer" XXII).

The oldest archeological traces are those of the Middle Paleolithic Mousterian culture. The characteristic artifacts of this culture with Taubachien affinities were recovered from layers XXb (assemblage with mammoth tooth), XIX, XVII, and XVI (assemblage with hyena mandible). The materials from layer XVb could be associated with the southeast Charentian; the inventory recovered from layer XIII comprises mainly small, often microlithic tools, shaped with retouch typical for Mousterian assemblages with denticulate tools. Upper Paleolithic artifacts were recovered in layers XI, VIII, V, and IIIa and the pit XXII. Moreover, a number of items were found in a secondary deposit of heterogeneous origin. The state of preservation of the artifacts varies. A large number of the artifacts found in the lower layers were smoothed by water. The edges of many artifacts in cultural horizons or layers were mechanically damaged and some of them "retouched" as a result of trampling and sedimentation processes, a phenomenon known as pseudo-retouch (McBrearty *et al.* 1998: 124–5).

A number of distinct technical and typological features have been identified in association with the

Mousterian assemblages from Obłazowa Cave. There are no distinct traces of the **Levallois** technique; the discoidal core technique was employed with the majority of the tools. The above observations are also relevant with respect to flakes, none of which displays all the Levallois elements (Boëda 1995: 45–61; Sellet 1995: 31–3). There are very few faceted or diagonal platforms.

Side scrapers and, to a lesser extent, small backed knives predominate. This is why certain similarities should be sought among the Middle Paleolithic materials found across the landscape of modern Hungary, Moravia, and Slovakia. In particular, the inventory from layer 11 in Külna Cave in Moravia, known as Taubachien (Valoch 1988: 73–80; 1996) includes a number of comparable items. Many elements of this inventory bear some resemblance to the assemblage discovered in the Bojnicka Cave or at Ganovce in Slovakia, connected with the Carpathian facies of the Mousterian culture characterized by a significant miniaturization of tools, classified by J. Svoboda as "les industries de petites dimensions" and dated to the Eemian interglacial or the beginning of the last glaciation (Svoboda 1984: 178).

Based on the recent discoveries conducted by L. Kaminská at the Hôrka-Ondrej site in North Slovakia, an inventory characteristic for the previously mentioned Taubachien has been found in the 12th layer in the C1 section. It has been dated to the Eemian Interglacial (Kaminská 2005: 83–4). In Poland similar materials were recently described from the Biśnik Cave from levels 11–8 recognized as archeological complexes B/C, D, and probably E (Cyrek 2002: 50–1, 68, Table 5).

The style of inventory known from the Hungarian site in Tata represents yet another Middle Paleolithic variety similar to those recovered from Mousterian layers at Obłazowa, in particular those found in layer XVb. The method of retouching, shape, and size of the tools are similar. The lack of distinct characteristics of the Levallois technique is noteworthy. The backed knives identified in the Mousterian assemblages at Obłazowa cannot be directly related to the knives/side scrapers of the Tata type. An artifact recovered from layer XVb with a cortical back and a pointed tip is the closest to this type although in this case we are evidently dealing with a Tata side scraper (Vertes 1964: 153, Figure 16). Many concave side scrapers in the inventory from the Erd, another site in Hungary, that is, "racloir simple concave" (Gabori-Csank 1968: Table XXIX:7) are similar to the artifacts from the assemblage recovered from layer XVb at Obłazowa (e.g., Figure 12:4), including a blunt natural back.

When we examine the Mousterian sequence from Obłazowa, we may point to the stylistic similarity of the materials found in the lower XXb–XVI layers, which can be related to the finds of the Taubachien types as well as to a number of differences observed in layers XVb and XIII.

It appears that the materials from layer XVb could be dated to the southeast Charentian (Gabori-Csank 1968: 182; Gabori 1976: 77). The preponderance of side scrapers with distinct typological features substantiates such an assumption. Some of them were shaped with a multiseries, steep, or partially steep gradual retouch. However, insufficient evidence has been found to confirm that the Levallois technique was employed. The southeast Charentian assemblages are dated to the Brörup or the beginning of the II Pleniglacial (Gabori-Csank 1968: 110; Kozłowski and Kozłowski 1977: 87–93, map 8). The more numerous artifact assemblage recovered from layer XIII comprises mainly small, often microlithic tools, shaped with denticulate retouch. For this reason we would rather classify it to the Mousterian culture with denticulate tools whose inventories often do not contain either backed points or typical core tools, as well as backed knives, points, and side scrapers (Bordes 1963, 1968: 102–103; Farizay and Jaubert 1994: 172; Thiébaut 2005). A bifacial tool similar to that recovered at Obłazowa has been found in the Sesselfelsgrotte E layer among items with denticulate retouch and small side scrapers. This layer is the youngest Mousterian on this site dated to 37 700 ± 1000 BP (Freund 1984; Weissmüller 1996: 145, Figure 9) whose position in the stratigraphy appears to be similar to be the location of layer XIII in Obłazowa Cave, at the Mousterian sequence in this cave.

For this chapter, the raw material aspects should be initially discussed. The spectrum of raw materials in layer XIX is representative of Middle Paleolithic raw material utilization (Table 14.1).

It is remarkable that radiolarites, as the local rock material, predominate in all categories of the inventory. Only obsidian from the Zemplin-Prešov Upland (East Slovakia) or Bükk Mountains (northeastern Hungary) has been noted as a long distance import.

The materials recovered in layer XI have distinctive features of the Middle Paleolithic tradition, for example in the form of a backed knife, made with the core technique, and a flake side scraper. However, artifacts with characteristic features of the Upper Paleolithic predominate. The most important find from the perspective of culture chronology is a small leaf point with barely visible tang, similar to some Szeletian

Table 14.1 Quantitative and raw-material structure of inventory from layer XIX.

Inventory group	Raw material								
	Radiolarite				Radiolarite limestone	Obsidian	Sandstone	Quartz	Total
	Red	Green	Steel-Gray	Other					
Cores	2	2	–	–	–	–	–	–	4
Negative nodules	10	2	7	3	–	–	2	–	24
Flakes	226	91	62	11	11	5	–	2	408
Tools	69	12	13	9	1	1	–	7	112
Total	307	107	82	23	12	6	2	9	548

Table 14.2 Quantitative and raw-material structure of lithic inventory from layer XI.

Inventory group	Raw material						
	Radiolarite			Obsidian	Undetermined flint	Flysh sandstone	Total
	Red	Green	Steel-gray				
Cores	1	–	–	–	–	–	1
Flakes	16	2	5	1	1	–	25
Blades	3	–	–	–	–	–	3
Negative pieces	2	1	2	–	–	–	5
Tools							
Bifacial leaf point	4	1	–	–	1	–	6
Core backed knife	1	–	–	–	–	–	1
Micro-backed knife	1	–	–	–	–	–	1
Scraper	1	2	–	–	–	–	3
Segment	1	–	–	–	–	–	1
Burin	1	–	–	–	–	–	1
Retouched blade	5	–	–	–	–	–	5
Retouched flake	3	–	–	–	–	–	3
Raclette	2	–	–	–	–	–	2
Borer	2	–	–	–	–	–	2
Pieces of unidentified tool	2	1	–	–	–	–	3
Hamerstone /retouchers	–	–	–	–	–	4	4
Total	45	7	7	1	2	4	66

forms from Hungary (Vertes 1955; Allsworth-Jones 1977) and items from Moravian assemblages, for example from Modřice (Klíma 1957: 102) or Vedrovice V (Valoch 2000: 103, Figure 3: 9–11; Valoch et al. 1993). In the raw material structure of the assemblage of the stone artifacts the local Pieniny radiolarite predominates. Only one small fragment of a leaf point and a flake were made of undetermined flint, a gray-white material with dark gray flecks. One obsidian flake and four sandstone hammerstones/retouchers were present. The inventory quantifications are presented in Table 14.2.

The most important aspect of the Obłazowa sequence is layer VIII, the assemblage that included the boomerang, which has received considerable attention in numerous publications (Valde-Nowak 1991, 2000; Valde-Nowak *et al.* 2003). It has been suggested that this inventory is related to the Pavlovian culture, known from famous Moravian sites such as Dolní Vestonice (Absolon and Klíma 1977; Svoboda 1994; Škrdla *et al.* 1995).

It should be noted that the inventory of layer VIII contained a wide range of raw material, which includes chocolate (Świeciechów) and Jurassic Cracow flint, originating mainly from northern regions (Table 14.3). The proportion of local radiolarite is, by contrast, negligible. Mountain crystal, most probably from the neighboring Magura Spiska Mountains (northern Slovakia) was also present and is known from other Pavlovian sites (Škrdla *et al.* 1995: 175). The presence of one artifact made of the Volhynian flint reinforces the overall impression of a broad range of raw materials. All undetermined flint artifacts (all are patinated) in this assemblage provide evidence of long distance import, as flint beds are absent in the North Carpathians.

How can we explain this phenomenon of a relatively rich raw material spectrum in so small an inventory? The utilized flints were brought from distances over 300 km and we conclude that such materials would have been very valuable. In addressing this question, the general characteristics of other elements of the assemblage should also be considered.

The substantial proportion of raw materials from the northern regions may indicate the Obłazowa assemblage dates to the early phase in the development of the Pavlovian period, traces of which have not as yet been discovered elsewhere in Poland (Kozłowski 1996: 14). In inventories dating from the same period, J.K. Kozłowski (1987) classified 76% of the raw material as flint of southern Poland. This may suggest the penetration of this area during the oldest phase of the Pavlovian, as reflected at Dolní Vestonice I and II and in Willendorf layer 5. The relatively large quantity of imported raw materials contrasts with all other layers of the Obłazowa sequence.

In this assemblage artifacts made of organic material are of special importance. Two horn wedges similar to those known from flint mines of different time periods, bone perforators, arctic fox tooth pendants, and two *Conus* shells represent very rare elements of Upper Paleolithic inventories.

The truly unique artifact found at Obłazowa, which has achieved international fame, is an intact boomerang fashioned on a mammoth tusk made by cutting the tusk along the longitudinal axis of the tooth and polishing the inner surface. The outer surface of the tusk, which shows numerous traces that might have appeared during the life of the animal, was not altered by Paleolithic man. The inner surface displays arch-shaped dentine rings and longitudinal lines. Delicate, slightly oblique cuts on both sides at the end of the object may be connected with a period of use for the boomerang. The re-examination of remains of red pigment (Łaptaś and Pazzkowski 2003) has confirmed the possibility that the applied dye was a kind of hematite. To complete the inventory of finds from layer VIII, it is necessary to mention the presence of human bones: two fragments of different distal phalanxes (comp. Gleń-Haduch 2003).

The inventory of the artifacts recovered from layer VIII is unique. No production waste, for example, related to the repeated production of flint artifacts, was present. The material seems to have been selected. The majority of the artifacts can be regarded as items of great value to prehistoric man. Therefore, this location is evaluated as one of special importance during the layer VIII occupation.

There are several hypotheses regarding the history of this site. One should consider that the site might have served as a stopping point and foothold for extensive penetration or even a sanctuary located far away from permanently inhabited areas. The two horn mining tools and the abundant deposits of Pieniny radiolarite in the vicinity of the cave are probably interrelated. However, the importance of this discovery would be underestimated if we linked the deposits of layer VIII only to mining activities. Radiolarite mines are not yet known in the vicinity of Obłazowa. Only two of the cores from layer VIII are made of local radiolarite. Moreover, radiolarite cores have limited technical parameters, being worked with no preparation of the platform and sides. Analyzing the position of many discovered objects within the cultural contexts, we have the impression that the horn tools were selected and deposited in the cave.

A very interesting circle-like construction of boulders (quartzite and granite Tatra boulders) found in layer VIII that must have been carried to the cave from the bed of the River Pra-Białka should be mentioned. Similar provenence can be postulated in the case of the sandstone plate found inside the circle. In the center

Table 14.3 Quantitative and raw-material structure of inventory from layer VIII.

Inventory group	Raw material									
	Radiolarite			Other	Flint				Rock crystal	Total
	Red	Green	Steel-gray		Chocolate	Świeciechów	Volhynian	Undetermined		
Pre-core	–	–	–	–	–	1	–	–	–	1
Cores	2	–	–	–	1	–	–	1	–	4
Flakes	4	12	1	3	–	–	–	10	–	30
Blades	2	–	1	–	–	1	–	–	–	4
Negative nodules	1	1	–	–	–	–	–	–	–	2
Tools	4	3	–	1	–	–	1	1	1	11
Total	13	16	2	4	1	2	1	12	1	52

of this stone circle the boomerang was discovered as well as a number of other objects of great significance to prehistoric people. One of the artifacts is a regular core made of patinated flint, probably from southern Poland. A pre-core of Świeciechów flint, a finger-like core (or splintered piece) and refitted bladelet made of chocolate flint and a human bone were also found. Within a 1.5 m² radius there were discovered the horn wedges, a collection of broken sandstone plates used as grinding stones or in stone tool production, and other artifacts. This association is fascinating but is also hard to interpret.

Artifacts from the upper layers V and IIIc in the Obłazowa sequence are less abundant. We cannot determine the cultural affiliation of layer V; it may date to the late phase of the Upper Paleolithic or even the Late Paleolithic – Epi-Gravettian or Magdalenian. The materials in secondary deposits, called the Aurignacian pit or "layer" XXII, are scarce but in fact typical for the Epi- or Late Aurignacian variety (comp.: Valoch *et al.* 1985; Richter 1987; Oliva 1993; Bachner *et al.* 1996; Kozłowski 1996; Street and Terberger 2000).

It is important to emphasize once again that the relatively high numbers of imported raw materials in layer VIII contrast with all other layers in the Obłazowa sequence. With the exception of the concentration in layer VIII, the excavated Paleolithic assemblages from cultural layers in Obłazowa represent typical inventories from hunting sites in caves. However, the intensity of settlement dynamics in the case of selected inventories can be discussed. For instance layers XXb and XVI had low densities of artifacts concentrated around the mammoth tooth (XXb) and hyena mandibles (XVI). In both cases we suppose that Mousterian occupants were coming for brief periods just after the phases of high water in the vicinity of Obłazowa Rock or possibly for short episodes between river floods in the area. The assemblages from layers XIX and XVII are relatively rich and show a fragment of larger settled area. The finds from layer XVb are connected with a thin level of cave sediments, situated on the top of a series of water-deposited sediments. The characters of these sediments suggest a period of warm and dry climatic conditions. The traces of human occupation are undisturbed. We are quite sure that layer XVb represents a settlement floor in the sense of a cultural level. The situation in layer XIII is different: we cannot document exactly the cultural horizon but the stratigraphic position of the artifacts reveals a cultural presence.

Layer XI is another settlement floor, one that was marked by red-brown sediments that were easily distinguished in many sections of the cave, despite the complicated configuration of this layer. The lithic inventory confirms the suggestion that we have here remnants of a hunting camp, resembling also some aspects characteristic of killing sites. All leaf points are broken, perhaps after use as spear points. The best example for this interpretation was the refitting of two pieces from a broken leaf point that may reflect the effect of successful hunting and a consequent need for tool rehabilitation.

The former hypothesis pointed to an important role of Obłazowa Cave as an element of transmission process of East Gravettian cultural elements from the Danube basin lying in a northeastern direction (Bárta 1966: 212; Kozłowksi 1983: 78, Figure 19). In fact this site is situated between a larger concentration of settlements in the Danube region (Bánesz 1961; Bárta 1970, 1980; Gabori-Csank 1970; Absolon and Klíma 1977; Klíma 1983; Oliva 1988) as in the Cracow area in the upper Vistula river basin (Chmielewski 1975; Kozłowski and Kozłowski 1977; 1993).

The finds of layer VIII are of a special importance for several reasons. Papers that were published after the initial phase of artifact analysis mainly focused on the discovery of the oldest boomerang in the world (Valde-Nowak *et al.* 1987, 2003; Valde-Nowak 1991) and presented the interpretation of finds in layer VIII as more or less coming from a typical hunting camp, perhaps seasonal, situated far from base camps (e.g., Moravia). The inventory should therefore reflect a special kind of weapon (boomerang), important stone tools like end scrapers, and some flint pre-cores and cores, planned as a raw material reservoir for flint tool rejuvenation prior to the next hunting foray. The horn core wedges may be explained as tools useful in mining activities such as radiolarite extraction in the vicinity of Obłazowa. Additional objects – two shells, three fox tooth pendants, a bone bead, and perforator – complete the unusual assemblage. Following long term analysis and reflection on all elements of the concentration in layer VIII, including cave stratigraphy and broader archeological contexts for the Obłazowa sequences, a new interpretation of this discovery will be presented in this chapter.

A distal phalanx from the left hand was found close to the boomerang; another previously unpublished human phalanx lay not far from the central concentration. It is argued that the red ochre and artistic/symbolic ornament explain this discovery. Some important observations were made during the stratigraphic analyses. The Pavlovian inhabitants delimited the area of this construction and made a pit about 2 m deep in the entrance area. As a result, a pebble circle was situated on the platform higher that the people who entered the cave, as the level exceeds the height of man. The anthropogenic disturbance, observed in the stratigraphy during some seasons of excavations, was ultimately removed as the excavation progressed.

All mentioned elements show the exceptional character of the inventory of layer VIII. The human bones may reflect only a symbolic grave. However, this interpretation has no good analogies. We argue instead that a ceremonial locus was situated here. The nature or character of the ceremony that took place here cannot be clearly defined. On the one hand, it should be emphasized once again that we have a concentration of relatively rare artifacts of apparent value that are well-preserved and in some cases covered with red ochre pigment. By contrast, more commonly encountered cultural artifacts, such as animal bone fragments and larger debitage, were absent. The presence of portions of two human fingers with an absence of other parts of the human skeleton has no analogies in other Paleolithic assemblages. It should be remembered that the cave sediments at Obłazowa were completely screened through fine mesh sieves (less than 1 mm).

We postulate either some practices like shamanism and suppose that Paleolithic man intentionally cut fingers as a ritual (for instance initiation) or, less probably, that the bones reflect accidental injuries or a medical "operation" (e.g., Clottes and Courtin 1995: 63–79). This supposition is supported when we consider explanations of hands in Paleolithic art in the Franco-Cantabrian region as cut or damaged, as first described at Gargas (Carthaillac and Breuil 1910). In this context the skeletons of Murzak-Koba in Crimea that lack portions of fingers should be mentioned (Bibikov 1940; Žirov 1940; Grünberg 2000: 342). The Obłazowa discovery may therefore provide a context for understanding the Gargas phenomenon.

At the very least, the finds of parts of human fingers in the context of the early Gravettian should be related to the older discussion about the explanation of the pictures of positive and negative presentation of hands, often with incomplete fingers. Such images are well-known from Franco-Cantabrian caves such as Gargas, Pech Merle, Cosquer, and Chauvet (Clottes and Courtin 1995; Clottes 2001). The chronology of the complex

from Obłazowa layer VIII – around 30 000 years BP – is certainly relevant. The left human distal thumb phalanx itself was dated by AMS analysis at Oxford to 31 100 ± 550 years BP (Housley 2003). Some cultural features suggest the older Gravettian, specifically early Pavlovian culture, as previously discussed.

The chronology of the Pavlovian in Obłazowa is now the oldest Pavlovian in the whole range of this culture. The timing therefore roughly coincides with the earliest occupations in the Chauvet Cave, dated to the interval 30 000–32 200 years BP or the later Aurginacian. However, many indications in Chauvet – especially the hands – may reflect the early Gravettian (Clottes 2001: 195). In addition, the chronology of other cave hand paintings, which would be estimated between 28 800 and 27 700 year BP, may coincide with the older Gravettian (Clottes and Courtin 1995: 78; Valladas et al. 2001: 980–5 and Table 2; Amormino 2000). The best example is Cosquer Cave, where three AMS dates from different samples (in two cases from the same hand picture) revealed ages over 27 700 BP. Therefore, the chronology of the oldest Gravettian/initial Pavlovian in central Europe (Obłazowa, Dolní Vestonice) is exactly contemporary with the initial Gravettian or transitional Aurignacian/Gravettian in Franco-Cantabria.

However, issues beyond those of chronology await further explication. It is hard to explain why well-excavated, rich sites from the Moravian, Austrian, Ukranian, or Russian Upper Paleolithic have not yielded either complete or fragmentary boomerangs. These sites have yielded numerous mammoth bones, including ivory, as well as artifacts made from these materials. One can assume theoretically that for ordinary hunting practices wooden boomerangs were normally used and have not survived at Paleolithic sites. Excavations at European Mesolithic and younger sites have recovered boomerangs made of wood. The ivory example from Obłazowa may have had special, possibly magical, significance, and great value.

CONCLUSION

Hłomcza and Obłazowa are two Paleolithic settlements in the North Carpathians that are situated far from their respective cultural contexts. The distance to the nearest contemporary sites can be estimated as about 300 km. The former site is an open air dwelling site with traces of a tent with a stone pavement, typical for the Magdalenian. The second site, Obłazowa, is a multicomponent cave shelter that was occupied over a long period. In both cases the local beds of stone raw materials provided the primary sources for exploitation. Only the deposit of special, probably ceremonial character found in Obłazowa level VIII shows the exceptional nature of the raw materials that were brought from far distant sources, such as chocolate (Świeciechów) and Volhynian flints. The special character of this deposit is interpreted as reflecting aspects of shamanism that would serve to emphasize the great value of materials transported long distances.

The Carpathians area should no longer be regarded as a "desert" of siliceous rocks. The technical and typological features and quality of all artifacts made from the local stone materials – radiolarite at Obłazowa and Bircza flint at Hłomcza – are the same as those reflected in materials imported from long distances. We therefore do not observe modifications caused by use of different kinds of lithic raw materials.

REFERENCES

Absolon, K. and Klima, B. (1977) *Predmosti. Ein Mammutjägerplatz in Mähren.* Archeologicky Ustav ČSAV v Brne, Brno.

Allsworth-Jones, P. (1977) Szeleta Cave, the excavations of 1928, and the Cambridge Archaeological Museum Collection. *Acta Archaeologica Carpathica* 18, 5–38.

Amormino, V. (2000) L'art paleolithique et le carbone 14. *L'Anthropologie* 104, 373–81.

Bachner, M., Mateiciucova, I., and Trnka, G. (1996) Die Spätaurignacien-Station Albendorf im Pulkautal. NÖ, In: Svoboda, J. (ed.) *Paleolithic in the Middle Danube Region, Anniversary Volume to Bohuslav Klima*, Archeologicky Ustav ČSAV v Brne, Brno, pp. 93–119.

Bánesz, L. (1961) Prehlad paleolitu Východného Slovenska. *Slovenská Archeólogia* 9, 33–48.

Bárta, J. (1966) *Slovensko v staršej a strednej dobe kamennej.* Slovenska Akademia Vied, Bratislava.

Bárta, J. (1970) Zur Problematik der Höhlensiedlungen in den Slowakischen Karpaten. *Acta Archaeologica Carpathica* 2, 5–33.

Bárta, J. (1980) *Vyznamne paleoliticke lokality na strednom a zapadnom Slovensku.* Archeologický ústav SAV Nitra, Nitra.

Bibikov, S.N. (1940) Grot Murzak-Koba – novaja pozdnepaleolitičeskaja stojanka v Krymu. *Sovetskaja archeologija* 5, 159–78.

Boëda, E. (1995) Levallois: A volumetric construction, methods, a technique. In: Dibble, H.L. and Bar-Yosef, O. (eds) *The Definition and Interpretation of Levallois Technology*,
Monographs in World Archaeology 23, Prehistory Press, Madison, Wisconsin, pp. 41–68.

Bordes, F. (1963) Le Mousterien a Denticules. *Archeolovski Vestnik* 13–14, 43–9.

Bordes, F. (1968) *The Old Stone Age*, World University Library, Weidenfeld & Nicolson Ltd., London.

Cartaillhac, E. and Breuil A.H. (1910) Gargas, Cne D'Aventinian (Hautes-Pyrénées), *L'Anthropologie XXI*, p. 129–50.

Chmielewski, W. (1975) Paleolit środkowy i górny. In: Hensel, W. (ed.) *Prahistoria ziem polskich* I, Ossolineum, Wrocław-Warszawa-Kraków-Gdańsk, pp. 9–158.

Clottes, J. (ed.) (2001) *La Grotte Chauvet, L'art. Des origines*, Seuil, Paris.

Clottes, J. and Courtin, J. (1995) *Grotte Cosquer bei Marseille. Eine im Meer versunkene Bilderhöhle.* Jan Thorbecke Verlag, Sigmaringen.

Cyrek, K. (2002) *Jaskinia Biśnik. Rekonstrukcja zasiedlenia jaskini na tle zmian środowiska przyrodniczego.* Wydawnictwo Uniwersytetu Mikołaja Kopernika, Toruń.

Demars, P. and Laurent, P. (1989) *Types d'outils lithiques du paléolithique supérieur en Europe.* Éditions du Centre National de la Recherche Scientifique (CNRS), Paris.

Farizy, C. and Jaubert, J. (1994) Comparaisons régionales et synthèse. In: Farizy, C., David, F., and Jaubert, J. (eds) *Hommes et bisons du Paléolithique moyen à Mauran (Haute-Garonne)*, XXXᵉ supplément à "Gallia Préhistoire," Éditions du Centre National de la Recherche Scientifique (CNRS), Paris, pp. 169–75.

Freund, G. (1984) Die Sesselfelsgrotte im unteren Altmühltal. Ein Beitrag zur Gliederung der *letzten Eiszeit*, Führer zu archäologischen Denkmälern in Deutschland 6, Regensburg-Kelheim-Straubing II, Theiss Verlag, Stuttgart, pp. 79–89.

Gabori, M. (1976) *Les civilisations de paléolithique moyen entre les Alpes et L'oural.* Akademiai Kiado, Budapest.

Gabori-Csank, V. (1968) *La station du paléolithique moyen d'Érd-Hongrie.* Akademiai Kiado, Budapest.

Gabori-Csank, V. (1970) C14 dates of the Hungarian Palaeolithic. *Acta Archaeologica Academiae Scienciarum Hungaricae* 22, 3–11.

Gleń-Haduch, E. (2003) Human remains. In: Valde-Nowak, P., Nadachowski, A., and Madeyska, T. (eds) *Obłazowa Cave. Human Activity, Stratigraphy and Palaeoenvironment.* Institute of Archaeology and Ethnology Polish Academy of Sciences – Cracow Branch, Kraków, pp. 89–90.

Grünberg, J.M. (2000) *Mesolithische Bestattungen in Europa. Ein Beitrag zur vergleichenden Gräberkunde.* Teil 2 – Katalog, Internationale Archäologie 40, Marie Leidorf GmbH, Rahden/Westf.

Housley, R. (2003) Radiocarbon dating. In: Valde-Nowak, P., Nadachowski, A. and Madeyska, T. (eds) *Obłazowa Cave. Human Activity, Stratigraphy and Palaeoenvironment.* Institute of Archaeology and Ethnology Polish Academy of Science – Cracow Branch, Kraków, pp. 81–5.

Kaminská, L. (2005) *Hôrka-Ondrej. Osídlenie spišských travertínov v staršej dobe kamennej.* Archeologický ústav Slovenskej Akademii Vied Nitra, Košice.

Klíma, B. (1957) Übersicht über die jüngsten paläolithischen Forschungen in Mähren. *Quartär* 9, pp. 85–130.

Klíma, B. (1983) *Dolní Věstonice. Tábořište lovců mamutů, Památníky naši minulosti 12.* Nakladatelství Československé Akademie Věd, Praha.

Kozłowski, J.K. (1983) Le paléolithique supérieur en Pologne. *L'Anthropologie* 87, 49–82.

Kozłowski, J.K. (1987) Changes in raw material economy of the Gravettian technocomplex in Northern Central Europe. In: Biro, K. (ed.) *International Conference on Prehistoric Flint Mining and Raw Material Identification in the Carpathian Basin*, 2, KMI Rota, Budapest, pp. 65–80.

Kozłowski, J.K. (1996) The Danubian Gravettian as seen from the Northern perspective. In: Svoboda, J. (ed.) *Paleolithic in the Middle Danube Region*, Archeologicky ustav AV ČR, Brno, pp. 11–22.

Kozłowski, J.K. and Kozłowski, S.K. (1977) *Epoka kamienia na ziemiach polskich.* Państwowe Wydawnictwo Naukowe, Warszawa.

Kozłowski, J.K. and Kozłowski, S.K. (1993) *Le Paléolithique en Pologne*, Serie "Préhistoire d'Europe" 2, Éditions Jerome Millon, Grenoble.

Kozłowski, S.K. and Sachse-Kozłowska, E. (1993) Magdalenian family from the Maszycka Cave. *Jahrbuch des Römisch-Germanisches Zentralmuseum Mainz* 40, 115–205.

Łanczont, M., Madeyska, T., and Kusiak, J. (2000) Late Pleistocene natural environment of the San River Valley near the Hłomcza archeologiczal site. *Acta Archaeologica Carpathica* 35, 33–48.

Łaptaś, A. and Paszkowski, M. (2003) Ferruginous pigments. In: Valde-Nowak, P., Nadachowski, A., and Madeyska, T. (eds) *Obłazowa Cave. Human Activity, Stratigraphy and Palaeoenvironment*, Institute of Archaeology and Ethnology Polish Academy of Sciences – Cracow Branch, Kraków, pp. 78–79.

McBrearty, M.S., Bishop, L., Plummer, T., Dewar, R., and Conard, N. (1998) Tools underfoot: Human trampling as an agent of lithic artifact edge modification. *American Antiquity* 63(1), 108–29.

Oliva, M. (1988) A Gravettian site with mammoth-bone dwelling in Milovice (Southern Moravia). *Anthropologia* 26(2), 105–12.

Oliva, M. (1993) Aurignacian in Moravia. In: Knecht, H., Pike-Tay, A., and White, R. (eds) *Before Lascaux. The Complex Record of the Early Upper Paleolithic*, CRC Press, Boca Raton, Ann Arbor, London, Tokyo, pp. 35–56.

Richter, J. (1987) Jungpaläolithische Funde aus Breitenbach, Kr. Zeitz im Germanischen Nationalmuseum Nürnberg. *Quartär* 37/38, 63–96.

Sellet, F. (1995) Levallois or not Levallois: Does it really matter? Learning from an African Case. In: Dibble, H.L. and Bar-Yosef, O. (eds) *The Definition and Interpretation of Levallois Technology*, Monographs in World Archaeology 23, Prehistory Press, Madison, Wisconsin, pp. 25–39.

Street, M. and Terberger, T. (2000) The last Pleniglacial and the human settlement of Central Europe: New information from the Rhineland site of Wiesbaden-Igstadt. *Antiquity* 73, 259–72.

Svoboda, J. (1984) Cadre chronologique et tendances évolutives du paléolithique Tchécoslovaque. Essai de synthèse. *L'Anthropologie* 88(2), 169–92.

Svoboda, J. (1994) Středni paleolit. In: Svoboda, J., Czudek, T., Havlíček, P., Ložek, V., Macoun, J., Prichistal, A., Svobodova, H., and Vlček, E. (eds) *Paleolit Moravy a Slezska*, Archeologický ustav ČR, Brno, pp. 76–93.

Škrdla, P., Cilek, V., and Prichystal, A. (1995) Dolní Věstonice III, Excavations 1993–1995. In: Svoboda, J. (ed.) *Paleolithic in the Middle Danube Region*, Archeologicky ustav ČR, Brno, pp. 173–90.

Thiébaut, C. (2005) *Denticulate Mousterian: Variability or techno-economic diversity?* PhD Thesis in Université d'Aix-Marseille I – Université de Provence UFR Archéologie et Histoire de l'art, Marseille (unpublished disertation).

Valde-Nowak, P. (1991) Studies in Pleistocene settlement in the Polish Carpathians. *Antiquity* 65, 593–606.

Valde-Nowak, P. (1995) Stone sources from the North-Carpathian province in the Stone and Early Bronze Ages. *Archaeologia Polona* 33, 111–8.

Valde-Nowak, P. (2000) The boomerang from Oblazowa and its prehistoric context. In: C. Bellier, C., Cattellain, P., Otte, M. (eds) *La Chasse dans la Prehistoire. Hunting in Prehistory*, Actes du Colloque international de Treignes, 3–7 octobre 1990, CNRS, Liège, pp. 88–94.

Valde-Nowak, P., Nadachowski, A., and Madeyska T. (eds) (2003) *Obłazowa Cave. Human Activity, Stratigraphy and Palaeoenvironment.* Institute of Archaeology and Ethnology, Kraków.

Valde-Nowak, P., Nadachowski, A., and Wolsan, M. (1987) Upper Palaeolithic boomerang made of a mammoth tusk in southern Poland. *Nature* 329, 436–38.

Valladas, H., Tisnérot-Laborde, N., Cachier, H., Arnold, M., Berlando de Quiròs, F., Cabrena-Vadés, V., Clottes, J., Courtin, J., Fortea-Pérez, C., and Gonzàles-Sainz, C. (2001) Radiocarbon AMS Dates for Paleolithic cave paintings. *Radiocarbon* 43/2B, 977–86.

Valoch, K. (1988) *Die Erforschung der Kulna-Höhle 1961–1976.* Morevske Museum/Anthropos Institut, Brno.

Valoch, K. (1996) *Le paléolithique en Tchéquie et en Slovaquie.* Éditions Jerome Millon Grenoble.

Valoch, K. (2000) L'histoire de la connaissance et les questions des industries paleolithiques à pointes foliacées sur le territoire de l'ancienne Tschécoslovaquie. *Prehistoria* 1, 95–107.

Valoch, K., Oliva, M., Havlíček, J., Karašek, P., and Smolikova, L. (1985) Das Frühaurignacien von Vedrovice II und Kuparovice I in Südmähren. *Anthropozoikum* 16, 107–203.

Valoch, K., Koči, A., Mook, W.G., Opravil, E., van der Plicht, J., Smoliková, L., and Weber, Z. (1993) Vedrovice V, eine Siedlung des Szeletien in Südmähren. *Quartär* 43–44, 7–93.

Vencl, S. (1995) *Hostim. Magdalenian in Bohemia*. Archeologický ústav Akademie Věd České Republiky, Prague.

Vertes, L. (1955) Neue Ausgrabungen und paläolithische Funde in der Höhle Istalloskö. *Acta Archaeologica Hungarica* 5, 111–31.

Vertes, L. (1964) Die Ausgrabung und die archäologischen Funde. In: Vertesz L. (ed.) *Tata, Eine mittelpaläolithische Travertin-Siedlung in Ungarn*. Akademiai Kiado, Budapest, pp. 133–249.

Weissmüller, W. (1996) Evaluating the incompleteness of Middle Palaeolithic silex inventories. *Quaternaria Nova* VI, 127–48.

Žirov, E.V. (1940) Kostjaki iz grota Murzak-Koba. *Sovetskaja archeologija* 5, 179–86.

Chapter 15

RAW MATERIAL ECONOMY AND TECHNOLOGICAL ORGANIZATION AT SOLVIEUX, FRANCE

Linda T. Grimm[1] and Todd Koetje[2]

[1]Department of Anthropology, Oberlin College
[2]Department of Anthropology, Western Washington University

ABSTRACT

In the Middle Isle Valley of southwestern France, the Upper Paleolithic inhabitants of Solvieux exploited readily available and abundant chert resources. Predominant black and brown Senonian cherts are accompanied by other varieties such as chalcedonys, jaspers, and Bergeracois types. Focusing on three industries that spanned 15 000 years of occupation at this open-air locality, we demonstrate much overlap in the raw materials used while also revealing considerable diversity in the organization of core technology and tool production. Differences in raw material utilization contribute to our understanding of changes in land use patterns over time.

INTRODUCTION

A preliminary analysis of patterns in raw material acquisition and use in three Upper Paleolithic levels from the open-air site of Solvieux, (Dordogne), France, establishes a foundation for understanding patterns of change in land and resource use through time in the Middle Isle Valley, and integrating these with broader

Lithic Materials and Paleolithic Societies, 1st edition. Edited by B. Adams and B.S. Blades. ©2009 Blackwell Publishing. ISBN 978-1-4051-6837-3.

patterns throughout the Aquitaine region. Excavated during the 1960s and 1970s (Sackett 1999), Solvieux preserves evidence of successive occupations from Middle Paleolithic to Middle Magdalenian times. The three levels under consideration here span a period of roughly 15 000 years and concern early, middle, and late manifestations of the regional Upper Paleolithic. The earliest is Level IV, or the Beauronnian (Taranik 1977), a typologically idiosyncratic horizon of probable Early Upper Paleolithic date that may bear some relationship to regional Aurignacian industries. It was recovered from deep deposits in western Solvieux and is represented by close to 7000 lithic artifacts. Level M is a Raysse Perigordian industry dating to around 24 000 BP (Sackett 1999: 5) that comes from the eastern part of Solvieux. This deposit was buried about a meter below the modern surface and is represented by a lithic inventory of more than 5000 chert artifacts. Couche A belongs to the Early Magdalenian, or Badegoulian, and falls within the time range of 15 000–18 000 BP (Sackett 1999: 5). Our study of Couche A is based on a sample of just under 3000 pieces of debitage recovered from a narrow strip on the western edge of Locality 3 in central Solvieux. All the other debitage from this vast exposure was lost in a fire set by an arsonist during the initial years of the Solvieux Project (Sackett 1999: 82). Though organic remains are not preserved in Solvieux's acidic soils, lithic artifactual remains, in the form of stone tools, debitage, rocks, and rock tools are numerous. The present analysis is restricted to the chipping debris, or debitage, that constitutes the bulk of knapped chert materials recovered in each deposit.

Following a brief summary of the broad technological patterns that characterize the three components of the study sample, we turn to the matter of source identification of chert materials in the Beauronnian, Late Perigordian, and Early Magdalenian. Similarities and differences between and among the three are explored within a quantitative framework that provides a basis for assessing the ways in which dimensions of lithic variability co-vary over space and time. We end by summarizing our present thinking about raw material and land use patterns in the Middle Isle Valley over the 15 000 year period covered by our study.

While there is much overlap in the raw materials used in the three levels, there is considerable diversity in the organization of core technology and tool production. Refitting provides significant information about technological organization in Levels IV and M (Grimm and Koetje 1992; Grimm 2000) whereas only limited success has been achieved with this method in the case of the Early Magdalenian material. We assume this is due in some degree to the fact that we are working with a small sample from the edge of a vast exposure. At the same time, our sample numbers in the thousands and is representative of the full range of debitage elements. It is also possible that the low number of refits is related to greater intensity of core reduction in the Early Magdalenian in comparison to the other two components of our sample. Whether this is due to an emphasis on flake as opposed to blade production in

Couche A, or a reflection of differences in settlement size and duration cannot be determined at the present time. Whatever the case, it is clear that this settlement was very different from the short term hunting camps of the Beauronnian and Upper Perigordian of Levels IV and M.

LEVEL IV

One tool class, called Beauronnian truncations (Taranik 1977), made on robust blades by direct percussion, dominates the Level IV assemblage. The core technology that is employed in blade production makes only modest use of crested blades, relying instead on the natural ridges of oval-shaped nodules to open and preform cores. This technique is especially marked in its application to large nodules of brown chert as shown in Figure 15.1.[1] In this ensemble, 47 pieces, including 9 tools, have been refitted to an abandoned core. In a pattern typical of the Beauronnian, the initial striking platform was established by the removal of a large flake that truncated the entire end of the nodule. Black chert nodules were smaller-sized and more commonly selected than the brown variety. At the same time, the large brown nodules are distinctive enough in both their size and well-preserved cortex to suggest some special selection criteria associated with their acquisition. Whether this reflects a choice on the part of an individual knapper or a characteristic of the Beauronnian industry as a whole awaits future investigation. In any event, the local origin of all of this core material seems strongly indicated by both the large amount of waste that accompanied their reduction in

Figure 15.1 Beauronnian Refitted Core.

Level IV as well as by their widespread occurrence in all cultural horizons at Solvieux.

LEVEL M

Level M cores are, on average, half the size and twice as productive of blade supports as Beauronnian ones, due, no doubt, to the extensive use of the crested blade and punch techniques. Gravette points and Raysse burins constitute the distinctive typological elements of this industry. There is more variability in the chert types found as well as signs of systematic thermal alteration in the form of luster, color changes, and blanks that are long, straight and thin (Grimm 1983).[2]

COUCHE A

The Early Magdalenians made minimal use of the crested blade technique in production of the raclettes that are characteristic of this industry, though core rejuvenation elements are far more common in the Couche A debitage we have examined than in that from levels M and IV. Some form of punch technique was used along with what can be gauged as a well-controlled and fairly widespread thermal alteration of the raw material.

RAW MATERIAL TYPES

With this background, we can now turn to a systematic examination of similarities and differences in raw material among the three industries in our sample. Cherts ranging from good to excellent in quality were abundant and easily obtained by the prehistoric inhabitants of the Middle Isle Valley. Access to workable stone was evidently a primary factor affecting settlement location, since otherwise attractive situations like rock shelters apparently went unoccupied if good cherts were not readily available (Texier 1986). Chert raw materials are associated in the region with a range of Upper Cretaceous contexts where they are found both *in situ* in calcareous formations and, more commonly, in colluvial deposits on slopes and in the riverbeds of the Isle and its tributaries (Texier 1986). Four limestone formations – Coniacian, Santonian, Campanian, and Maestrichtian (from earliest to most

recent) – together constitute the Senonian Epoch of the late Cretaceous and provide the geological framework for the majority of cherts utilized by prehistoric inhabitants (Seronie-Vivien and Seronie-Vivien 1987). Our methodology of chert classification involved standard macroscopic identification based on color, structure, and inclusions. It was complemented, in specific cases, by examination of microfossils using a binocular microscope. Lithic artifacts from Levels IV and M could be confidently identified as to source since they lack the white patination that commonly obscures colors and features on Paleolithic cherts. Patination on Couche A cherts ranges from moderate to heavy, but this did not interfere to any serious degree with the identification of the main groups used in our analysis. In the process of relating raw material to known geological sources, we utilized the substantial published literature (Demars 1982; Chadelle 1983; Larick 1983; Seronie-Vivien and Seronie-Vivien 1987; Geneste and Rigaud 1989; Blades 2001) and consulted one-on-one with experts.[3] Since we are working at this point in time from a type collection, we could not systematically address a broader ranger of chert characteristics. Seronie-Vivien and Seronie-Vivien (1987) provide a detailed geological analysis of Aquitaine cherts that includes useful guidelines for flint sourcing research. Our analysis is structured in terms of three major chert groupings, designated Senonian, Bergeracois, and Isle Valley Translucent (IVT), as described below.

Senonian cherts, our first group, constitute the bulk of the chipped stone recovered at Solvieux and, in this regard, Solvieux's archeological horizons are typical of those found throughout the Perigord (Bricker 1975; LeBrun-Ricalens 1996; Lucas 2000; Nespoulet 2000; Blades 2001). By convention, archeologists working in this region lump Santonian and Campanian cherts together into a Senonian category (their epoch name), which is subdivided, on the basis of color, into gray/black and brown/beige variants (Demars 1982; Chadelle 1983; Lucas 2000: 36; Blades 2001: 72). Sheer abundance and ready accessibility appear to have set the stage for a substantial accumulation of both black and beige varieties of Senonian chert in the tool inventories and manufacturing debris of all three industries examined here. Nodules of Senonian chert are amorphous in shape and of variable size, ranging from small cobbles to boulders. Fossil microfauna such as bryozoans and forams are common in these cherts (Seronie-Vivien and Seronie-Vivien 1987). Cortex is variable in thickness, smoothness, and in the nature of the transition from cortex to body. The chert body usually has a differentiated core and periphery, the latter of which is pure silica and often of fine grain. In fact, the periphery is usually the most workable part of the core since calcareous inclusions are common and constitute the Senonian's weak spot from a flintknapper's perspective (Larick 1983). Color range is broad within the divisions of black and brown and does not seem to be correlated with geological origin or workability, although the beige varieties are generally viewed as coarser (Blades 2001: 72).

Our second type is designated IVT, a fine-grained, highly translucent material with good technological properties. We created this category because it seemed to subsume a range of variation we could not identify in other research on chert sources. The chert body is homogeneous in grain and lacks calcareous or other inclusions. Cortex is thick and rough in some aspects with an abrupt boundary between the cortex and the chert body. Color ranges from tan to dark brown. One black example from Level M shows signs of intense heating on its cortex and probably belongs to this type. A source for this material is not known although one sample was collected from a small road cut in the tributary valley of the Crempsoutie to the east of Mussidan on a survey one of us conducted a number of years ago.[4] Nodule shape is globular to slightly lenticular and ranges from large cobbles to small boulders. It may be possible to subdivide this into several types once more is known about its geological context or it is related to a documented source. In any case it does not include chalcedenous materials or the so-called pointed chert that is associated at Solvieux exclusively with Middle Magdalenian levels (Sackett 1999: 136; see also Demars 1982: 87).

The third type, designated Bergeracois, occurs in all three levels and is a distinctive chert raw material that takes its name from a well-documented source area just northeast of the town of Bergerac on the Dordogne River (Demars 1982; Seronie-Vivien and Seronie-Vivien 1987; Blades 2001). It occurs as nodules in secondary geological contexts and has been recovered in numerous archeological sites throughout the Perigord (Bricker 1975; LeBrun-Ricalens 1996; Lucas 2000; Nespoulet 2000). where it is most frequently characterized as a non-local material. This high quality chert is typified by colorful banding in the chert body that occurs in combinations of lavender, yellow, pale blue, brown, tan, and red colors that cut through the core of the nodule (Larick 1983). Nodules themselves are

regular in shape, either globular or lenticular in form, and range from cobble to boulder-sized. Cortex is thin and smooth with an abrupt transition between the cortex and the chert body. Demars (1982: 88) reports the presence of a gray variety in secondary deposits on the plateau near Neuvic, *ca.* 10 km upstream from Solvieux. Significantly, Bergeracois chert is distinguished from other Maestrichtian cherts by the presence of *Orbitoides media* fossils, a type of foram that is associated specifically with the uppermost part of this final Cretaceous limestone formation (Demars 1982: 87–8; Seronie-Vivien and Seronie-Vivien 1987: 79–80 and plates 37 and 38; Lucas 2000: 37; Blades 2001: 74).

We have identified both the classic multicolor and gray varieties of Bergeracois chert in the Early Magdalenian at Solvieux where they occur in small but nonetheless significant quantities.[6] They were knapped on the site and are represented by tools, cores, and associated debitage. Moreover, based on the presence of *Orbitoides media*, we recognize these cherts in both Levels IV and M where their occurrence is more incidental compared to the Early Magdalenian. A single moderate sized nodule was worked in Level IV with a coarse but homogeneous chert body with no inclusions. Greenish in color, unlike either of the Classic Bergeracois types mentioned above, it possesses a very eroded exterior. This nodule, extensively refitted, includes a core, numerous tools, flakes, blades, and chipping debris. A second core from this nodule was apparently carried away since it was not found during excavation. Similarly, in Level M, a lenticular nodule of tan colored chert with smooth white cortex was knapped. Many debitage and tool products were recovered and refitted from this nodule. While the color is not typical, the presence of *O. media*, again, provides a basis for classifying it as Bergeracois. A second group from Level M of unrefitted flakes and blades, in a high quality gold colored chert, contains diagnostic *O. media*, so, was included in the group. Time will tell whether these minority types will hold up as true representatives of Bergeracois chert.

Our use of the term "Bergeracois" is consistent with the work of researchers who have sought to accurately tie archeological materials to geological sources (Demars 1982; Chadelle 1983; Larick 1983; Seronie-Vivien and Seronie-Vivien 1987 and others). We emphasize that our present usage departs both from our own past practice as well as that of the Solvieux Project, where brown cherts have been identified as Bergerac(ois) and attributed to the Maestrichtian

(Sackett 1999: 135). In our view, most if not all of these cherts are standard examples of beige Senonian varieties that are common across the Perigord (Urbano and Cole 2005). Certainly, we have not identified *Orbitoides media* in any we have examined. Gaussen's (1980: 88) report of classic Bergeracois nodules eroding out of Maestrichtian-derived deposits along the Isle near Mussidan, is tantalizing but inconclusive without microfaunal verification. Yet it is well established that *O. media* are generally absent from the early Maestrichtian, so an association between some brown Solvieux cherts that lack *O. media* and these earlier deposits remains a possibility. Indeed, Seronie-Vivien and Seronie-Vivien (1987: 79) reports non-Bergeracois type cherts from Maestrichtian deposits in the nearby valley of the Crempse. However this ultimately develops, the vast majority of brown / beige cherts from Solvieux are clearly still in the Senonian category.

This clarification is important when it comes to understanding changing patterns in raw material acquisition and use in the three levels being studied. Bergeracois cherts clearly played a much more substantial role in the raw material economy of Early Magdalenian people than they did for either Beauronnian or Upper Perigordian folk. This remains true even if the Bergeracois cherts are largely local, obtained from sources in the Middle Isle Valley. The previously mentioned reports of Bergeracois chert at Sourzac (Gaussen 1980: 88), the Valley of the Crempse (Demars 1982: 87; Seronie-Vivien and Seronie-Vivien 1987: 79), and the plateau above Neuvic (Demars 1982: 88) suggest that some Early Magdalenian sources were more distant (across the river), and located in different biogeographical zones (uplands), in contrast to local valley slopes and bottoms exploited by the Level IV and M inhabitants. More distant sources such as the classic ones found northeast of Bergerac may have been used by Magdalenians whose patterns of landscape use were distinct from those of Solvieux's earlier inhabitants. These patterns remain a matter for future investigation.

Non-local cherts and related materials (e.g., a flake of smoky quartz from Level M) are present in small amounts in all three levels. Included here are several flakes of jasper in Level IV that may derive from well documented but somewhat distant upstream localities, and a handful of translucent materials in Level M that no doubt belong with the chalcedonies that are well known throughout the Perigord. No cores of chalcedony were recovered in the Level M excavation but

the presence of multiple burin spalls indicates some on-site re-tooling activity. Such materials may have been available locally even though modern sources are not known in the vicinity of Solvieux. Gaussen (1980) indicates that chalcedony is common in many of the open-air sites he excavated in the Middle Isle Valley, and the lithic inventory at Guillassou is dominated by this raw material type. Demars appears to confirm this in reports of both jasper and chalcedony finds near Mussidan (1982: 90). Yellow-colored cherts from Couche A may include examples of the oft-mentioned Gavaudun chert (Chadelle 1983; Blades 2001: 75). Additional research will be needed to clarify these associations.

QUANTITATIVE ANALYSIS OF SOLVIEUX RAW MATERIAL DISTRIBUTIONS

Here we discuss some basic patterns in the raw material distributions among three levels at Solvieux. Table 15.1 presents the distribution of blank and material types by Level. Examining the distribution of blanks, it is clear that Levels IV and M are strongly similar in overall blank proportions, while Couche A contains many more flakes relative to other blank types, and includes markedly fewer spalls. These patterns

Table 15.1 Blank and material type by level (within level percentage).

	Level		
	IV (n = 5925)	A (n = 2865)	M (n = 5677)
Blank type			
Blade	37.44	20.98	38.91
Flake	49.21	71.41	44.78
Spall	9.25	3.86	12.84
Core	0.87	1.04	1.27
Chunk	3.24	2.72	2.20
Material type			
Bergerac	0.02	35.74	0.82
Black Senonians	63.44	32.79	54.67
Brown Senonians	36.46	26.12	43.59
IVT	0	5.97	0.91

Table 15.2 Material and blank types by level (within level percentages).

	Blade	Flake	Spall	Core	Chunk
Level A					
Bergeracs	35.97	36.66	32.47	13.33	8.96
Black Senonians	32.67	30.89	37.66	70.00	64.18
Brown Senonians	25.41	26.40	28.57	6.67	20.90
IVT	5.94	6.05	1.30	10.00	5.97
Total	**788**	**2682**	**145**	**39**	**102**
Level M					
Bergeracs	0.81	1.13	0.16	0.00	0.00
Black Senonians	57.58	50.02	60.22	69.44	47.41
Brown Senonians	40.71	48.37	38.21	29.17	50.86
IVT	0.90	0.48	1.42	1.39	1.72
Total	**2209**	**2542**	**729**	**72**	**125**
Level IV					
Bergeracs	0.00	0.03	0.00	0.00	0.00
Black Senonians	62.89	65.11	49.49	76.00	71.73
Brown Senonians	36.97	34.82	44.18	24.00	28.27
IVT	0.00	0.00	0.00	0.00	0.00
Total	**2242**	**2947**	**554**	**52**	**194**

are nicely consistent with the general technological features of the Aurignacian/EUP, Perigordian and Badegoulian (or Initial Magdalenian) lithic industries, respectively, that the assemblages represent.

Table 15.1 also shows that when raw material types are grouped generally as Bergeracois types, Black Senonian types, Brown Senonian types, IVT types, and the few unidentified raw material types that make up about 1% of each assemblage are ignored, Levels IV and M are again similar in being dominated by Black and Brown Senonian types. Couche A, in contrast, is numerically dominated by Bergeracois types, but the overall distribution is nearly even among the three major raw material groups. Black Senonian types account for about twice as much of the assemblage in Level IV as do Brown Senonian types. Bergeracois and IVT types are absent or nearly so from Level IV, and present at only about 1% each in Level M. Bergeracois types are the most common in Couche A, where IVT types make up almost 6% of the assemblage. The sample from each level is large enough so that these differences clearly cannot be attributed to stochastic effects.

Table 15.2 breaks down the blank types by raw material group and Level. In Level IV, the Black Senonians dominate all blank types, usually by about 2:1, with the exception of spalls where the Black and Brown Senonian groups are approximately even, and cores, where the Black Senonians reach almost 3:1 over the Brown group. In Level M, Black Senonians also tend to predominate, especially in cores, but generally not so disproportionately, and not at all in flakes. In both these levels there are proportionately far more Black Senonian cores than either flakes or blades, suggesting selective removal of the Black Senonian products. This pattern is also true of Couche A, which shows even more extreme differences in the opposite direction of proportions of Bergeracois and Brown Senonian types. More than a third of the debitage is from Bergeracois types, but only 10% of the cores, suggesting differential transport of the cores, with perhaps only the most thoroughly exhausted ending up at the site, implying a very different pattern of resource use and transport during this period. These suggestions, although intriguing, must be qualified by the very small sample of horizontal space that the material is derived from, compared to the vastly larger area of Couche A whose data was destroyed by fire.

If we look at mean blade and flake size by raw material and Level, blanks in Level IV show equal dimensions across raw material groups, with the exception of a single Bergeracois flake (which can be ignored). Blanks in Level M show more variation. Mean blade length in the Bergeracois ($n = 17$) and IVT ($n = 19$) groups is significantly longer than in the Black and Brown Senonian ($n = 2059$) groups. Oddly, perhaps, mean flake length on Bergeracois types ($n = 28$) is significantly shorter than the others, but mean IVT flake length ($n = 12$) is significantly longer than among the Senonian groups. Given the very large differences in sample size, this seems as likely to be a sample size effect as much as anything else, but the differences are statistically significant at $p < 0.01$. Couche A shows a very different pattern. Brown Senonian blades ($n = 154$) are the only standout, being significantly ($p < 0.01$) longer than any other combination. All other material and blank combinations show no significant differences.

Cortex was coded in five categories: (1) none, (2) up to 25% of the surface, (3) 26–50%, (4) 51–75%, and (5) 76–100%. Again, the breakdown by Level shows some intriguing differences. In Level IV all the

Bergeracois pieces are without cortex, whereas *ca.* 60% of the Brown Senonian and 40% of the Black Senonian pieces are without cortex. Level M shows the same fundamental pattern, but the differences are much less pronounced, and there are some Bergeracois pieces with cortex. IVTs, show a cortex distribution similar to the Brown Senonians. Couche A again shows very different patterns. Nearly 30% of Bergeracois pieces show some cortex, while the Brown Senonians and IVTs are more strongly dominated by pieces absent cortex, and more than half of the Black Senonians show some cortex. Even given the subjective nature of this classification, these differences are stark.

Using blade and flake thickness as a proxy for size, and sorting by amount of cortex and Level (Table 15.3) shows that where present, Bergeracois and IVT type blanks are always thinner compared to the other material types as the amount of cortex increases, but that there is no fixed relationship between the Brown and Black Senonian groups aside from the trend towards thinner blanks as cortex decreases. Analysis of variance among the material categories and cortex groups by level shows that these differences are all statistically significant ($p < 0.01$).

Consistently across all the levels Black Senonian raw material types show good evidence for substantial

Table 15.3   Average flake and blade thickness (cm) by level, material type, and cortex group.

	Cortex				
Material	**None**	**1–25%**	**26–50%**	**50–75%**	**75–100%**
Level A					
Bergerac ($n = 993$)	0.62	0.83	0.85	0.80	0.77
Black Senonians ($n = 851$)	0.67	0.93	1.12	1.13	1.28
Brown Senonians ($n = 712$)	0.77	0.99	0.94	1.19	0.95
IVT ($n = 164$)	0.84	0.79	0.91	0.68	0.72
Level M					
Bergerac ($n = 45$)	0.52	1.00	0.00	0.00	0.40
Black Senonians ($n = 2436$)	0.50	0.66	0.80	0.81	0.90
Brown Senonians ($n = 2059$)	0.56	0.72	0.86	1.12	0.91
IVT ($n = 31$)	0.57	0.70	0.68	0.00	0.53
Level IV					
Bergerac ($n = 0$)	0.00	0.00	0.00	0.00	0.00
Black Senonians ($n = 3295$)	0.54	0.74	0.95	1.05	1.15
Brown Senonians ($n = 1836$)	0.56	0.74	0.87	0.79	0.87
IVT ($n = 0$)	0.00	0.00	0.00	0.00	0.00

amounts of on-site reduction, and often preferential use for blade blanks. Only in Couche A does the selection of Bergeracois raw material types surpass the Black Senonian types as blade blanks, and those differences are small. Bergeracois, in contrast, consistently shows proportionately fewer blanks with cortex in all categories, even in Couche A where it is the predominant material used for blades and flakes. Brown Senonian types also consistently show a smaller proportion of cortex bearing blanks, as do the IVT types, despite their overall rarity. This pattern, combined with the consistently small blanks as cortex percentage increases in Bergeracois and IVT raw materials, suggest strongly that frequently they are being reduced off-site, although not far enough away to eliminate all primary reduction, and/or reduction of cores on the site itself. Changes in cortex percentages on flakes and blades, especially from almost no cortex on Bergeracois materials in Levels IV and M, to percentages more typical of the other materials in Couche A, imply substantial changes in procurement and transport patterns. Combining these patterns with data from refitting, as discussed above, suggests that Levels M and IV can be seen as palimpsests of numerous short-term uses of the site (Grimm and Koetje 1992) that involved intensive processing of primarily local cherts. Couche A, however, seems to represent a distinct type of occupation at Solvieux, in which there is extensive evidence for utilization of non-local raw materials coupled with more off-site processing. This is a conclusion that is not compromised by sampling problems with the Couche A materials.

CONCLUSIONS

While the story at Solvieux is one of overwhelmingly local raw material utilization, even at this fairly coarse level, there are obvious differences in the choices being made for particular materials and products. The Solvieux artifacts also demonstrate that there is substantial variation in the patterns of raw material distribution and use on a regional level, which can be particularly well contrasted with the patterns recovered at nearby "classic" cave and rockshelter sites that show very different evidence and proportions of extra-local exotics. For example, many Aurignacian industries exhibit more variability in raw materials types as well as strong evidence of long distant transport in comparison to the Beauronnian (Morala 1984; LeBrun-Ricalens 1996; Lucas 2000; Blades 2001). Indeed, the raw material use patterns we describe here reinforce conclusions from typological studies (Sackett 2001) and further underscore the uniqueness of the Beauronnian industry in comparison to both the Aurignacian and Chatelperronian. To us this suggests that we are just beginning to understand the complexity of both lithic resource utilization and human movement in this area during the Upper Paleolithic, and that whatever else might be said, the story is not a simple one.

A range of issues will be relevant in future research on raw material use patterns at Solvieux, and these investigations will be more efficacious with additional field surveys in the Middle Isle Valley. The possibility that superimposed palimpsests might have an homogenizing effect on data and mask variability that may be manifested in short-term habitations has long been recognized (Binford 1982; Grimm and Koetje 1992; Blades 2001: 179). This is a potential problem for our analysis that may be reflected in some of the quantitative results reported here since all three of our study samples may conform, in slightly different ways, to this situation. For example, a recent re-analysis of Level M (Grimm *et al.* 2006) reinterprets its spatial distribution to represent palimpsests of overlapping, short-term hunting camps in the center of the Level M exposure in eastern Solvieux; also, stratigraphic evidence from western Solvieux suggests that disturbance of the Beauronnian deposit, especially in the area of the probable hearth feature, is due to solifluction; lastly, Couche A, which separates into multiple lenses at the northern limit of its extent, varies in thickness between 15 and 20 cm in some areas and up to 30 cm in others (Sackett 1999: 81–2) and thus could easily incorporate multiple habitations. Insight into these issues of variability will come from future analyses that concentrate specifically on the refitted data from these levels. Finally, methodological and theoretical frameworks that focus attention on the social contexts of tool production and use in relation to particular raw materials will, no doubt, contribute to our growing understanding of technological change over time (Grimm 2000; Fisher 2006).

ENDNOTES

1 Without permission, Tankersley (2002: 166) published an image of the refitted core shown in Plate (1), having obtained it, apparently, from a mutual acquaintance. Inexplicably, he identifies the ensemble as Clovis-like, Solutrean artifacts from the Iberian Peninsula.

2 In Larick's view (1983: 87) Raysse burins – with their multiple, flat-faced, tiny removals – and Solutrean points, are the best tool candidates for heat treatment in the Upper Paleolithic. Grimm thinks the fine, bifacial thinning that is seen on the butts of Gravette armatures would be similarly facilitated by the use of this technique.

3 Grimm met individually with Brooke Blades and Roy Larick in 2003 and 2004, respectively, to review a type collection and discuss how to best relate Solvieux cherts to standard geological sources. Grimm gratefully acknowledges their guidance while taking full responsibility for any errors in implementation. Digital images of Solvieux chert types can be viewed on Grimm's website at http://www.oberlin.edu/anthropo/MainPages/faculty.html

4 Seronie-Vivien and Seronie-Vivien (1987: 79) describe a type of Maestrichtian chert (type c-2) from the same area but our IVT type seems to be distinguished from it by its highly translucent character.

REFERENCES

Binford, L.R. (1982) The Archaeology of place. *Journal of Anthropological Archaeology* 1, 5–31.

Blades, B. (2001) *Aurignacian Lithic Economy: Ecological Perspectives from Southwestern France.* Kluwer Academic/Plenum Publishers, New York.

Bricker, H.M. (1975) Provenience of flint used for the manufacture of tools at the Abri Pataud. In: Movius, H.L. Jr., (ed.) *Excavations at the Abri Pataud*, American School of Prehistoric Research Bulletin 30, 194–7.

Chadelle, J.-P. (1983) *Technologie et utilization du silex au Périgordien supérieur: l'exemple de la couche VII du Flageolet I.* Memoire en vue de l'obtention du diplome de École des Hautes Études en Sciences Sociales, Toulouse.

Demars, P.-Y. (1982) *L'Utilisation du Silex au Paléolithique Supérieur: Choix, Approvisionnement, Circulation.* Cahiers du Quaternaire No. 5. CNRS, Paris.

Fisher, L.E. (2006) Blades and microliths: Changing contexts of tool production from Magdalenian to Early Mesolithic in southern Germany. *Journal of Anthropological Archaeology* 25, 226–38.

Gaussen, J. (1980) *Le Paleolithque Supérieur de Plein Air en Perigord. XIVe Supplement à Gallia Préhistoire.* Editions du CNRS, Paris.

Geneste, J.-M. and Rigaud, J.-Ph. (1989) Matières premières lithiques et occupation d'espace. In: Laville, H., (ed.) *Variations des Paléomilieux et Peuplement Préhistorique*, CNRS, Paris; *Cahiers du Quaternaire* 3, 205–18.

Grimm, L. (1983) Systematic heat treatment of flint in the Upper Perigordian at Solvieux. *Paper Presented at the 48th Annual Meeting*, Society for American Archaeology, Pittsburgh, PA.

Grimm, L. (2000) Apprentice flintknapping: Relating material culture and social practice in the Upper Paleolithic.

In: Derevenski, S.J., (ed.) *Children and Material Culture*, Routledge, New York, pp. 53–71.

Grimm, L. and Koetje, T. (1992) Spatial patterns in the Upper Perigordian at Solvieux: Implications for activity reconstruction. In: Hofman, R. and J. Enloe, (eds) *Piecing Together the Past*, BAR International Series 578, Oxford, British Archaeological Reports, pp. 264–86.

Grimm, L., Seidel, A., and Koetje, T. (2006) Ties that bind: chronological applications of lithic refitting. *Paper Presented at the 71st Annual Meeting of the Society for American Archaeology*, San Juan, P.R.

Larick, R. (1983) *The Circulation of Solutrean Foliate Point Cherts: Residential Mobility in the Perigord.* PhD Dissertation, Department of Anthropology, S.U.N.Y. Binghamton.

LeBrun-Ricalens, F. (1996) L'occupation d'Hui à Beauville, Lot-et-Garonne: Activities domestiques en plein air à l'Aurignacien ancien. In: *La Vie Préhistorique.* Éditions Faton, pp. 80–5.

Lucas, G. (2000) *Les Industries Lithiques du Flageolet I (Dordogne): approche économique, technologique, fonctionnelle et analyse spatiale.* Thèse de Doctorat, Préhistoire et Géologie du Quaternaire, Université de Bordeaux 1, 2 Vols.

Morala, A. (1984) *Périgordien et Aurignacien en Haut-Agenais: Etude d'ensemblers lithiques Memoire en vue de l'obtention du diplome de École des Hautes.* Études en Sciences Sociales, Toulouse.

Nespoulet, R. (2000) Le Gravettian final de l'abri Pataud, Les Eyzies-de-Tayac, Dordogne. *L'Anthropologie* 104, 63–120.

Sackett, J. (1999) *The Archaeology of Solvieux.* Institute of Archaeology, UCLA, Los Angeles, Monumenta Archaeologica 19.

Sackett, J. (2001) The Beauronnian: An early Upper Paleolithic industry from Solvieux. In: Hays, M.A. and Thacker, P.T. (eds) *Questioning the Answers*, BAR International Series 1005, 65–77.

Seronie-Vivien, M. and Seronie-Vivien, M.R. (1987) *Les Silex du Mésozoïque nord-aquitain: approche géologique de l'étude des silex pour servir à la recherche préhistorique.* Bordeaux: Supplement au tome XV du Bulletin de la Société Linéenne de Bordeaux).

Taranik (Grimm), L. (1977) *The Beauronnian at Solvieux.* PhD Thesis, Department of Anthropology, UCLA.

Tankersley, K. (2002) *In Search of Ice Age Americans.* Gibbs, Smith Publisher, Salt Lake City.

Texier, J.-P. (1986) L'occupation Paleolithique du basin de l'Isle: ses relations avec les resources en matieres premieres et la geomorphologie. In: Laville, H. (ed.) *Variations des Paléomilieux et Peuplement Préhistorique*, CNRS, Paris; *Cahiers du Quaternaire* 3, 119–22.

Urbano, L. and Cole, S. (2005) Chertman: A model to simulate the procurement of lithic raw material assemblages for Paleolithic archeological sites using a random agent. http://des.memphis.edu/lurbano/vpython/Research/chertman/chertman.html (accessed October 10, 2006).

Part IV

HOMINID COGNITION, ADAPTATION, AND CULTURAL CHRONOLOGY

Chapter 16

INFERRING ASPECTS OF ACHEULEAN SOCIALITY AND COGNITION FROM LITHIC TECHNOLOGY

Ceri Shipton[1], *Michael D. Petraglia*[1], *and Katragadda Paddayya*[2]

[1]Leverhulme Centre for Human Evolutionary Studies, Department of Biological Anthropology, University of Cambridge
[2]Department of Archaeology, Deccan College, India

ABSTRACT

In this chapter we use data on Acheulean biface technology from the Hunsgi-Baichbal valley to suggest some socio-cognitive capabilities of Acheulean hominins. We examine raw material selection and transfer distances for ten sites in the valley and we analyze metric measurements on bifaces from the site of Isampur Quarry. Our data portray hominins who were flexible in their approach to biface manufacture, who transmitted biface technology through imitation, who were capable of creating and differentiating two distinct tool types, and who planned the creation and use of bifaces for the day ahead.

INTRODUCTION

This chapter aims to outline some of the cognitive and communicative capabilities of Acheulean hominins using the metric dimensions of Acheulean tools, raw

Lithic Materials and Paleolithic Societies, 1st edition. Edited by B. Adams and B.S. Blades, ©2009 Blackwell Publishing. ISBN 978-1-4051-6837-3.

material selection, and the transfer of artifacts across the landscape. The homogeneity of the Acheulean is perhaps its most striking feature, as the same basic tool types are manufactured for well over a million years in East Africa, while they are also found as far afield as North Wales and South India. The stasis of the Acheulean has led researchers to write-off these hominins as lacking the capacity to innovate (Mithen 1996), while some even go so far as to suggest that biface form is the product of motor habit pattern and not intentional design (Noble and Davidson 1996). This chapter will surmise that the communicative and cognitive powers of Acheulean hominins are considerably more sophisticated than is traditionally believed.

The Acheulean phenomenon

The Acheulean is a technocomplex which first appears in the East African Great Rift Valley around 1.7 million years ago (hereafter mya) (Roche and Kibunjia 1994). Its diagnostic lithics are handaxes: tear-drop shaped bifaces; and cleavers: bifaces with broad bits as their tips. Experimental, archeological, and micro-wear studies have shown that the principal function of handaxes was in all likelihood butchery (Jones 1980; Mitchell 1997; Pitts and Roberts 1997; Roberts and Parfitt 1999), although they may have been used on vegetable matter as well (Keeley 1980). The function of cleavers is more elusive, as their name suggests they may also have been involved in butchery, given the broad bits at their tips it seems their function was to slice or cleave some material, be it animal, plant matter, or both.

The oldest non-African Acheulean site, 'Ubeidiya, occurs in the Jordan Valley at 1.4 mya (Goren-Inbar and Saragusti 1996). Traces of this initial dispersal of the Acheulean have also been found in India, where the site of Isampur Quarry has been preliminarily dated to 1.1 mya (Blackwell *et al.* 2001). After 600 kya it is possible to discern a new facies of the Acheulean (Wynn 2002), associated with the species *Homo heidelbergensis*. This Late Acheulean is characterized by highly symmetrical artifacts (Wynn 2002), widespread use of prepared core techniques (Clark 1994; Debono and Goren-Inbar 2001; White and Ashton 2003), and the earliest tentative evidence for symbolism (Oakley 1981; Wymer 1982; Bednarik 1995, 1996; Mania 1990). For the first time the Acheulean is found in Europe during this period, even as far north as Britain (Gamble 1995; Pitts and Roberts 1997). During the 1.25 million years or so of the Acheulean, hominin brain size undergoes its biggest expansion, increasing from around 850 cc in species such as *Homo ergaster*, to 1300 cc in species such as *Homo heidelbergensis* (Ruff *et al.* 1997), with body size remaining largely the same. While this trend is by no means uniform, there nonetheless appears to be a marked evolutionary

trajectory for increasing encephalization during the Acheulean. Understanding the behavior of Acheulean hominins is therefore crucial to understanding human cognitive evolution.

The Indian peninsula

The geography of peninsula India is dominated by the Deccan Plateau, a vast upland area stretching from west to east and constituting the majority of the geological provinces. The peninsula shield is comprised of cratonic blocks, ranging in age from Archaean to Proterozoic, overlain by the younger cover of Deccan volcanic rocks (Korisettar 2004). Fluvio-lacustrine formations have further shaped the landscape during the quaternary. While the coast is very wet, inland peninsula India is in the rain shadow of the Ghat Mountains which run continuously along its western edge and more sporadically along the eastern side. Rainfall is seasonal coming with the monsoons, which vary greatly in intensity making the area drought prone. Peninsula India is thus left high and dry, giving it a very similar climate to East Africa, this, in all likelihood, is the principal explanation for the extremely dense occupation of Peninsula India during the Acheulean. In the Vindhayan range on the border of Peninsula India, 94 Acheulian sites were found in an area of 175 km² (Jacobson 1985), in the Ghataprabha and Malaprabha basins, tributaries of the Peninsula's largest river, the Krishna, 74 Lower Paleolithic sites have been identified (Pappu and Deo 1994), while in the Kortallayar basin in Tamil Nadu in the southeast around 200 Acheulean localities have been found (Pappu 1996).

THE HUNSGI AND BAICHBAL VALLEYS

The Hunsgi-Baichbal basin is located in the physiographical region of the Deccan Plateau and in the political district of Gulbarga, Karnataka, roughly in the center of peninsula India (Paddayya 1982). The basin is of Tertiary age covering an area of *ca.* 500 km² and consists of the two valleys of the Hunsgi and Baichbal rivers, separated by a narrow remnant of a shale-limestone plateau. The basin is flanked on its southern and western sides by shale-limestone plateaux and capped by volcanic basalts of the Deccan Trap. Seep springs emanate from the junctions of shale and limestone beds, forming shallow perennial streams. Thick travertine deposits indicate that the seep springs were present in the Middle Pleistocene (Szabo *et al.* 1990). The valleys are located in a semi-arid tract of peninsular India, with a mean annual precipitation of 600 mm largely falling during the monsoon season. The presence of perennial springs may have permitted and encouraged human occupation in the basin since the earliest times (Paddayya *et al.* 2002).

Over 200 Acheulean localities have been found in the basin making it one of the densest Lower Paaeolithic occupation areas in India (Paddayya *et al.* 2002). In accordance with the overall South Asian pattern (Korisettar and Petraglia 1993; Pappu and Deo 1994), Acheulean sites invariably occur close to ancient springs and paleo-streams. The Acheulean sites appear to be Middle Pleistocene in age, with some exceeding the uranium series maximum range of 350 kya (Szabo *et al.* 1990) and with a preliminary date of 1.1 mya for the Isampur Quarry site. The 10 sites used in this analysis are: Isampur Quarry (see below); Mudnur VIII – a possible cache site where several large bifaces were found with no other artifacts in the vicinity; Yediyapur I, Yediyapur IV and Yediyapur VI – where bifaces were commonly produced on quartzite

flakes as in the majority of South Asian Acheulean industries; Hunsgi II and Gulbal II – where limestone was used to make bifaces; Hunsgi V – a very large site yielding around 150 bifaces made on limestone flakes; Fatephur V – a collection of multiple surface localities, probably varying considerably in age; and Mudnur X – an assemblage of small, extensively flaked bifaces, which may described as Late Acheulean.

Isampur Quarry

The construction of an irrigation canal in the Hunsgi valley involved the removal of 2–3 m of brown and black silt topsoil from the Isampur Quarry locality, exposing the upper levels of the archeology. The site occurs on a siliceous limestone pediment on the outer edge of a paleodrainage tract (Paddayya *et al.* 2002). During the Pleistocene the limestone beds would have occurred on the surface and a spring existed at the locality (Paddayya *et al.* 2000). The limestone bedrock weathers in a predictable way leaving joint-bounded bedding plane surfaces of thicknesses varying from 2 to 20 cm (see Figure 16.1). The weathered form of

Figure 16.1   The bedrock at Isampur Quarry. Note the variation in thickness of the slabs in the upper part of the picture. Slab thickness ranges from 20–165 mm (*Source*: Shipton 2003).

the limestone makes it easy to procure and its highly siliceous nature gives it excellent Hertzian flaking properties, so that it is an ideal material for tool manufacture (Petraglia *et al.* 1999). Hominins may have been drawn to the site by the overlapping resources of the limestone bedrock and the perennial water supply (Petraglia *et al.* 1999).

The limestone bedrock was extracted at Isampur by smashing away a piece of a slab at the intersection of natural joints. This would then allow the slab to be pried up for further manipulation, evidence for which activity is provided by the fact that many limestone slabs at Isampur were found inclined at 20° to 45° rather than their natural horizontal position (Petraglia *et al.* 1999). Based on observations of modern limestone quarries and experimental reduction studies it is inferred that extracting the bedrock slabs at Isampur would have required multiple individuals to work cooperatively (Petraglia *et al.* 2005).

Limestone cores in various stages of working, a variety of hammerstones and a large amount of debitage pieces of various sizes indicate that the site was a lithic reduction area (Paddayya *et al.* 1999). Over 15 500 artifacts have been recovered from the site from various stages of manufacture allowing for a reconstruction of the reduction sequence. Experimental work has shown that handaxes were made by reducing thinner slabs of bedrock (see Figure 16.2), while cleavers were made from large side-struck flakes taken off thicker slabs (see Figure 16.3) (Petraglia *et al.* 1999, 2005).

INTER-SITE ANALYSIS AND RESULTS

Here we will analyze data from the 10 sites detailed above with a particular focus on Isampur Quarry. Three different analyses will be used: metric measurements of the bifaces from Isampur Quarry; raw material selection at all 10 sites, and raw material transfer distances for all 10 sites in the sample.

Biface metrics

The following metric measurements were taken in millimeters on the Isampur biface assemblage from drawings of the bifaces: maximum length, maximum breadth, maximum thickness, butt width, mid-width,

Figure 16.2   An early stage handaxe from Isampur Quarry. The original surface of the bedrock slab from which it was made is retained in the center of the artifact on both surfaces. Handaxes were made at this site by reducing the surface area of thinner slabs to leave the tear-drop outline (*Source*: Shipton 2003).

and tip width. The number of flake scars on each surface of the Isampur bifaces was also counted.

In studies of East African Acheulean bifaces it has been found that butt breadth is the most conservative of dimensions (Crompton and Gowlett 1993; Noll 2000). In order to determine if this pattern occurs at Isampur all four breadth measurements were compared on four different measures of variation: the standard error of the mean; the variance, the range, and the coefficient of variation (see Table 16.1).

Longer East African bifaces tend to be relatively narrower than shorter ones (Isaac 1977; Clark and Kurashina 1979; Crompton and Gowlett 1993; Noll 2000), with the breakpoint in isometry occurring at around 160 mm. Maximum length was plotted against

Figure 16.3 A core from Isampur Quarry displaying a large side-struck flake scar (*Source*: Shipton 2003).

Table 16.1 A comparison of biface breadth variation.

	Max. breadth	Tip width	Mid width	Butt width
Standard error of mean	2.798	3.028	2.79	2.072
Variance	665.461	779.393	661.662	365.038
Range	137	163	117	82
Coefficient of variation	0.302	0.399	0.268	0.242

Source: Shipton (2003).

maximum breadth and a Lowess regression line was fitted to see the relationship between the major biface dimensions at Isampur (see Figure 16.4). Figure 16.4 shows that longer Isampur bifaces are relatively narrower than shorter ones with the breakpoint in isometry occurring at 160 mm, thus replicating the East African pattern.

To measure the extent of standardization within the Isampur assemblage, correlations were determined between the major dimensions of the bifaces. Regression analysis showed that maximum length was highly correlated both with maximum breadth and width, the relationships both having p-values of less than 0.001. Maximum breadth and maximum

thickness were not significantly related at all however, with a p-value of 0.264.

Using multivariate statistics it is demonstrated that handaxes and cleavers are distinct tool types. A discriminant functions analysis was conducted using all of the metric variables taken from the Isampur Quarry bifaces and the flake scar count, in order to determine if it is possible to differentiate handaxes and cleavers using these variables. Since there are two groups one function was extracted with an eigenvalue of 1.475. A Wilks' Lambda value of 0.404 was obtained for this function, which has a p-value of less then 0.001. This shows that the population means for handaxes and cleavers are significantly different, so that they can be referred to as two separate

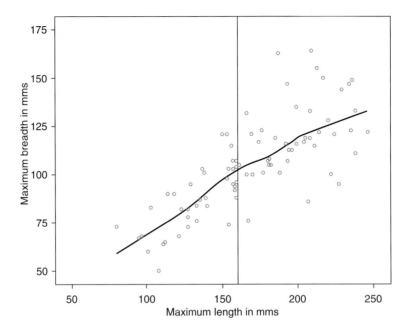

Figure 16.4   Maximum length plotted against maximum breadth for Isampur bifaces. The reference line is at 160 mm, note the breakpoint in isometry which occurs at this point (*Source*: Shipton 2003).

Table 16.2   Structure matrix multiplied by the functions at group centroids (given in parentheses) for handaxes and cleavers.

	Structure matrix	Handaxes (–0.841)	Cleavers (1.681)
Tip width	0.56	–0.47096	0.94136
Max. thickness	–0.396	0.333036	–0.665676
Max. length	–0.159	0.133719	–0.267279
Max. breadth	0.143	–0.120263	0.240383
Butt width	–0.123	0.103443	–0.206763
Flake scar count	0.087	–0.073167	0.146247
Mid width	0.049	0.041209	0.082369

Source: Shipton (2003).

tool types. The structure matrix (Table 16.2) confirms that handaxes and cleavers are principally differentiated by tip width, with the functions at group centroids illustrating that cleavers have much wider tips, a difference which is directly attributable to their differing functions. The second most powerful discriminating variable is maximum thickness, with handaxes being thicker than cleavers. This difference arises because handaxes are made on slabs while cleavers are made on flakes, which tend to be thinner.

Raw material selection

Bifaces, cores, and hammerstones across the entire basin were broken down by raw material type to determine if hominins were targeting different raw materials for different purposes (see Figures 16.5 and 16.6). Figure 16.5 illustrates that over 70% of bifaces are made on limestone, while the rest are made on chert, granite, dolerite, and quartzite. As would be expected the cores that bifaces are made from mirror this pattern

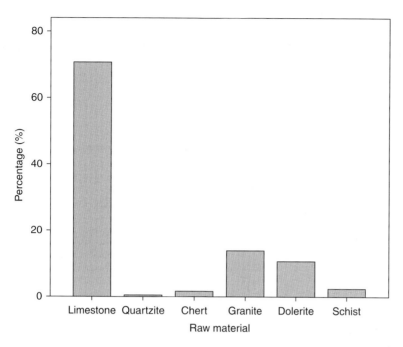

Figure 16.5 Bifaces by raw material for all 10 sites in the sample (*Source*: Shipton 2003).

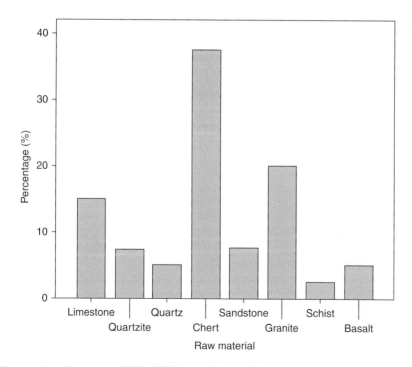

Figure 16.6 Hammerstones by raw material for all 10 sites in the sample (*Source*: Shipton 2003).

closely. Hammerstones by contrast, which perform a different function, have a remarkably different raw material distribution, with chert predominating and two raw materials: sandstone and basalt, exclusively used as hammerstones (see Figure 16.6).

Focusing on Trench 1 at Isampur this analysis determined if hominins were targeting limestone slabs of different thickness for different purposes. In this analysis the thicknesses of handaxes that preserve cortex parallel to the bedding plane on both their dorsal and ventral surfaces, were compared with the thicknesses of cores and natural slabs. Figure 16.7 illustrates that slabs of under 40 mm thickness are generally left by hominins, the great majority of those between 40–88 mm thickness are used for manufacturing handaxes, while those thicker than 88 mm are used as cores. Kolmogorov-Smirnov tests showed that the data were not normally distributed, therefore Mann-Whitney U tests were performed to test for significant differences between the group means. The results showed that the difference between the thickness of slabs used for handaxes

and those used as cores was significant with a *p*-value less than 0.001. The difference between slabs used as handaxes and those left unworked was also significant with a *p*-value of 0.014. To create a large side-struck flake suitable for cleaver manufacture requires the thickest slabs, so it is these which have been used as cores. Medium thickness slabs are good for whittling down to leave a handaxe effective at cutting, while the thinnest slabs are liable to break during manufacture therefore they tended to be ignored by hominins.

Raw material transfer distances

Recently several studies have recognized the potential significance of raw material transfer distances in reconstructing hominin behavior (Feblot-Augustins 1997; Marwick 2003). Adapted from Paddayya *et al.* (1999) and using the published literature on the basin (Paddayya 1982, 1989, 1999), a map was created showing the geology of the basin with the 10 sites in

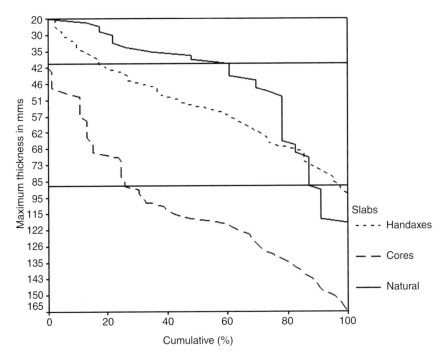

Figure 16.7 Cumulative percentages of the thickness of natural slabs, slabs used to make handaxes, and those used as cores. Note that most remaining natural slabs are under 40 mms thick, most handaxes are between 40–88 mms thick and most cores are thicker than 88 mms (*Source*: Shipton 2003).

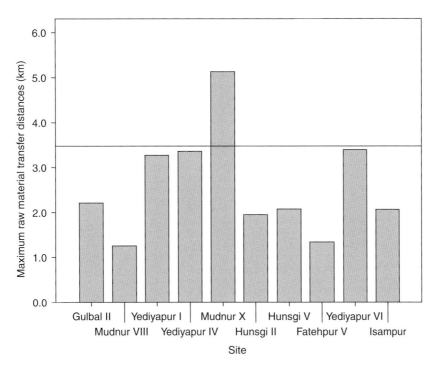

Figure 16.8   Maximum raw material transfer distances in kilometers by site. The reference line is at 3.5 km (*Source*: Shipton 2003).

the analysis plotted on it (see Figure 12.6 in Petraglia *et al.* 2005). From this map, as-the-crow-flies distances were calculated between each site and the relevant raw material sources. Accurately pinpointing a raw material source for a particular tool is extremely difficult without expensive and time-consuming chemical analyses and it would be fair to say that such studies are in their infancy. These measurements are inevitably underestimates as they assume a straight-line trajectory for the movement of tools, they assume the nearest geological source was the one used and they assume that the underlying geology outcrops everywhere. Despite these limitations, this and other raw material studies are interesting for the comparative data they contain.

No raw materials were used at any site that were not locally available in the same valley. Mudnur X, the one discernibly Late Acheulean site in the sample, stands out clearly above the rest for both mean and maximum transport distances (see Figure 16.8). Bifaces were transported much further across the landscape than either cores or hammers, suggesting functional differences in landscape use (see Figure 16.9).

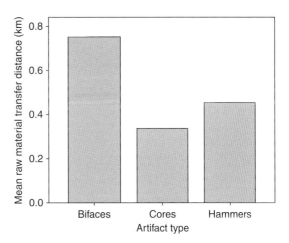

Figure 16.9   Mean raw material transfer distances in kilometers, by artifact type (*Source*: Shipton 2003).

DISCUSSION

As illustrated in Table 16.1, butt breadth is the most conservative of all breadth measurements at Isampur on any index of variation, thus reproducing the East African pattern. It has been suggested that this pattern may result from constraints of hand size (Gowlett and Crompton 1994).

Figure 16.4 shows that longer bifaces tend to be relatively narrower than shorter ones with the breakpoint in isometry occurring at 160 mm as at Olorgesailie (Noll 2000). It has been claimed that this pattern simply reflects different degrees of rejuvenation (McPherron 2000), however, given the low levels of rejuvenation in certain Acheulean assemblages (Petraglia et al. 1999), this explanation seems highly unlikely. A more plausible explanation is that width was sacrificed in longer bifaces in order to keep the weight at a wieldly level, as weight increases as the cube of the linear dimensions (Crompton and Gowlett 1993). The pattern of long, narrow and short, fat bifaces may then be the result of a conscious effort on the part of hominins. Overall there is considerable similarity in the relationship between length and width and the conservative nature of butt breadth between Isampur and the East African Acheulean sites. Such consistent patterning on this global scale is indicative of not only tool manufacturing norms, but also probably a highly robust mechanism.

Length–breadth and length–thickness relationships are standardized at Isampur, while the breadth–thickness relationship is not. In order to manufacture Late Acheulean bifaces, which often have congruent symmetry in three dimensions, it is necessary to hold in mind three perspectives that cannot be viewed simultaneously (Wynn 1993). Translated into sociality this ability implies the capacity of each individual to consider the perspective of several other hominins at a time. Since the breadth–thickness relationship is not standardized at Isampur, it may be that in the earlier Acheulean, hominins could only keep in mind a single spatial relationship at a time and by inference were not capable of multiple order theory of mind.

It is suggested that Acheulean hominins followed rigid patterns of behavior due to limitations of cognitive fluidity (Mithen 1996). The raw material selection data for the Hunsgi-Baichbal valleys (Figures 16.5 and 16.6); shows that far from simply bashing two stones together, hominins had a sophisticated enough knowledge of material properties to be able to seek different raw materials depending on their intended function. The overall diversity of raw materials used by hominins indicates a flexibility among Acheulean hominins far greater than that which they are usually accorded.

It has been argued that the distinction between handaxes and cleavers is artificial and that they in fact represent extremes along a continuum of bifacial variation (Dibble 1989; Ambrose 2001). That the discriminant analysis produced significantly different population means for the two tool types indicates that handaxes and cleavers are in fact two separate tool types. Some researchers contest whether Acheulean bifaces are indeed finished tools in themselves, or whether in fact they are merely cores, flaked according to a rigid motor pattern (Noble and Davidson 1996; McPherron 2000). Given the variety of raw materials and manufacturing techniques used to produce similar shaped artifacts the world over (McBrearty 2001), we find such an explanation extremely unlikely. While most cleavers at Isampur were made on flakes, a few were made on slabs, so that the mental template of cleaver form must have existed independently of manufacturing technique in the hominin mind. That cleavers made on slabs exist at the same site where the dominant blank type is a flake reiterates the cognitive fluidity of the hominin mind, as different manufacturing procedures could be applied to create different tool types.

The selection of particular slab thicknesses at Isampur for different purposes shows that even as hominins approached the bedrock they knew what they wanted the finished tool to be (see Figure 16.7). Since handaxes and cleavers have contrasting manufacturing methods, we can also say that hominins had the planning capacity to envision an extremely complex motor sequence.

Raw material transfer distances are informative of hominin planning and communication, as moving across the landscape in a group to specific target locations would require a communally understood knowledge of the goal. The average maximum transport distance for chimpanzees is 0.5 km (Boesch and Boesch-Achermann 2000), whereas of the Acheulean sites in the sample none has maximum transport distances less than 1 km (see Figure 16.8). Acheulean planning depth can thus be said to exceed that of chimpanzees, although it is still limited in comparison to the huge transport distances achieved by later hominins (Feblot-Augustins 1997). Figure 16.9 illustrates a dichotomy between manufacturing and subsistence behavior, which may be attributed to the differing predictabilities of the resources: geological sources do not move about the landscape whereas animal prey does. The acquisition of raw materials can therefore be planned whereas the use of bifaces is likely to occur more opportunistically. While bifaces generally travel further than cores or hammers, the mean transport distance is still less than a kilometer (see Figure 16.9), suggesting hominin foraging is highly localized. Since hominins seldom carry bifaces further than 3.5 km during their use lifetimes this implies that hominins are attached to a specific range. The spatially predictable manufacturing resources (cores and hammers) are rarely transported further than 2 km, consistent with the hypothesis that the planning depth of Acheulean hominins was usually confined to a few hours ahead (Gamble 1999), though the spatial abundance of limestone in the Hunsgi and Baichbal valleys may have played a role. The Late Acheulean site of Mudnur X stands out above the rest as having longer maximum raw material transfer distances (see Figure 16.8). This is not due to increases in foraging range, but it is instead the spatially predictable items for manufacture that are being transported these greater distances. This suggests that Late Acheulean hominins may have had a deeper capacity for planning and coordinating group activity than earlier hominins.

CONCLUSION

A technocomplex that lasted for 1.25 myrs must have had a robust means of reproduction. Far from the mental templates of Acheulean tools being genetically hardwired into hominin minds, we would argue that the homogeneity of the Acheulean is evidence of a robust form of social transmission, such as true imitation. True imitation requires the imitator to understand not only the goal and the means used to achieve it, but also an understanding that the observed party has chosen to use that particular means based on their understanding of the situation (Tomasello and Call 1997). Imitation therefore requires some level of theory of mind (Meltzoff 1995). It has recently been demonstrated that the neurological substrate for imitation, known as the mirror neurone system, is found in Broca's area (Rizzolatti and Arbib 1998). Endocranial studies have demonstrated that Broca's area was present in something like its modern anatomical form in some of the earliest members of our genus (Tobias 1987; Holloway 1996). We suggest that from the outset Acheulean social interactions were characterized

by a propensity for true imitation and it is this, rather than any genetic predisposition to produce bifaces, which has produced the remarkable homogeneity of the Acheulean.

On the basis of the presence of discrete tool types it has been proposed that only Upper Paleolithic hominins had the cognitive capacity for lexicon (Dibble 1989). However here we have demonstrated that handaxes and cleavers are indeed distinct tool types with contrasting shapes and manufacturing methods. Therefore we would suggest that Acheulean hominins were capable of communicating concepts of a lexical specificity, although we remain agnostic as to the issue of whether these words were actually verbalized or whether they were gestural.

Manufacturing techniques are seen to exist independent of a specific final product, as occasionally cleavers at Isampur were made on the slab blanks predominantly used to make handaxes. This mixing and matching of concepts points to a great flexibility in Acheulean cognition, further testimony for which is provided by the wide variety of raw materials which hominins were prepared to use to make their tools. In selectionist terms such flexibility translates into adaptability, and it would have been an invaluable asset and probably a vital prerequisite for the global scale colonization achieved by Acheulean hominins.

The long term planning capacity of Acheulean hominins is indicated by the transfer of raw materials across the landscape. In the Hunsgi-Baichbal basin we can see that Acheulean hominins only used raw materials that were locally available suggesting that hominin landscape-use was localized, or that the need for tools could not be anticipated. This accords with the suggestion that the rhythms of life in the Acheulean were confined to a single day (Gamble 1999). The Late Acheulean site of Mudnur X has been seen to have notably higher raw material transfer distances, which we would suggest is tentative evidence of an increase in planning capacity during the Acheulean. Indeed in the European Late Acheulean raw material transfer distances are known to exceed 80 km (Feblot-Augustins 1997). We would suggest that the behavioral innovations which occur during the Acheulean, mentioned at the beginning of this chapter, are indicative of vast changes in the cognitive abilities of hominins during the Acheulean, which can only be confirmed or refuted by further research. However, far from earlier Acheulean hominins being automatons, here we have demonstrated that they possessed great flexibility, understanding of raw material properties, hierarchical short-term planning, and an ultra-social propensity for imitation.

ACKNOWLEDGEMENTS

The authors would like to thank Richa Jhaldyial for help in fieldwork and for providing some information on the geology of the Hunsgi-Baichbal Valley. The Allen, Meek and Read scholarship; the Prehistoric Society Tessa and Mortimer Wheeler fund; the Smithsonian Human Origins Program; the Leakey Foundation; the University Grants Commission, New Delhi; and Deccan College, Pune, are all gratefully thanked for their financial support. We thank the Government of India and the Archaeological Survey of India for permission to conduct this research and the American Institute of Indian Studies for facilitating our work.

REFERENCES

Ambrose, S.H. (2001) Paleolithic technology and human evolution. *Science* 291, 1748–53.

Bednarik, R.G. (1995) Concept-mediated marking in the Lower Paleolithic. *Current Anthropology* 36, 605–16.

Bednarik, R.G. (1996) The cupules on Chief's Rock, Auditorium Cave, Bhimbetka. *The Artefact* 19, 63–72.

Blackwell, B.A.B., Fevrier, S., Blickstein, J.I.B., Paddayya, K., Petraglia, M.D., Jhaldiyal, R., and Skinner, A.R. (2001) ESR dating of an Acheulian Quarry site at Isampur, India. *Journal of Human Evolution* 40, A3 (abstract).

Boesch, C. and Boesch-Achermann, H. (2000) *The Chimpanzees of the Tai Forest: Behavioural Ecology and Evolution*. Oxford University Press, Oxford.

Clark, J.D. (1994) The Acheulian industrial complex in Africa and elsewhere. In: Corrucini, R.S. and Ciochon, R.L. (eds) *Integrative Pathways to the Past*, Prentice-Hall, Englewood Cliffs, NJ, pp. 451–70.

Clark, J.D. and Kurashina, H. (1979) An analysis of Earlier Stone Age bifaces from Gadeb (Locality 8E), northern Bale highlands, Ethiopia. *South African Archaeological Bulletin* 34, 93–109.

Crompton, R.H. and Gowlett, J.A.J. (1993) Allometry and multi-dimensional form in Acheulean bifaces from Kilombe, Kenya. *Journal of Human Evolution* 25, 175–99.

Debono, H. and Goren-Inbar, N. (2001) Note on a Link between Acheulian Handaxes and Levallois Method. *Journal of the Israel Prehistoric Society* 31, 9–23.

Dibble, H. (1989) The implications of stone tool types for the presence of language during the Lower and Middle Palaeolithic. In: Mellars, P.A. and Stringer, C. (eds) *The Human Revolution: Behavioural And Biological Perspectives On The Origins Of Modern Humans*, Princeton University Press, Princeton, NJ, pp. 415–32.

Feblot-Augustins, J. (1997) *La circulation des matieres premieres au Paleolithique*, ERAUL 75, Liege.

Gamble, C. (1995) The earliest occupation of Europe: The environmental background. In: Roebroeks, W. and van Kolksholten, T. (eds) *The Earliest Occupation of Europe*, European Science Foundation and University of Leiden, Leiden, pp. 279–95.

Gamble, C. (1999) *The Palaeolithic Societies of Europe*. Cambridge University Press, Cambridge.

Goren-Inbar, N. and Saragusti, I. (1996) An Acheulian biface Assemblage from the Site of Gesher Benot Ya'aqov, Israel: Indications of African Affinities. *Journal of Field Archaeology* 23, 15–30.

Gowlett, J.A.J. and Crompton, R.H. (1994) Kariandusi: Acheulean morphology and the question of allometry. *African Archaeological Review* 12, 1–40.

Holloway, R.A. (1996) Evolution of the human brain. In: Lock, A. and Peters, C.R. (eds) *Handbook of Symbolic Evolution*, Oxford University Press, Oxford, pp. 74–125.

Isaac, G. (1977) *Olorgesailie: The Archaeology of a Middle Pleistocene Lake Basin in Kenya*. Chicago University Press, Chicago.

Jacobson, J. (1985) Acheulian surface sites in central India. In: Misra, V.N. and Bellwood, P. (eds) *Recent Advances in Indo-Pacific Prehistory*, Oxford-IBH, Delhi, pp. 49–57.

Jhaldiyal, R. (1997) *Formation Processes of the Prehistoric Sites in the Hunsgi and Baichbal Basins, Gulbarga District, Karnataka*. Unpublished Ph.D. dissertation, Deccan College, Pune.

Jones, P.R. (1980) Experimental butchery with modern stone tools and its relevance for Palaeolithic archaeology. *World Archaeology* 12, 153–75.

Keeley, L.H. (1980) *Experimental Determination of Stone Tool Uses: A Microwear Analysis*. University of Chicago Press, Chicago.

Korisettar, R. (2004) *Geoarchaeology of the Purana and Gondwana basins of peninsula India: peripheral or paramount*. Presidential Address, Archaeology Section, 65th session of the Indian History Congress, Mahatma Jyotiba Phule Rohilkand University, Bareilly, Uttar Pradesh, India.

Korisettar, R. and Petraglia, M.D. (1993) Explorations in the Malaprabha Valley, Karnataka. *Man and Environment*, 18, 43–8.

Mania, D. (1990) Auf den Spuren des Urmenschen: Die Funde von Bilzingsleben, Theiss, Berlin.

Marwick, B. (2003) Pleistocene exchange networks as evidence for the evolution of language. *Cambridge Archaeological Journal* 13, 67–81.

McBrearty, S. (2001) The Middle Pleistocene of east Africa. In: Barham, L. and Robson-Brown, K. (eds) *Human Roots: Africa and Asia in the Middle Pleistocene*, Western Academic & Specialist Press Limited, Bristol.

McPherron, S.P. (2000) Handaxes as a measure of the mental capabilities of early hominids. *Journal of Archaeological Science* 27, 655–63.

Meltzoff, A. (1995) Understanding the intentions of others: Re-enactment of intended acts by 18 month old children. *Developmental Psychology* 31, 838–50.

Mitchell, J.C. (1997) Quantitative image analysis of lithic microwear on flint handaxes. *USA Microscopy and Analysis* 26, 15–17.

Mithen, S.J. (1996) *Prehistory of the Mind: A Search for the Origins of Art, Religion and Science*. Phoenix, London.

Noble, W. and Davidson, I. (1996) *Human Evolution, Language and Mind: A Psychological and Archaeological Enquiry*. Cambridge University Press, Cambridge.

Noll, M.P. (2000) *Components of Acheulean Lithic Assemblage Variability at Olorgesailie, Kenya*. Unpublished Ph.D. dissertation, University of Illinois, Urbana-Champaign.

Oakley, K.P. (1981). Emergence of higher thought 3.0–0.2 Ma BP. *Philosophical Transactions of the Royal Society London B* 292, 205–11.

Paddayya, K. (1982) *The Acheulian Culture of the Hunsgi Valley (Peninsular India): A Settlement System Perspective*. Deccan College, Pune.

Paddayya, K. (1989) The Acheulian culture localities along the Fatehpur Nullah, Baichbal Valley, Karnataka (Peninsular India). In: Kenoyer, J.M. (ed.) *Old Problems and New Perspectives in the Archaeology of South Asia*, University of Wisconsin Press, Madison, Wisconsin, pp. 21–8.

Paddayya, K., Jhaldiyal, R., and Petraglia, M.D. (1999) Geoarchaeology of the Acheulian Workshop at Isampur, Hunsgi Valley, Karnataka. *Man and Environment* 24, 95–100.

Paddayya, K., Jhaldiyal, R., and Petraglia, M.D. (2000) The significance of the Acheulian Site of Isampur, Karnataka, in the Lower Palaeolithic of India. *Puratattva* 30, 1–27.

Paddayya, K., Jhaldiyal, R., and Petraglia, M.D. (2001) Further field studies at the Lower Palaeolithic Site of Isampur, Karnataka. *Puratattva* 31, 8–15.

Paddayya K., Blackwell, B.A.B., Jhaldiyal, R., *et al.* (2002) Recent findings on the Acheulian of the Hunsgi and Baichbal valleys, Karnataka, with special reference to the Isampur excavation and its dating. *Current Science* 83, 5.

Pappu, R.S. and Deo, S.G. (1994) *Man-land Relationship During Palaeolithic Times in the Kaladgi Basin, Karnataka*. Deccan College, Pune.

Pappu, S. (1996) Reinvestigation of the prehistoric archaeological record in the Kortallyar basin, Tamil Nadu. *Man and Environment* 21, 1–23.

Petraglia, M.D., LaPorta, P., and Paddayya, K. (1999) The first Acheulean quarry in India: Stone tool manufacture, biface morphology, and behaviours. *Journal of Anthropological Research* 55, 39–70.

Petraglia, M.D., Shipton, C., and Paddayya, K. (2005) Life and Mind in the Acheulean: A case study from India. In: Gamble, C. and Porr, M. (eds) *The Hominid Individual in Context: Archaeological investigations of Lower and Middle Palaeolithic Landscapes, Locales and Artefacts*, Routledge, London, pp. 197–219.

Pitts, M. and Roberts, M. (1997) *Fairweather Eden: Life in Britain Half a Million Years Ago as Revealed by the Excavations at Boxgrove*. Random House, London.

Rizzolatti, G. and Arbib, M.A. (1998) Language within our grasp. *Trends in Neuroscience* 21, 188–94.

Roberts, M.B. and Parfitt, S.A. (1999) *Boxgrove: A Middle Pleistocene Hominid Site at Eartham Quarry, Boxgrove, West Sussex*. English Heritage, London.

Roche, H. and Kibunjia, M. (1994) Les sites archeologiques plio-pleistocenes de la Formation de Nachukui, West Turkana, Kenya. *Comptes Rendus de l'Academie des sciences de Paris* 318 serie II, 1145–51.

Ruff, C.B., Trinkhaus, E., and Holliday, T. (1997) Body mass and encephalisation in Pleistocene *Homo*. *Nature* 387, 173–6.

Shipton, C.B.K. (2003) *Sociality and Cognition in the Acheulean: A Case Study on the Hunsgi-Baichbal Basin, Karnataka, India*. Unpublished M.Phil thesis, University of Cambridge, Cambridge.

Szabo, B.J., McKinney, C., Dalbey, T.S. and Paddayya, K. (1990) On the Age of the Acheulian Culture of the Hunsgi-Baichbal Valleys, Peninsular India. *Bulletin of the Deccan College Post-Graduate and Research Institute* 50, 317–21.

Tobias, P.V. (1987) The brain of *Homo habilis*: A new level of organisation in cerebral evolution. *Journal of Human Evolution* 16, 741–61.

Tomasello, M. and Call, J. (1997) *Primate Cognition*. Oxford University Press, Oxford.

White, M.J. and Ashton, N.M. (2003) Lower Palaeolithic core technology and the origins of the Levallois method in NW Europe. *Current Anthropology* 44(4), 598–609.

Wymer, J. (1982) *The Palaeolithic Age*. St. Martin's, New York.

Wynn, T. (1993) Two developments in the mind of early *Homo*. *Journal of Anthropological Archaeology* 12, 299–322.

Wynn, T. (2002) Archaeology and cognitive evolution. *Behavioural and Brain Sciences* 25, 4.

Chapter 17

QUINA PROCUREMENT AND TOOL PRODUCTION

Peter Hiscock[1], Alain Turq[2], Jean-Philippe Faivre[3], and *Laurence Bourguignon[4]*

[1]School of Archaeology and Anthropology, Australian National University
[2]Musèe National de Prèhistoire des Eyzies
[3]IPGQ-PACEA UMR 5199 du CNRS; Université de Bordeaux
[4]INRAP Direction interrégionale Grand-Sud-Ouest

ABSTRACT

Quina Mousterian lithic variability is explicated as a consequence of interactions between the flexible Quina technology and the economic context in which it was employed. Quina technology was expressed somewhat differently in each assemblage as the needs of people in specific contexts led them to emphasize different possibilities of Quina flake and core forms. Variation in Quina tools was linked to blanks produced by the distinctive core reduction strategy as well as the intensity with which tools were maintained and transported. However, despite variation that exists in Quina assemblages, it is still clear that this expresses a common technological and economic theme. Quina strategies represent particular solutions to the problems of provisioning Neanderthal foragers in the Dordogne during the Middle Paleolithic. Transporting highly extendable packages of stone, usually in the form of thick flakes from which knapper's could remove large numbers of flakes for prolonged periods, either to maintain/create working edges and/or to produce the flakes themselves, formed the central strategy of Quina behavior. The nature of core reduction and flake retouching appears to have been efficient and appropriate for producing these packages, and the pattern and directionality of material transfers indicates the structure of economic needs rather than the intellectual capacity of Neanderthals. Nevertheless, the existence of a "Quina system" which integrated a variety of procurement and technological practices to create an economic structure that was suited to a specific environmental/economic context may indicate the presence of planning in Neanderthal foragers who developed and employed it.

Lithic Materials and Paleolithic Societies, 1st edition. Edited by B. Adams and B.S. Blades, ©2009 Blackwell Publishing. ISBN 978-1-4051-6837-3.

Neanderthals are an extinct form of hominid, closely related to modern humans, who lived in Europe and the Middle East during the Pleistocene, until about 30 000 years ago (Green *et al.* 2006). The nature of their cultural systems has been much debated, and views about them have often been polarized around questions of their cognitive ability. Negative evaluations of Neanderthal cognition have been embedded in arguments about what they could not do: they were not capable of hunting large game, lacked capacity for long-term planning and organization (including social activities), lacked even the foresight of their own mortality, and did not have complex language (e.g., Binford 1989; Gargett 1989; Lieberman 1989, 1991; Stinger and Gamble 1993; Noble and Davidson 1996). More positive evaluations have been offered, but in the absence of representational art and elaborate bone and antler technologies researchers still concluded that Neanderthal were cognitively slightly less complex than modern humans (see Clark and Lindly 1989; Lindly and Clark 1990; Mellars 1996). Researchers have also employed evidence for the manufacture and transportation of stone tools as a way to reflect on the cognitive and social situation of Neanderthals, and particularly the notion that future planning was not a key characteristic of their lives. For example, studies of implement assemblages have been used to debate the existence and nature of regular tool designs in the European Middle Paleolithic (e.g., Binford 1989; Mellars 1996), while the scale at which stone was redistributed has been used as evidence of the nature of inter-group social interactions and the language/cognitive capacity of Neanderthals (e.g., Marwick 2003). While these debates have been productive we argue that interpretations of Neanderthal life are best built from an examination of the range of material residues of their behavior systems, rather than from evidence for single aspects of behavior treated in isolation. For instance, we suggest that the level of planning possessed by prehistoric hominids is difficult to assess simply from the maximum distance stones traveled. As Kuhn (2004: 446) explained, there is no need to interpret differences in the scale of raw material circulation in terms of cognitive capacities, including language use, since the transportation of stone is explicable as an economic response to patterns of mobility and site use. Distance to source is an imprecise measure of cognitive evolution and by itself cannot signal the nature of economies and social networks in the Paleolithic. Rather than base

reconstructions of Middle Paleolithic life-ways on a single trait of that kind, it is desirable to employ information on the patterns of material movements and the regularity of associated transportation and production behavior to yield a more coherent depiction of the ancient behavioral system. The task of reconstructing the behavior and cognitive patterns of tool production and use is difficult, but it begins to be possible in regions such as the Perigord where research has gradually developed robust descriptions of stone procurement and reduction.

In the Perigord current interpretations of Mousterian assemblages reflect the history of archeological research (Figure 17.1). Before exploring the central point of this chapter it is valuable to briefly review the research which led to the identification and description of the Mousterian. The Lower and Middle Paleolithic of Europe covered considerable time and space. Confronted with the diversity and volume of evidence it was necessary to create structures with which to understand these ancient periods; structures which reflected the intellectual perspectives of the researchers and depended on typological principals of tool analysis. In 1930 Denis Peyrony (1930: 176) advanced the culture hypothesis, that industrial differences signaled groups with distinct identities, a proposition later reprised and developed by François Bordes (1953, 1961, 1973) and Maurice Bourgon (Bordes and Bourgon 1951). Still later a different interpretative model explaining variability in the Mousterian was presented by Lewis and Sally Binford (1966, 1969, 1973, 1983). Their model attributed temporal and spatial patterns of tools to different functional contexts rather than to cultural differences (in the narrow sense of distinctive identities and traditions). From the same time Paul Mellars (1965, 1969, 1970, 1988, 1989, 1996) proposed, on the basis of evidence from key archeological sites such as Combe Grenal, Pech de l'Azé, and Moustier, that there was a sequence of three Mousterian facies: Ferrassie, Quina, and Mousterian of Acheulian Tradition. He took each of these industrial units to be distinct cultural entities which succeeded each other over time, in that order. Beginning in the 1980s Harold Dibble (1984, 1987a,b, 1988a,b, 1991, 1995) developed the idea that the diversity of typological tool forms within the Mousterian was the result of different degrees of reduction; an idea since pursued by other researchers (e.g., Rolland 1988; Rolland and Dibble 1990; Gordon 1993; Holdaway *et al.* 1996).

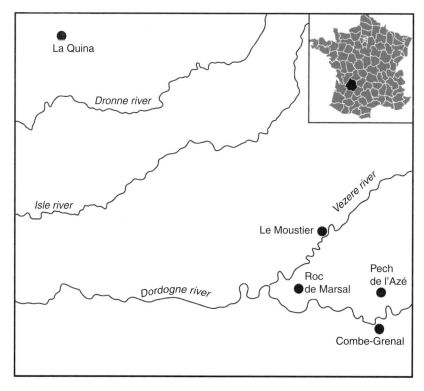

Figure 17.1   Map of the Dordogne, France, showing the location of sites mentioned in the text and the major rivers of the Periogord.

In France lithic analyses have proceeded along different lines. Since the 1950s they have increasingly focused on technology, especially the concept of the *chaîne opératoire* (Bordes 1947; Leroi-Gourhan 1956, 1966; Tixier 1978; Tixier *et al.* 1980), and since the 1980s French analysts have pursued studies of modes of lithic production in preference to typological studies (e.g., Geneste 1985; Boëda 1986). These studies were often pursued by researchers replicating ancient technologies, such as Tixier, Bordes, and Pelegrin. In this approach the sequence of technical events creating the artifacts was constructed and integrated with lithic sourcing studies to yield a new understanding of geographical dimensions to past economies (Turq 1977; Demars 1980; Geneste 1985). This step towards a techno-economic characterization of lithic assemblages was seen as providing a powerful framework for analyzing lithic material, not as privileging a particular explanatory model. The outcome of this approach has been to shift archeological concerns away from industrial classifications towards detailed explorations of the constraints and conditions in which prehistoric techno-economic choices were made and repeated. This is the viewpoint employed here.

Investigations of the preserved artifact assemblages have identified regional and chronological differences in hominid behavior. For more than a century archeologists have recognized a series of different Mousterian archeological patterns, and attempts to understand what each pattern means in terms of Neanderthal life-ways continue. This chapter examines one such pattern, the Quina Mousterian of the Dordorgne in France, and presents an interpretation of lithic procurement and production as it is now known.

Quina assemblages are conventionally described as having very high frequencies of thick scrapers with convex retouched edges, notably transverse scrapers and those with distinctive "Quina" retouch and little evidence of Levallois core reduction (Bordes 1953,

1972, 2002; Turq 2000: 24). These lithic assemblages are consistently associated with reindeer-dominated faunal assemblages and pollen sets with little arboreal signal (Chase 1986). This indicates that they were created during colder, periglacial conditions, probably the final part of OIS4 and the start of OIS3, but perhaps also at earlier cold phases (Bordes et al. 1966; Mellars 1969; Dibble and Rolland 1992; Turq 1992, 2000: 312). However the absolute chronology for Quina industries is currently poorly defined and Quina assemblages are not necessarily synchronous at all sites in the Perigord.

MATERIAL PROCUREMENT

The Perigord region, dominated by the valleys of the Dordogne and Vezere Rivers, contains a remarkable variety of flints. Middle Paleolithic hominids living in the region employed these flints as materials on which to make tools, and transported artifacts as they moved across the landscape. It is possible to identify the general source of many flint artifacts found in archeological sites because there is a well-defined pattern of geological outcrops and the flints which come from each source are often distinctive. The Perigord is the northeastern part of the Aquitaine Basin; it contains flint-bearing limestone formations of different ages (Jurassic, Cretaceous, and Tertiary), through which the main valleys have been cut. Fortunately the appearance of flint from each geological stratum is often visually distinguishable, and can be identified from inspections of color, texture, fossiliferous inclusions, cortex, and other features (e.g., Demars 1980; Morala 1983; Seronie-Vivien and Seronie-Vivien 1987; Turq 2000, 2005). Analyses of the sources of artifacts in archeological assemblages have identified patterns of Mousterian rock procurement and transport across the region (Geneste 1988, 1989; Turq 1988, 1989, 1992). These patterns can be summarized as follows:
• Flint was readily available on the ground surface, as colluvial or solifluction deposits adjoining flint-rich outcrops and deposits in the beds of rivers, and was probably collected without the need to excavate it from bedrock. Throughout most of the region these sources were for practical purposes unlimited (Turq 1988).
• In most assemblages the artifacts were made on flint obtained from the local environment, typically 95–99% come from localities within a 5 km radius of the site. In a flint rich landscape the use of locally

abundant stone resources, perhaps embedded in other economic activities, would have been an expected consequence of foraging.
• Small numbers of artifacts were carried much further, from sources up to 100 km away to the site at which they were discarded. The specimens transported over these distances were typically made only of high quality flint and they were retouched, an indication of careful selection of objects to be transported and maintained, and perhaps of rapid, logistical movements of people carrying tools (Geneste 1989) or even exchanges between neighboring groups of hominids.
• There is no evidence for transportation of stone more than 100 km in the Dordogne, even though greater transport distances are known elsewhere in Europe during the Middle Paleolithic (Féblot-Augustins 1993: 215). While this has sometimes been seen as an indication that Neanderthals in the Dordogne did not transfer material though exchange systems, it is more plausible that the maximum transfer distance is merely a reflection of the richness of this landscape in flint and that trading flint made little economic sense (Mellars 1996).
• Furthermore, the movement of material across the landscape is not random or haphazard. While some source areas were not only the origin of flint transported elsewhere but also contained sites that received non-local flint, other source areas gave material to but did not receive it from the surrounding region. These patterns reveal not only a complex set of movements across this landscape but also that the redistribution of raw material was repeatedly carried out according to a regular system which existed for millennia.
• Artifacts that were transported substantial distances, more than 20 km, were mainly flake blanks and tools that had great potential for reworking. Cores were not regularly relocated, but were reduced close to their occurrences to produce portable flakes. This process involved testing nodules by removing a few flakes and then transporting them to sites where they could be reduced. It was the distinctive Quina flake blanks that were carried over long distances and intensively resharpened and recycled (Meignen 1988).

These observations are not consistent with a simple, opportunistic pattern of resource exploitation. Instead, they indicate a highly structured system of resource use that involved regular, well-defined selectivity of items to be transported, and repeated transfers of blanks and tools across the landscape in the same directions. The most plausible interpretation of material redistributions in this region is that they were a consequence of a system of seasonally semi-permanent occupation in the Perigord, with seasonal forays to other regions and perhaps occasional exchanges with neighboring groups (Turq 2000). The regularity of archeological evidence for transportation of flint indicates that this foraging pattern, and the system for procuring and redistributing rock, was relatively stable over a long period. Additionally the system was well adjusted to supplying tools in the environmental context in which foragers were operating. A similar image of an elaborate and well-structured economic system has been obtained from investigations into the ways in which flakes for transportation were removed from cores.

FLAKE PRODUCTION

The flake production typically associated with Quina Mousterian assemblages differed from the Levallois and discoidal strategies associated with some other Mousterian industries. Quina core reduction technologies often involved the flaking of large, thick flakes from cobbles in a way that allowed cortex to be retained on many of the dorsal surfaces of many flakes, a process that has sometimes been described as "clactonian." Many of the flakes removed from Quina cores were also asymmetrical in cross-section. The regularity with which thick, asymmetrical flakes bearing cortex along a margin were created indicates the existence of systems of core reduction.

The shapes and dimensions of raw materials imported into sites varied. Some were rounded nodules, others were angular, thermally fractured pieces, while others were humanly manufactured flakes. The consequence of this variety of material forms was the production of heterogeneous collections of large flakes which share the same morphological characteristics and result from similar strategies of core reduction. Diversity in core forms and flake products stems from not only differences in the initial raw material but also from the knapping actions applied to each piece of rock being worked. However, the variation in core forms masks strong similarities in core reduction. To understand patterning in the variable forms found in assemblages, archeologists such as Turq (1988, 1989, 1992, 2000) and Bourguignon (1997) examined the knapping actions and patterns of fracture: the selection of hard hammers, the use of the hammers, the direction

in which blows were applied, and the arrangement of blows to different surfaces as alternative or successive. Examinations of the angles between surfaces and their effect on the symmetry of artifacts obtained (Bourguignon 1997) has led researchers to conclude that Quina core reduction obeys a specific logic of volumetric production, producing thick flakes with cortex, resembling Clactonian systems of core reduction (Ashton et al. 1992, 1998; Ashton 2000), trifacial systems (Boëda 1997) and SSDA (*systéme par surface de débitage alternée*) (Forestier 1993). This basic format of this core reduction is illustrated in Figure 17.2a.

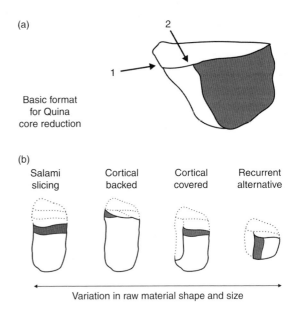

Figure 17.2 Summary of Quina core reduction. (a) Schematic cross-section of the basic format of Quina flake removal from cores, (b) Representations of inferred variation displayed in Quina core reduction [based on proposals by Turq (1988, 1989) and Bourguignon (1997)], and (c) Drawings of two cores from Combe Grenal Layer 17 (Drawings by Jean-Philippe Faivre).

Turq (1988, 1989, 1992, 2000) described some of the variations of this Quina core reduction system, singling out three strategies which together accounted for the range of flake forms observed in excavated assemblages such as Roc de Marsal where raw material occurred in elongate blocks. These strategies are schematically illustrated in Figure 17.2b, and can be summarized as follows:

• The first is the application of a series of powerful blows to the center of the natural surface of a block to remove a series of parallel flakes that progressively truncate one end of the nodule. Turq (1988, 1989, 2000), following Bourlon (1907) and Cheynier (1953), described this metaphorically as a "salami slicing" strategy, and it yielded large, thick, flakes with cortical platforms and steep cortical edges on both lateral margins.

• The second scheme also involves the application of blows to the natural surface of cobbles, but in this case they are positioned towards one side of the platform, to take advantage of prominent ridges on the core face formed by the junction of large flake scars and unflaked cortical surfaces. The result was large, thick flakes with cortical platforms that were asymmetrical in transverse cross-section and had cortex only on the steep angled lateral margin, a pattern which Turq referred to as a "cortical-backed" strategy.

• The third scheme involved stripping cortex from the platform as well as portions of the core face with a series of large flakes removed along the nodule length as well as across its end, prior to again focusing blows down ridges created by the junction of cortical and flaked surfaces. The flakes produced were often also large and thick, often asymmetrical and with cortex on one margin, but typically with a conchoidal platform. Turq (1992) named this a "covering" strategy for producing flakes with cortical backs.

Other schemes of Quina core reduction have also been proposed. For example, Bourguignon (1997) described procedures in which flakes were alternately struck from two or three platforms (Figure 17.2b). This system is suited to blocks of raw material that are roughly equi-dimensional, and could maintain similar core shapes during reduction. The flakes produced are again thick with large platforms, and with cortex on a lateral margin or distal end.

Although the strategies defined by Turq and Bourguignon represent varied procedures of core reduction they share several features. The flakes produced are often thick, with large platforms, substantial amounts of cortex on the dorsal face, and frequently have an asymmetrical transverse cross-section in which the cortex is located on the steep angled margin and the opposing low angled margin is formed by two conchoidal surfaces (Turq 2000: 318). Typical examples are presented in Figure 17.3a. As discussed below it was this conchoidal margin that was typically retouched (Turq 2000: 314). These patterns are extremely regular. For example, at sites such as Roc de Marsal and Combe Grenal as much as 70–80% of retouched flakes have cortex on their dorsal surface and the majority were often made on flakes between 1 cm and 3 cm thick, and rarely thinner than 0.7 cm. The consistency with which flakes of this form were struck from cores contributed to the standardization of tools being produced.

The core reduction strategy applied in particular cases was probably chosen to be appropriate to the size

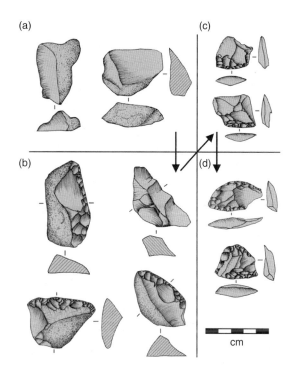

Figure 17.3 Examples of Quina flake selection and retouching from Combe Grenal (layers 17 and 20). Arrows show the sequence of reduction. (a) Unretouched flakes removed from cores, (b) Similar flakes with retouching, (c) Smaller flakes struck from retouched flakes of the kind portrayed in b, and (d) Small flakes from c that have subsequently been retouched. (Drawings by Jean-Philippe Faivre.)

and shape of the raw material nodules at hand. For instance, the salami slicing version was suited to slab- and cylinder-shaped blocks and was less commonly employed on more spherically-shaped blocks, where other strategies such as cortical-backed, covering, and recurrent alternative strategies would have been more successful. Figure 17.2b schematically depicts one dimension of the relationship between core reduction strategy and raw material form. Examples of discarded cores are shown in Figure 17.2c.

Another indication of the adjustment of reduction strategy to circumstances is the production of tools blanks from very large retouched flakes rather than cores, a reduction stage Bourguignon *et al.* (2004) term "Production 2". Reduction of large flakes to generate flakes which were used as tools and/or retouched took place in several ways. One approach was the use of a Kombewa technique (Tixier and Turq 1999), which involved striking the proximal end to remove a flake which had the bulb of the original flake on its dorsal surface. The resulting flakes typically have asymmetrical cross-sections, conforming to the pattern of Quina flakes struck from cores, although the asymmetry is often longitudinal rather than transverse. A Kombewa technique is particularly suited to stone materials with irregular form, such as the local chalcedony used to manufacture many artifacts at Combe Grenal, where it was emphasized in preference to some of the core reduction strategies (Turq 1989: 251). Other strategies for striking tool blanks from large flakes have also been observed in Quina industries, such as by notching scraper edges, has been suggested Bourguignon *et al.* (2004), revealing that there were multiple options for the production of blanks from large flakes. These options form part of what has been described as a "ramified" production system in which there are not only multiple strategies within a production system but also recursion of operations at several points in the production history, so that flakes serving as tool blanks were produced on cores, on flakes removed from cores, and so on (Bourguignon *et al.* 2004) – see Figure 17.3.

Quina blank production not only employed regular flaking strategies that were matched to the forms of raw material available, it also appears to have been highly efficient. One expression of flaking efficiency is the amount of core mass converted into tools. Although this is difficult to accurately measure, because tools are difficult to identify even with detailed studies of wear/residues, Turq (1992: 76) provides evidence that indicates efficiency was high in Quina industries in the Dordogne. His experiments suggested that Quina strategies yielded between twice and six times as many suitable blanks as the Levallois techniques by allowing successive removals of tool blanks without intensive core preparation between them. Furthermore, he concluded that between half and three-quarters of flakes were retouched into implements, a frequency that indicates a high rate of converting core mass into tool mass. The efficiency and economy implied by those patterns is also reflected in the treatment of tools.

QUINA IMPLEMENTS

Quina assemblages are dominated by dorsally retouched "scrapers" with a single, long retouched edge (Figure 17.3b). Approximately 40–90% of the typologically defined tools are single-edged scrapers. These scrapers have traditionally been separated into two groups: single-edged side scraper with retouch on one lateral margin (Bordes types 9–11), and transverse scrapers which have retouch across the distal end of the flake (Bordes types 22–24). Transverse forms are common in Quina assemblages, making up between a quarter and a half of all single-edged, dorsally retouched scrapers. Other implements are often recovered from Quina levels in small proportions, such as notched scrapers (Bordes types 42 and 43), abrupt pieces (Bordes type 48), and double and convergent scrapers (Bordes types 12–17, 18–21), but it is the numerical dominance and distinctiveness of single scrapers that are noteworthy of Quina retouching.

Quina scrapers typically have a long retouched edge, created by blows applied to the ventral face and removing scars from the dorsal face (Figure 17.3). In many instances these dorsal scars were created with direct soft-hammer percussion using antler, bone, and perhaps wood, and the resulting retouched surfaces contain regular invasive scars which expand away from the impact point and often terminate with small steps or hinges (Turq 2000: 331). Scars of different lengths can frequently be observed, with shorter, later ones superimposed over longer, earlier ones creating a pattern that has been described as four or more "lines" or layers of retouch forming the profile of a retouched edge (Figure 17.4). The expression of these characteristics on retouched edges reflect the interaction of several factors, including the thickness of the flake and the amount of retouch it had received (Verjux and Rousseau 1986). The majority of Quina scrapers have

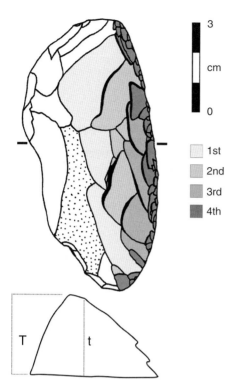

Figure 17.4 Illustration of the multiple "lines" of retouch on Quina scrapers, after Verjux and Rousseau (1986: 406), with the cross-section showing the measurement of the Geometric Index of Unifacial Reduction (GIUR).

a retouched edge which is long relative to the size of the flake on which it has been made, and the retouched edge is often orientated parallel to the long axis of the specimen, irrespective of the location of the platform. The length of retouch and its location on each specimen varies in relation to the size and shape of the blank, but in most cases there was a clear preference for a large, regular, retouched edge that was smoothly curving in plan shape (Bordes 1961a; Dibble 1987a; Turq 2000: 314). This preference appears to reflect cultural rules of tool design.

The functions of these tools has not been established, many more studies of residue and use-wear are required, but it is likely that they could have been multifunctional, providing edges for cutting and scraping of a variety of organic materials including hides, meat, and wood. In a detailed examination of side scrapers with Quina retouch at the site of La Quina Hardy (2004: 556) showed that they had been used on a variety of materials including plant, animal, hard and soft materials (see also Hardy *et al.* 1997). Associations of scrapers with red deer and reindeer at Combe Grenal (Chase 1986) may also hint at the functional roles of these tools. Some of these scrapers were probably repeatedly resharpened, a proposition that has been supported by several studies (e.g., Dibble, 1984, 1987a,b, 1988a, 1988b 1995; Lenoir 1986; Meignen 1988). The multiple layers of retouch scars, as well as other characteristics such as different levels of patina on layers of retouched scars, have led researchers to the conclusion that some Quina scrapers were intensively retouched on successive occasions, and that the retouch changed the shape of the retouched edge.

In a series of papers Dibble (1984, 1987a,b, 1988a,b, 1995) and others (e.g., Gordon 1993; Holdaway *et al.* 1996) hypothesized that this reworking of scrapers was responsible for the differences between the typological classes, and that much of the typological distinction between industries is a consequence of the degree of resharpening to which tools had been subjected. Dibble and Rolland (1992: 17) concluded that in Quina assemblages the dominant scrapers were often intensively retouched and that the characteristics of Quina industries were a consequence of higher levels of tool maintenance that occurred in Denticulate or Typical industries. These conclusions were based primarily on the evidence Dibble presented for differences in the size of platforms relative to the size of remaining ventral surfaces on retouched flakes. Platforms of transverse and convergent scrapers were relatively bigger than those on single scrapers, a pattern that Dibble (1995: 319) argued resulted from them having lost more of their ventral surface through retouching and therefore being more intensively reduced. In a similar way he interpreted the relatively large average platform size of scrapers in Quina assemblages as an indication that they were more heavily retouched than scrapers in Typical or Denticulate Mousterian assemblages.

The hypothesis that differences between typological categories of Quina scrapers, and the distinctiveness of Quina industries, resulted principally from the extent of retouching is not supported by other analyses. Two kinds of observations have demonstrated that intensity of reduction is probably a minor factor in the creation of typological patterning in the Mousterian. The first is a study of scrapers from a Quina layer (21) at Combe Grenal, using a more direct and precise method of measuring the intensity of retouching: a version of Kuhn's (1990) geometric index of unifacial reduction. By comparing the height of retouched scars (t) to the thickness of an implement (T), as shown in Figure 17.4, it is possible to estimate the proportion of original flake mass that had been removed during retouching (see Hiscock and Clarkson 2005). The evidence showed that traditional implement types were not well-correlated with levels of retouching, and that each typological group contained specimens with both high and low levels of reduction (Hiscock and Clarkson 2008). Consequently the typologically different scrapers were largely a product of alternative retouching strategies; in general they were not a series of sequential stages in a single reduction process but a more complex and somewhat parallel reduction sequence of the kind shown in Figure 17.5. Direct measurement of the

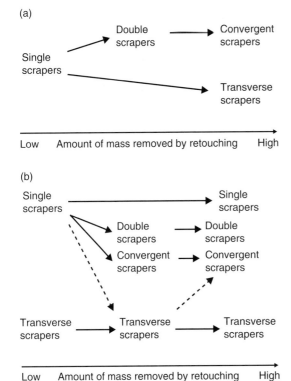

Figure 17.5 (a) Representation of the staged reduction model proposed by Dibble (based on Dibble 1987b: 115). (b) Representation of the reduction history of Quina scraper types at Combe Grenal proposed by Hiscock and Clarkson (2008).

extent of retouch in this manner is yet to be repeated in other Dordogne assemblages but the same conclusion had previously been reached by Turq (1989), who concluded that there was a morphological distinction between lateral and transverse scrapers in the Quina industries, and that this was a consequence of the particular relationship of implement types with blank form (see also Mellars 1996).

The second observation that is critical to understanding variations in implement types is that different retouch strategies appear to be associated with different flake blanks, and blank form had a strong and demonstrable influence on the distribution and intensity of retouch on many specimens. This proposition has been advanced for Mousterian assemblages in France (e.g., Turq 1989, 1992) and more broadly in Europe (e.g., Kuhn 1992). A number of relationships between flake blanks and retouch have been observed, including:

- Retouch was frequently positioned opposite a steep cortical margin, sometimes called a "natural back" (Turq 2000: 314).
- The position of retouch creating the long edge characteristic of Quina scrapers reflected the plan shape of the flake. On short but wide flakes the optimal way to create a single long retouched edge is to retouch the distal end; whereas on long flakes the creation of a long retouched edge requires retouch to be positioned primarily on one of the lateral margins. This observation implies the relative frequency of lateral and transverse scrapers in Quina assemblages is a result of the form of the available flake blanks (Mellars 1992).

Together these relationships explain much of the typological variation in Quina assemblages. They also potentially explain the patterns of platform size that Dibble had observed. Transverse and lateral scrapers were made on different blanks (Turq 1988, 1989, 1992; Lenoir 1990), and experiments by Turq indicated that the short, wide flakes on which transverse scrapers were made typically had larger platforms, and hence their characteristics reflect the nature of the blanks selected rather than extent of scraper reduction.

However, even if reduction intensity does not account for differences between implement types there is little doubt that some scrapers were extensively reduced, as already mentioned. Multiple factors appear to underlay the variation displayed in scraper reduction. Given Quina scrapers were predominantly designed with a single long edge, the asymmetry of flake cross-section had a powerful effect on the potential of each specimen

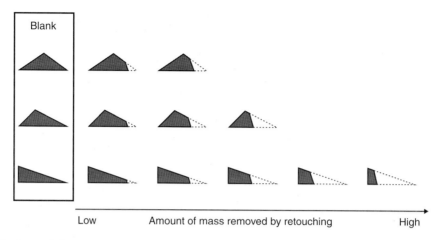

Figure 17.6 Notional illustration of the relationship of blank cross-section and extent of reduction for dorsally retouched Quina scrapers (after Turq 1992: Figure 6.2).

for extended reduction (Turq 1992). Thick flakes with highly asymmetrical cross-sections could receive more retouch without breaking than those with symmetrical cross-sections, as shown in Figure 17.6. Economic context was also a factor conditioning the extent to which the potential of blanks for reduction was exploited. For instance, evidence of tool recycling is more pronounced in assemblages from localities without easy access to abundant stone sources (Turq 2000: 343). Since the design of Quina scrapers is extremely suited to prolonged maintenance and recycling it is not surprising that in some economic contexts tools were more extensively reduced. The design of Quina scrapers and their extended maintenance, together with their dominance in assemblages, are all indications of the importance of these tools in Neanderthal subsistence activities.

THE QUINA SYSTEM

Lithic variability in Quina assemblages can be understood as a consequence of interactions between the flexible Quina technology and the context in which it was employed. All Quina knapping products offered their maker multiple opportunities. For instance, flakes could have been used or struck like a core to yield new flakes, as per Figure 17.3. Another example is the multipurpose quality of Quina retouched flakes,

which were transformed into different forms including scrapers, denticulates, or notches (Bourguignon et al. 2004). These possibilities reflect the way similar reduction processes permitted objects to be treated in different ways: as a tool, as a core, and so on. The range of artifact forms observed archeologically was expanded as knappers responded to the diversity of functional contexts represented in different environmental settings and different situations of occupation (cave, long-term specialized camp, open-air). The technology was expressed somewhat differently in each assemblage as the needs of people in each of context led them to emphasize different possibilities of the Quina forms. Variation in Quina tools was also an outcome of the intensity with which tools were maintained and transported. Some tools were substantially reduced, what (Binford 1977, 1979) called "curated" (see Bamforth 1986; Shott 1989, 1995; Kuhn 1994, 1996; Hiscock 2006). The most well-known examples of this are Quina scrapers or the heavily reduced bifacial pieces at the La Quina site, which have been successively and extensively modified through multiple stages, maintained over long periods and moved to distant places. These extensively maintained artifacts represent only one component of Quina assemblages. Whether occupations were short-term or long-term, most material was obtained locally, and artifacts made on these local materials were often employed in an expedient manner, used, and discarded locally.

Consequently, assemblage composition also reflects the different balance between maintained or expediently used tools in different contexts.

The technological emphasis revealed in different assemblages can be understood in terms of the length of occupation and economic context of each site. Quina production activities were suited to mobile foragers who transported personal equipment, comparable with Binford's (1979) concept of "personal gear" although the toolkits were dominated by general purpose tools in the sense of Kuhn (1995) and many objects had high potential for transformation and multifunctionality described above. This system provided tools or materials for flake production, depending on the needs of people in each context. Transporting and maintaining tools was an expression of the strategy Kuhn (1992, 1995) called provisioning of individuals. But as Costamagno *et al.* (2006) recently illustrated at Marillac, transported personal gear was usually minimal and foragers employed locally available lithic materials to supply their specific needs at each site. In many contexts, especially during long-term occupations when site activities were diversified, the consequence of minimal personal gear was that the needs for tools were efficiently fulfilled by the exploitation of local raw materials, comparable to the strategy Kuhn called provisioning of place. The use of local materials resulted in the introduction into assemblages of blocks of material, cores, and large flakes with the potential to be reduced. The contribution to assemblages of maintained or expedient tools, local or non-local materials, personal gear, or site provisioning varied with different residence and mobility patterns as well as with different functional contexts. This process underlies variation between Quina assemblages.

Acknowledging the variation that exists in Quina assemblages, it is still clear that they express a common technological and economic theme. The Quina system of artifact production and material transport functioned to supply stone for use as tools where and when it was required. Quina strategies represent particular solutions to the problems of provisioning Neanderthal foragers in the Dordogne during the Middle Paleolithic. Transporting highly extendable packages of stone, usually in the form of thick flakes from which knapper's could remove large numbers of flakes for prolonged periods, either to maintain/create working edges and/or to produce the flakes themselves, formed the central strategy of Quina behavior. The nature of core reduction and flake retouching appears to have been efficient and appropriate for producing these packages, and the pattern and directionality of material transfers indicates the structure of economic needs rather than the intellectual capacity of Neanderthals. Consideration of this system raises several implications.

Inevitably the difficult question about the articulation of Quina retouching strategies and core production strategies is: which was the principal factor in the development of this system of behavior? Was the Quina approach to retouching merely a consequence of dealing with the blanks being produced, an argument that implies core reduction was the primary cause of the system, or was the approach to core reduction refined as a means of producing blanks appropriate for creating the preferred retouched implements? In addressing this question of causation Turq (1988, 1989, 1992, 2000) cited the evidence that blanks were chosen to be appropriate to the tools produced, and hypothesized that Quina core reduction strategies were deliberate and preconceived attempts to create flakes suitable for manufacture of Quina tools. His model implies that the fundamental conceptual shift represented by Quina industries involved the reliance on and investment in retouching to create regularity in tool form, whereas in many Levallois-based industries initial flake (blank) production rather than retouching was the mechanism creating regularity. Turq's insight also implies that requirements of tool design were primary considerations in the adoption of patterns of core reduction, a proposition that implies a high level of planning, but in an economic context in which tools needed to be highly extendable, through their capacity to have edges maintained or alternatively to be recycled.

The result was that Quina production was not only both efficient and adaptable to different situations of raw material, as discussed above, but also yielded implements of a regular form which were suited to multiple tasks and could be intensively maintained and recycled. These tools, together with the production and supply activities that supported their production, provided the technological basis for exploitation of the region during harsh climatic conditions, when it seems that foragers were encountering different prey in different seasons and at different locations within their territory (Chase 1987, 1989). In this context, where the rigid use of activity-specific tools would frequently have required economically costly retooling and foraging uncertainty and may have made predictions of required toolkits dangerous because of the consequence of planning an inappropriate toolkit, a technological system dedicated to producing flexible, long lasting, multifunctional, portable tools might have been the better choice. In this context, the emphasis on highly extendable retouched tools, created on thick flakes, produced in a time-economizing way that did not tie foragers to particular outcrop areas, would also have been an advantageous strategy (see Turq 1992). In this way it is possible to recognize a "Quina system" which integrated a variety of procurement and technological practices to create an economic structure that was suited to the specific environmental/economic context in which hominids living in southwest France found themselves. This system was elaborate and coherent, traits which indicate the presence of planning in the Neanderthal foragers who developed and employed it.

REFERENCES

Ashton, N.M. (2000) The clactonian question: On the interpretation of Core-and-Flake Assemblages in the British Lower Paleolithic. *Journal of World Prehistory* 14, 1–63.

Ashton, N.M., Cook, J., Lewis, S.G., and Rose, J. (1992) *High Lodge: Excavations by G. de G. Sieveking 1962–68 and J. Cook 1988*. British Museum Press, London.

Ashton, N., Dean, P., Lewis, S.G., and Parfitt, S. (1998) *OP 125: Excavations at the Lower Palaeolithic Site of East Farm Barnham, Suffolk 1989–94*. British Museum, London.

Bamforth, D.B. (1986) Technological efficiency and tool curation. *American Antiquity* 51, 38–50.

Binford, L.R. (1973) Interassemblage variability – the Mousterian and the "functional" argument. In: Renfrew, C. (ed.) *The Explanation of Culture Change*, Duckworth, Surrey, pp. 227–54.

Binford, L.R. (1977) Forty-seven trips. In: Wright, R.S.V. (ed.) *Stone Tools as Cultural Markers: Change, Evolution and Complexity*, Humanities Press, Syndey, pp. 24–37.

Binford, L.R. (1979) Organization and formation processes: Looking at curated technologies. *Journal of Anthropological Research* 35, 255–73.

Binford, L.R. (1989) Isolating the transition to cultural adaptations: An organizational approach. In: Trinkaus, E. (ed.) *The Emergence of Modern Humans: Biocultural Adaptations in the Later Pleistocene*, Cambridge University Press, Cambridge, pp. 18–41.

Binford, L.R. and Binford, S.R. (1966) A preliminary analysis of functional variability in the Mousterian of Levallois Facies. *American Anthropologist* 68, 238–95.

Binford, L.-R. and Binford, S.-R. (1969) Stone tools and human behavior. *Scientific American* 220, 70–84.

Boëda, E. (1986) Approche technologique du concept Levallois et evaluation de son Champs d'Application: etude de trios gisements saaliens et weichseliens de la France septentrionale. Doctoral dissertation, University of Paris X.

Boëda, E. (1997) Technogenése de systémes de production lithique au Paléolithique inférieur et moyen en Europe occidentale et au Proche-Orient. Dissertation, University of Paris X.

Bordes, F. (1947) Étude comparative des différentes techniques de taille du silex et des roches dures. *L'Anthropologie* 51, 1–29.

Bordes, F. (1953) Essai de classification des industries "moustériennes." *Bulletin de la Société préhistorique française* 50, 457–66.

Bordes, F. (1961) *Typologie du Paléolithique inférieur et moyen*. Delmas, Bordeaux, 2 Vols. Publications de l'Institut de Préhistoire de l'Université de Bordeaux, Mémoire 1.

Bordes, F. (1972) *A Tale of Two Caves*. Harper and Row, New York.

Bordes, F. (2002) *Typologie du Paléolithique Ancien et Moyen*. CNRS Editions, Paris, France. Originally published 1961, Mémoires de l'Institut Préhistoriques de l'Université de Bordeaux 1, Bordeaux, Delmas.

Bordes, F. and Bourgon, M. (1951) Le complexe Moustérien: Moustériens, Levalloisien and Tayacien. *L'Anthropologie* 55, 1–23.

Bordes, F., Laville, H., and Paquereau, M.M. (1966) Observations sur le Pléistocène supérieur du gisement de Combe-Grenal (Dordogne). *Actes de la Sociéte Linnéene de Bordeaux* 10, 3–19.

Bourguignon, L. (1997) *Le Moustérien de type Quina: Définition d'une nou-velle entité technique*. Thèse de Doctorat de l'Université de Paris X, Nanterre.

Bourguignon, L., Faivre, J.-P., and Turq, A. (2004) Ramification des chaînes opératoires: une spécificité du Moustérien? *Paleo* 16, 37–48.

Bourlon, M. (1907) Débitage des rognons des silex en tranches paralléles. *Bulletin de la Société Préhistorique Française* 4, 330–2.

Chase, P.G. (1986) Relationships between Mousterian lithic and faunal assemblages at Combe Grenal. *Current Anthropology* 27, 69–71.

Cheynier, A. (1953) Stratigraphi de l'abri Lachaud et les cultures de bords abattus. *Archivo de Prehistoria Levantina* 4, 25–55.

Clark, G.A. and Lindly, J. (1989) The case for continuity: observations on the biocultural transition in Europe and Western Asia. In: Mellars, P. and Stringer, C. (eds) *The Human Revolution: Behavioural and Biological Perspectives on the Origins of Modern Humans*, Princeton University Press, Princeton, pp. 626–76.

Close, A. (1991) On the validity of Middle Paleolithic tool types: A test case from the Eastern Sahara. *Journal of Field Archaeology* 18, 256–64.

Costamagno, S., Meignen, L., Maureille, B., and Vandermeersch, B. (2006) Les Pradelles (Marillac-le-Franc, France): A Mousterian reindeer hunting camp? *Journal of Anthropological Archaeology* 25, 466–84.

Demars, P.-Y. (1980) *Les matières premières siliceuses utilisées au Paléolithique supérieur dans le basin de Brive.* Thesis, University of Bordeaux I.

Dibble, H.L. (1984) Interpreting typological variation of Middle Paleolithic scrapers: function, style, or sequence of reduction? *Journal of Field Archaeology* 11, 431–6.

Dibble, H.L. (1987a) Reduction sequences in the manufacture of Mousterian Implements of France. In: Soffer, O. (ed.) *The Pleistocene Old World Regional Perspectives*, Plenum Press, New York, pp. 33–45.

Dibble, H.L. (1987b) The interpretation of Middle Paleolithic scraper morphology. *American Antiquity* 52, 109–17.

Dibble, H.L. (1988a) Typological aspects of reduction and intensity of utilization of lithic resources in the French Mousterian. In: Dibble, H. and Montet-White, A. (eds) *Upper Pleistocene Prehistory of Western Eurasia*, University Museum, University of Pennsylvania, Philadelphia, pp. 181–94.

Dibble, H.L. (1988b) The interpretation of middle Paleolithic scraper reduction patterns. In: Dibble, H. (ed.) *L'Homme de Néandertal, Vol 4, La Technique*, Actes du Colloque International de Liége, L'Homme de Neandertal, pp. 49–58.

Dibble, H.L. (1991) Rebuttal to close. *Journal of Field Archaeology* 18, 264–9.

Dibble, H.L. (1995) Middle Paleolithic scraper reduction: Background, clarification, and review of evidence to date. *Journal of Archaeological Method and Theory* 2, 299–368.

Dibble, H.L. and Rolland, N. (1992) On assemblage variability in the Middle Paleolithic of Western Europe: History, perspectives, and a new synthesis. In: Dibble, H.L. and Mellars, P. (eds) *The Middle Paleolithic: Adaptation, Behavior, and Variability*, University Museum, University of Pennsylvania, Philadelphia, pp. 1–28.

Féblot-Augustins, J. (1993) Mobility strategies in the late Middle Palaeolithic of central Europe and western Europe: Elements of stability and variability. *Journal of Anthropological Archaeology* 12, 211–65.

Forestier, H. (1993) Le clactonien: mise en application d'une nouvelle méthode de débitage s'incrivant dans la variabilité des systèmes de production lithique au Paléolithique ancient. *Paleo* 5, 53–82.

Gargett, R.H. (1989) Grave shortcomings: The evidence for Neanderthal burial. *Current Anthropology* 30, 157–90.

Geneste, J.-M. (1985) *Analyse Lithique d'Industries Moustériennes du Périgord: une approche technologique du comportement des groups humains au Paléolithique moyen.* Doctoral dissertation, University of Bordeaux 1.

Geneste, J.-M. (1988) Les industries de la Grotte Vaufrey: technologie du débitage, économie et circulation de la matière première lithique. In: Rigaud, J.-P. (ed.) *La Grotte Vaufrey: Paléoenvironment, Chronologie, Activités Humaines*, Memoires de la Société Préhistorique Française, Paris, vol. 19, pp. 441–517.

Geneste, J.-M. (1989) Economie des resources lithiques dans le Moustérien du sud-ouest de la France. In: Otte, M. (ed.) *L'Homme de Néandertal. Vol. 6: La subsistance*, Etudes et Recherches Archéologiques de l'Université de Liége, Liege, pp. 75–97.

Gordon, D. (1993) Mousterian tool selection, reduction and discard at Ghar, Israel. *Journal of Field Archaeology* 20, 205–18.

Green, R.E., Krause, J., Ptak, S.E., et al. (2006) Analysis of one million base pairs of Neanderthal DNA. *Nature* 444, 330–6.

Hardy, B.L. (2004) Neanderthal behaviour and stone tool function at the Middle Palaeolithic site of La Quina, France. *Antiquity* 78(301), 547–65.

Hardy, B.L., Raman, R., and Raff, R.A. (1997) Recovery of mammalian DNA from Middle Paleolithic stone tools. *Journal of Archaeological Science* 24, 601–12.

Hiscock, P. (2006) Blunt and to the point: Changing technological strategies in Holocene Australia. In: Lilley, I. (ed.) *Archaeology in Oceania: Australia and the Pacific Islands*, Blackwell, Oxford, pp. 69–95.

Hiscock, P. and Clarkson, C. (2005) Experimental evaluation of Kuhn's Geometric Index of Reduction and the flat-flake problem. *Journal of Archaeological Science* 32, 1015–22.

Hiscock, P. and Clarkson, C. (2008) The construction of morphological diversity: A study of Mousterianimplement retouching at Combe Grenal. In: Andrefsky, W. (ed.) *Lithic Technology*, Cambridge University Press, Cambridge, pp. 106–35.

Holdaway, S., McPherron, S., and Roth, B. (1996) Notched tool reuse and raw material availability in French Middle Paleolithic sites. *American Antiquity* 61, 377–87.

Kuhn, S. (1990) A geometric index of reduction for unifacial stone tools. *Journal of Archaeological Science* 17, 585–93.

Kuhn, S. (1992) Blank morphology and reduction as determinants of Mousterian scraper morphology. *American Antiquity* 57, 115–28.

Kuhn, S. (1994) A formal approach to the design and assembly of mobile tool kits. *American Antiquity* 59, 426–42.

Kuhn, S. (1995) *Mousterian Lithic Technology.* Princeton University Press, Princeton.

Kuhn, S. (1996) The trouble with ham steaks: A reply to Morrow. *American Antiquity* 61, 591–5.

Kuhn, S. (2004) Upper Paleolithic raw material economies at Uçağızlı cave, Turkey. *Journal of Anthropological Archaeology* 23, 431–48.

Lenoir, M. (1986) Un mode de retouche "Quina" dans le Moustérien de Combe-Grenal (Domme, Dordogne). *Bulletin de la Société Anthropologique du Sud-ouest* 21(3), 153–60.

Lenoir, M. (1990) Un mode d'obtention de la retouche "Quina" dans le Moustérien de Combe-Grenal (Domme, Dordogne). *Bulletin de la Société Anthropologique du Sud-ouest* 21, 153–60.

Leroi-Gourhan, A. (1956) La galerie mousterienne de la grotte du Renne (Arcy-sur-Cure, Yonne). *Congres Prehistorique de France*, 15e session, 676–91.

Leroi-Gourhan, A. (1966) *La Prehistorie.* University of France Press, Paris.

Lieberman, P. (1989) The origins of some aspects of human language and cognition. In: Mellars, P. and Stringer, C. (eds) *The Human Revolution: Behavioural and Biological Perspectives on the Origins of Modern Humans*, Princeton University, Princeton, pp. 391–414.

Lieberman, P. (1991) *Uniquely Human: The Evolution of Speech, Thought, and Selfless Behaviour.* Harvard University Press, Cambridge.

Lindly, J.M. and Clark, G.A. (1990) Symbolism and modern human origins. *Current Anthropology* 31, 233–61.

Marwick, B. (2003) Pleistocene exchange networks as evidence for the evolution of language. *Cambridge Archaeological Journal* 13, 67–81.

Meignen, L. (1988) Un exemple de comportement technologique differential selon les matiéres premiéres: Marillac, couches 9 et 10. In: Otte, M. (ed.) *L'Homme de Néandertal. Vol. 4: La technique*, Etudes et Recherches Archéologiques de l'Université de Liége, Liege, pp. 71–9.

Mellars, P. (1965) Sequence and development of Mousterian traditions in south-west France. *Nature* 205, 626–7.

Mellars, P. (1969) The chronology of Mousterian Industries in the Perigord region. *Proceedings of the Prehistoric Society* 35, 134–71.

Mellars, P.-A. (1970) Some comments on the notion of "functional variability" in stone-tool assemblages. *World Archaeology* 2, 74–89.

Mellars, P. (1988) The chronology of the south-west French Mousterian: A review of the current debate. In: Otte, M. (ed.) *L'Homme de Néanderthal, Vol.4: La technique*, etudes et recherches Archéologiques de l'Université de Liége, Liege, pp. 97–120.

Mellars, P.-A. (1989) Chronologie du Moustérien du sud-ouest de la France : actualisation du débat. *L'Anthropologie* 94, 1–18.

Mellars, P. (1992) Technological change in the Mousterian of southwest France. In: Dibble, H.L. and Mellars, P. (eds) *The Middle Paleolithic: Adaptation, Behavior, and Variability*, University Museum, University of Pennsylvania, Philadelphia, pp. 29–43.

Mellars, P. (1996) *The Neanderthal Legacy.* Princeton University Press, New York.

Morala, A. (1983) A propos des matiéres premiéres lithiques en Haut-Agenais. *Bulletin de la Société Préhistorique Française* 80, 169.

Noble, W. and Davidson, I. (1996) *Human Evolution, Language and Mind: A Psychological and Archaeological Inquiry.* Cambridge University Press, Cambridge.

Peyrony, D. (1930) Le Moustier: ses gisements, ses industries, ses couches géologiques. *Revue Anthropologique* 40, 48–76, 155–76.

Rolland, N. (1988) Observations on some Middle Paleolithic time series in southern France. In: Dibble, H. and Montet-White, A. (eds) *Upper Pleistocene Prehistory of Western Eurasia*, University museum monograph 54 University of Pennsylvania, Philadelphia, pp. 161–80.

Rolland, N. and Dibble, H.L. (1990) A new synthesis of Middle Paleolithic assemblage variability. *American Antiquity* 55, 480–99.

Seronie-Vivien, M. and Seronie-Vivien, M.-R. (1987) Les silex du Mésozoïque nord-aquitain: approche géologique de l'etude du silex pour server à la recherché. *Bulletin de la Société Linnéenne de Bordeaux* 15, 1–135.

Shott, M.J. (1989) On tool-class use lives and the formation of archaeological assemblages. *American antiquity*, 54(1), 9–30.

Shott, M.J. (1995) How much is a scraper? Curation, use rates, and the formation of scraper assemblages. *Lithic Technology* 20, 53–72.

Stringer, C.B. and Gamble, C. (1993) *In Search of the Neanderthals: Solving the Puzzle of Human Origins.* Thames and Hudson, London.

Tixier, J. (1978) *Méthode pour l'Etude des Outillages lithiques: notice sur les travaux scientifiques de Jacques Tixier presentee en vue de grade de Docteur es Lettres.* University of Paris X.

Tixier, J., Inizan, M.-L., and Roche, H. (1980) *Préhistoire de la pierre taille´e 1: Terminologie et technologie.* Cercle de Recherches et d'E´ tudes Pre´historiques, Valbonne.

Tixier, J. and Turq, A. (1999) Kombewa et alii. *Paléo* 11, 135–43.

Turq, A. (1977) Première approche sur le Paléolithique moyen du gisement des Ardailloux, commune de Soturac (Lot). *Bulletin de la Société des Etudes du Lot* 48, 222–42.

Turq, A. (1988) Le Moustérien de type Quina du Roc de Marsal à Campangne (Dordogne): contexte stratigraphique, analyse lithologique et technologique. *Documents d'Archéologie Perigourdine (A.D.R.A.P.)* 3, 5–30.

Turq, A. (1989) Approche technologique et économique du facies Moustérien de type Quina: etude préliminaire. *Bulletin de la Société Préhistorique Française* 86, 244–56.

Turq, A. (1992) Raw material and technological studies of the Quina Mousterian in Perigord. In: Dibble, H.L. and Mellars, P. (eds) *The Middle Paleolithic. Adaptation, Behavior, and Variability*, University Museum, University of Pennsylvania, Philadelphia, pp. 75–85.

Turq, A. (2000) Paléolithique inférieur et moyen entre Dordogne et Lot. *Paléo* 2, 1–456.

Turq, A. (2005) Réflexions méthodologiques sur les etudes de matiéres premiéres lithiques. 1 – des lithothéques au materiel archéologique. *Paleo* 17, 111–31.

Verjux, C. and Rousseau, D.-D. (1986) La retouche Quina: une mise au point. *Bulletin de la Société Préhistorique Française* 11–12, 404–15.

Chapter 18

THE IMPACT OF LITHIC RAW MATERIAL QUALITY AND POST-DEPOSITIONAL PROCESSES ON CULTURAL/ CHRONOLOGICAL CLASSIFICATION: THE HUNGARIAN SZELETIAN CASE

Brian Adams

Public Service Archeology and Architecture Program, University of Illinois at Urbana-Champaign

ABSTRACT

The Szeletian industry is a key component of models of the Middle to Upper Paleolithic transition in Central Europe. The acculturation model proposes that the co-occurrence of Middle and Upper Paleolithic artifact types in Szeletian assemblages, including "archaic" types such as notches and denticulates, is the product of interactions between Neanderthals and modern humans. In this chapter, the role of the Szeletian in transitional studies is re-examined based on new radiocarbon dates, a consideration of lithic raw material quality, and post-depositional processes.

Lithic Materials and Paleolithic Societies, 1st edition. Edited by B. Adams and B.S. Blades, ©2009 Blackwell Publishing. ISBN 978-1-4051-6837-3.

INTRODUCTION

The lithic assemblages from Szeleta Cave in Hungary are a crucial component of discussions of the Middle to Upper Paleolithic "transition" in Central Europe. Based primarily on lithic typology, material from the site has been interpreted as either an example of *in-situ* cultural evolution or the product of interactions between Neanderthals and modern humans. The acculturation model seeks to explain the co-occurrence of Middle and Upper Paleolithic artifact types as evidence of such interactions (e.g., Allsworth-Jones 1986), while models of *in situ* evolution explain the Szeleta assemblages as the product of Middle Paleolithic groups actually in the process of evolving into the Upper Paleolithic (e.g., Vértes 1959, 1968; Gábori 1969). This chapter focuses on the acculturation model as it continues to have a strong influence on discussions of the Middle to Upper Paleolithic transition. In this chapter the influence of lithic raw material quality and post-depositional processes on tool morphology are considered and it is argued that hypothesized "archaic" types are not reliable cultural or temporal signatures.

SZELETA CAVE AND THE SZELETIAN

The Szeletian Industry is named after Szeleta Cave in the Bükk Mountains of northeast Hungary (Kadić 1916) (Figure 18.1). Seven stratigraphic layers are recognized at this site with material classified as Szeletian associated with Layers 3 through 6 (Figure 18.2). Two developmental phases are recognized at Szeleta. The Early Szeletian is primarily associated with Layer 3 and is characterized by crudely flaked, asymmetrical leaf points, while the Developed Szeletian is associated primarily with Layer 6 and produced more finely worked, thin, symmetrical leaf points (Figure 18.3).

THE SZELETIAN AND THE MIDDLE TO UPPER PALEOLITHIC TRANSITION

The acculturation model of the Szeletian was popularized by Philip Allsworth-Jones (1986) and contends that local populations of Neandertals with Middle Paleolithic material culture interacted with immigrating groups of anatomically modern *Homo sapiens* associated with Upper Paleolithic material culture

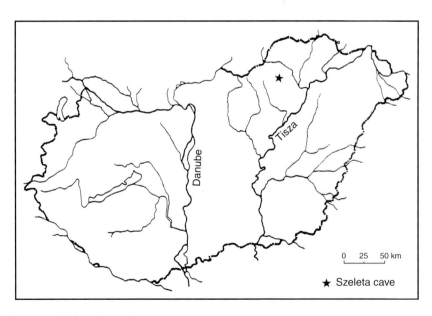

Figure 18.1 Location of Szeleta Cave in Hungary.

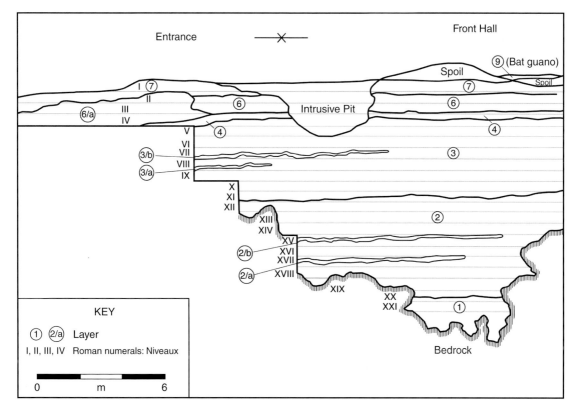

Figure 18.2 Detail of Szeleta Cave stratigraphy from the entrance and front hall, showing niveaux and layers (after Kadić 1916).

such as the Aurignacian. This resulted in the adoption of Upper Paleolithic technology by Neanderthals who began producing lithic assemblages such as the Szeletian characterized by blade core technology together with Middle Paleolithic tool types such as notches, denticulates, and side scrapers.

The acculturation model is based primarily on the typological classification of stone tools. According to this model, typical Middle Paleolithic tool types, which are commonly classified as "archaic" types (e.g., Gladilin and Demidenko 1989; Anickovich 1992; Zilhão and D'Errico 2003) occur together with "progressive" Upper Paleolithic types. In his analysis of the Szeletian material, Allsworth-Jones (1986) states that notched pieces and denticulates are the most abundant tools after leaf points, yet later suggests that these "tools" are most likely the products of cryoturbation rather than intentionally shaped artifacts. For Allsworth-Jones, it is ultimately the presence of side scrapers alongside Upper Paleolithic types that gives

the Szeletian its transitional "flavor". However, in regard to the presence of alleged "archaic" Mousterian tool types, such as side scrapers, in French and Spanish Solutrean assemblages, Straus (1978: 37) states that "... there are limited ways in which a flake of given dimensions can be retouched to form a desired tool shape, regardless of whether manufacture took place 50 000 or 20 000 years ago ...," and suggests that artifact function and/or raw material constraints must be taken into account.

Another component of the acculturation model is the attempted correlation of Szeletian assemblages with Neanderthal remains (Allsworth-Jones 1986; Kozlowski 1988). Straus (1993: xii) has criticized acculturation models as "... ad hoc non-explanation[s] ... because the term is simply declared without further analysis and argument as to specifically why and how Neanderthals might become (or need to become) acculturated ... to the ways of the Aurignacian ... 'invaders' ". This is especially true given

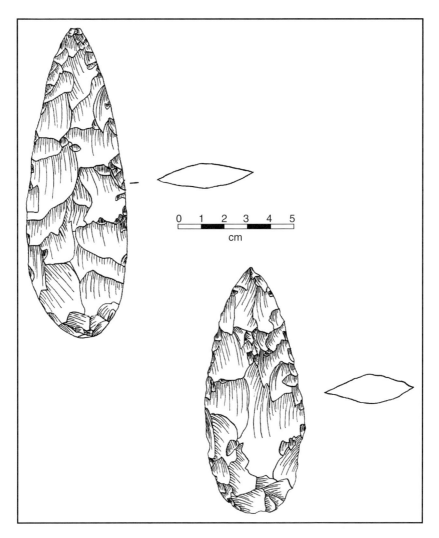

Figure 18.3   Szeleta Cave leaf points.

the success of the Neanderthal adaptation to life in Europe between approximately 130 000 and 35 000 years ago, a period of about 95 000 years! A detailed discussion of this argument is beyond the scope of this chapter, but suffice it to say that to date no Neanderthal remains have been recovered from Szeleta Cave or any other proposed Szeletian sites so that attribution of the Szeletian to Neanderthals is pure speculation.

Probably the most important event in the development of the acculturation argument was the publication of the first radiocarbon dates from the site in the 1960s by Lászlo Vértes (Vértes 1968). Based on three radiocarbon dates, Vértes argued that the Developed Szeletian dated to about 33 000 years ago while the Early Szeletian was approximately 43 000 years old, and it was on the basis of these three dates, together with typological data, that the Szeleta Cave material was assigned to the Middle to Upper Paleolithic transition period. However, uncertainty regarding the provenence of Vértes's dated samples casts doubt on their reliability.

THE SZELETIAN IN LIGHT OF RECENT RESEARCH

More recent work on the Szeleta Cave assemblages combined with results from new excavations at the site brings into question the validity of the acculturation argument. Based on the author's (Adams 1998) inspection of the Szeleta Cave material, the assemblages are described as follows. Retouched tools from the Early Szeletian assemblage are dominated by leaf points followed by retouched blades and bladelets, end scrapers, retouched flakes, and burins. Only four Middle Paleolithic types were recognized consisting of 2 denticulates, 1 side scraper, and 1 truncation. These types represent about 5% of the Early Szeletian assemblage. In addition, three split-based bone points are associated with the lower assemblage. The Developed Szeletian assemblage is also dominated by bifacial leaf points, followed by retouched blades and bladelets, burins, and end scrapers. Middle Paleolithic types were actually more abundant in the Developed Szeletian assemblage, accounting for about 13% of the assemblage. These consist of 3 convergent scrapers, 2 side scrapers, 1 denticulate, and 1 traverse scraper. Flake and blade debitage and flake and pyramidal blade cores occur in both assemblages. In brief, both the Early and Developed Szeletian material can be described as non-Levallois industries dominated by the production of bifacial tools and retouched blades/bladelets. However,

as Allsworth-Jones (1986) has emphasized, the Upper Paleolithic types appear to be poorly made, and Middle Paleolithic tool types represent relatively minor components of these assemblages.

New radiocarbon dates from Szeleta Cave secured by the author in cooperation with Dr. Árpád Ringer of Miskolc University, Hungary suggest that the site is younger than proposed by Vértes (Adams and Ringer 2004) (Table 18.1). These dates suggest that most of the material from Szeleta Cave dates to between approximately 20 000 to 30 000 years ago, and that deposits dating to approximately 40 000 year ago are approximately 10 000 years older than the Early Szeletian material. These new dates represent the most serious blow to the acculturation argument, as they suggest that the period of overlap between the Neanderthal and modern human presence in Central Europe was more restricted than previously believed, further suggesting a narrower window of opportunity for potential interactions between the two groups. Current data indicate that Neanderthal fossils in the northern Carpathian Basin region date to about 40 000 years ago at the latest, while *Homo sapiens sapiens* fossils date to around 35 000–30 000 years ago at the earliest (Valoch 1986; Svoboda and Simán 1989; Ringer 1990; Gábori-Csánk 1993; Svoboda *et al.* 1996). The currently available fossil data thus indicates a 5000 to 10 000 year gap between the disappearance of the Neanderthals and the appearance of modern humans in this region,

Table 18.1 New radiocarbon dates from Szeleta and Istállóskõ caves, Hungary.

Site	Date	Material	Stratigraphy	Archeology
Szeleta	22 107±130[a] (ISGS-A-0131)	Bone	Layer 6	"Developed Szeletian"
Szeleta	26 002±182[a] (ISGS-A-0189)	Charcoal	Layer 3	"Early Szeletian"
Szeleta	>25 200 (ISGS-4460)	Bone	Layer 3	"Early Szeletian"
Szeleta	11 761±62[a] (ISGS-A-0128)	Bone	Layer 3 (hearth)	"Early Szeletian"
Szeleta	13 885±71[a] (ISGS-A-0129)	Bone	Layer 3 (hearth)	"Early Szeletian"
Szeleta	42 960±860 (ISGS-4464)	Bone	Layer 2/3 boundary	Pre- "Early Szeletian"?
Istállóskõ	27 933±224[a] (ISGS-A-0186)	Bone	Layer 5/8[b]	Aurignacian II
Istállóskõ	31 608±295[a] (ISGS-A-0188)	Bone	Layer 5/8	Aurignacian II
Istállóskõ	29 035±237[a] (ISGS-A-0185)	Bone	Layer 7/9[b]	Aurignacian I/II
Istállóskõ	32 701±316[a] (ISGS-A-0187)	Bone	Layer 7/9	Aurignacian I level
Istállóskõ	33 101±512[a] (ISGS-A-0184)	Bone	Layer 7/9	Aurignacian I level

[a] AMS Date.

[b] Vértes (1955, Figure 18.3 and Plate LI) presents two profiles from his excavations at Istállóskõ Cave and has numbered analogous levels differently in each.

and the new dates presented here suggest that the assemblages from Szeleta, together with new dates from the Aurignacian site at nearby Istállóskő Cave are the product of anatomically *Homo sapiens sapiens* and not Neanderthals. However, two dates of approximately 28 000 and 29 000 years BP from Croatia indicate that isolated populations of Neanderthals may have continued to survive south of the Carpathian Basin, approximately 500 km south of the Hungarian Bükk Mountains (Smith *et al.* 1999; Karavani and Smith 2000).

IF THE SZELETIAN ISN'T OLD, WHY DOES IT LOOK OLD?

As argued above, recent research calls into question two of the main roots of the Szeletian as a transitional phenomenon. New dates suggest the Szeletian assemblages are not as old as originally believed, suggesting they were not produced by Neanderthals. Yet the issue of the general "crudeness" or "archaic" nature of the lithic assemblages remains to be addressed. It is proposed here that the presence of purported Middle Paleolithic or "archaic" tool types in the Szeletian collections can be explained by factors other than acculturation or *in situ* cultural evolution.

LITHIC RAW MATERIAL USE AT SZELETA CAVE

Due to an over-emphasis on the leaf point component of the Szeleta Cave assemblages, other aspects of the lithic inventories are ignored, and it is argued here that a full understanding of the assemblages requires consideration of lithic raw materials utilized at the site. Fortunately, a wealth of information exists regarding the lithic raw material stock utilized throughout prehistory in Hungary (Vértes and Tóth 1963; Allsworth-Jones 1986; Dobosi 1986, 1991; Simán 1986, 1991; Takács-Biró 1986a,b, and this volume; Biró and Dobosi 1991; Adams 1998; Markó *et al.* 2003). Nearly 90% of the artifacts from both the Early and Developed Szeletian assemblages are made from poor- to medium-quality local raw materials (Adams 1998). Both assemblages are dominated by felsitic quartz porphyry, which tends to be of medium-quality, followed by various poor- to medium-quality hydro- and limnoquartzites. In north Hungary, hydrothermal and limnic silicates occur within 15–60 km of Szeleta Cave, while felsitic quartz porphyry is of more limited distribution approximately 5 km to the south (Figure 18.4). It must be emphasized that raw material quality is variable even among these different types. The mechanical properties of these local lithic raw materials are highly

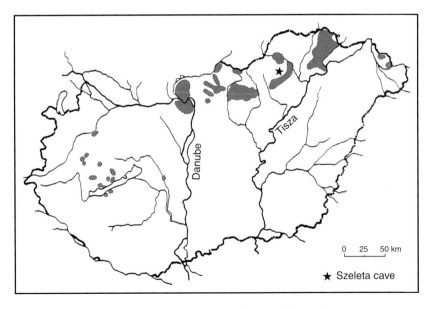

Figure 18.4   Hydrothermal and limnic silicite sources in Hungary (*Source*: Takács-Biró).

variable, especially with regard to homogeneity and isotropy due to imperfections such as fissures, cavities, impurities along bedding planes, fossil and crystal inclusions, etc. Clearly prehistoric knappers were able to produce finely worked leaf point from felsitic quartz porphyry. However, even this material exhibits inclusions which interfere with the otherwise laminar structure that is conducive to the production of thin bifaces (Simán 1986). Such inclusions can cause seemingly homogenous pieces of felsitic quartz porphyry to shatter unpredictably. The same is true for the hydro- and limnoquartzites, where quality of a single nodule can vary from very fine- to coarse-texture due to the presence of fossils, voids, and internal fissures. In short, the Szeleta Cave assemblages are made almost exclusively of medium- to poor-quality lithic materials that limit the production of standardized, typical tool forms. The influence of lithic raw material on artifact form is well documented (e.g., Vértes 1964; Gábori-Csánk 1968; Lischka 1969; Hayden 1980; Straus 1980; Dibble 1985, 1991; Barham 1987; Kretzoi and Dobosi 1990; Simek 1991; Tieu 1991; Reher and Frison 1991; Andrefsky 1994; Kuhn 1995; Amick and Mauldin 1997; Blades 2001). Poor quality raw material has been suggested as a partial explanation for the rarity of Lower Paleolithic handaxe cultures in eastern Asia and the general "simple" appearance of lithic industries in this area (Klein 1989; Schick and Toth 1993). Similarly, the use of more intractable raw material has been cited as a partial explanation for the production of crude Developed Oldowan bifaces and the contemporary, more refined early Acheulean axes in east Africa (Stiles 1979; Jones 1981; Schick and Toth 1993). Others working in northern Hungary have come to the similar conclusions. For example, the assemblage from Püspökhatvan in northern Hungary, excavated by Éva Csongrádi-Balogh and Viola Dobosi (1995), consists of both "archaic" and "Upper Paleolithic" characteristics. "Archaic" characteristics consist of crudely fashioned bifaces and large scrapers and burins, while "Upper Paleolithic" traits consist of smaller, more refined burins, scrapers, cores, tanged fragments, and blade production. Significantly, the bifaces are described as analogous to the rough types from Szeleta Cave. The Püspökhatvan assemblage is made primarily from local hydroquartzites which exhibit "faults", plant remains, and other inclusions. Based on typology alone, this assemblage could be classified as yet another "transitional" industry like the material from Szeleta Cave. However, the excavators attribute the "archaic"

attributes to poor raw material quality, while a C-14 date of about 28 000 years ago puts the material well within the range of the Upper Paleolithic and is close to the newly obtained dates from Szeleta Cave. Like the assemblage from Püspökhatvan, the "archaic" appearance of the Szeleta Cave assemblages can be explained by poor raw material quality. I suggest that in assemblages based on medium to poor quality cherts, the occurrence of "primitive" types is more likely the result of poor flaking quality rather than an indication of cultural "backwardness" or the retention of so-called "archaic" traits. In short, the raw material stock used for the bulk of the artifacts from Szeleta Cave lacks the homogeneity and, thus flaking predictability, of other materials in the region, such as obsidian from the Tokaj area or flints from southern Poland.

POST-DEPOSITIONAL CONSIDERATIONS

While raw material quality can help explain the occurrence of "archaic" types, the potential impact of post-depositional modification must also be considered. Like most Central European late Pleistocene cave sites, the faunal assemblage from Szeleta consists almost exclusively of cave bear remains, followed by other large species such as brown bear and cave hyena (Kadić 1916). The weight of an adult cave bear is estimated at close to 500 kg (Kurtén 1968, 1976), and continued trampling by such animals would have undoubtedly altered the appearance and distribution of lithic artifacts on the cave floor just as they crushed and dispersed their own remains in preparation for hibernation, as Gargett (1996) has demonstrated at Pod Hradem Cave in the Czech Republic. Indeed, recent experimental work by McBrearty et al. (1998) demonstrates that human trampling can produce edge damage that is easily mistaken for deliberately retouched Middle Paleolithic types such as notched and denticulated pieces. The authors observed that edge modification can be severe on artifacts trampled on fine-grained sediments, and one can easily envision the effect of repeated trampling by cave bears (and other cave occupants) on artifacts deposited on cave floors with coarse, gravelly substrates.

SUMMARY

In conclusion, the accuracy of a model is only as good as its component parts, and in the case of the Szeletian, models of the Middle to Upper Paleolithic transition have relied almost exclusively on typological studies of lithic artifacts. As I have argued here, other data exist that have not been incorporated into the currently popular models of the Szeletian material. These consist of lithic raw material quality and a consideration of the potential effect of non-hominid behavior on lithic artifact morphology. A consideration of these factors exposes the acculturation model as overly simplistic and emphasizes the weakness of arguments based on lithic typological analysis alone. In addition, new chronometric data from Szeleta Cave demonstrate the fallacy of equating lithic artifact crudeness with "archaic-ness." Recognition of the potential impact of lithic raw material quality and post-depositional processes on artifact morphology can help shift the focus of Paleolithic studies from purely cultural-chronological issues to questions of human adaptations and interactions.

REFERENCES

Adams, B. (1998) *The Middle to Upper Paleolithic Transition in Central Europe: The Record from the Bükk Mountain Region*. BAR International Series 693, Archaeopress, Oxford.

Adams, B. and Ringer, A. (2004) New C14 dates for the Hungarian Early Upper Palaeolithic. *Current Anthropology* 45(4), 541–51.

Allsworth-Jones, P. (1986) *The Szeletian*. Clarendon Press, Oxford.

Amick, D.S. and Mauldin, R.P. (1997) Effects of raw material on flake breakage patterns. *Lithic Technology* 22(1), 18–27.

Andrefsky, W. (1994) The geological occurrence of lithic material and stone tool production strategies. *Geoarchaeology* 9, 375–91.

Anikovich, M. (1992) Early upper paleolithic industries of Eastern Europe. *Journal of World Prehistory* 6(2), 205–45.

Barham, L.S. (1987) The bipolar technique in southern Africa: A replication experiment. *The South African Archaeological Bulletin* 42, 45–50.

Biró, K. and Dobosi, V. (1991) *Lithoteca Comparative Raw Material Collection of the Hungarian National Museum*. Hungarian National Museum, Budapest.

Blades, B. (2001) *Aurignacian Lithic Economy: Ecological Perspectives from Southwestern France*. Kluwer Academic/Plenum, New York.

Csongrádi-Balogh, É. and Dobosi, V.T. (1995) Palaeolithic settlement traces near Püspökhatvan. *Folia Archaeologica* XLIV, 37–59.

Dibble, H.L. (1985) Raw material variation in Levallois flake manufacture. *Current Anthropology* 26, 391–3.

Dibble, H.L. (1991) Local raw material exploitation and its effects on Lower and Middle Paleolithic Assemblage Variability. In: Montet-White, A. and Holen, S, (eds) *Raw Material Economies among Prehistoric Hunter-Gatherers*, University of Kansas, Publications in Anthropology 19. Lawrence, Kansas, pp. 33–47.

Dobosi, V. (1986, May 20–22) Raw material investigations on the finds of some Paleolithic sites in Hungary. In: Katalin, T.B. (ed.) *International Conference on Prehistoric Flint Mining and Lithic Raw Material Identification in the Carpathian Basin-Sümeg*, Magyar Nemzeti Muzeum, Budapest, pp. 249–55.

Dobosi, V. (1991) Economy and raw material: A case study of three Upper Paleolithic sites in Hungary. In: Montet-White, A. and Holen, S. (eds) *Raw Material Economies among Prehistoric Hunter-Gatherers*, University of Kansas, Publications in Anthropology 19. Lawrence, Kansas, pp. 197–203.

Gábori, M. (1969) Regionale Verbreitung Paläolithischer Kulturen Ungarns. *Acta Archaeologica Academiae Scientiarum Hungaricae* XXI, 155–65.

Gábori-Csánk, V. (1968) *La Station du Paléolithique Moyen d'Érd, Hongrie*. Akadémia Kiadó, Budapest.

Gábori-Csánk, V. (1993) *Le Jankovichien: Une Civilisation Paléolithique en Hongrie*. Etudes et Recherches Archéologiques de l'Université de Liège.

Gargett, R.H. (1996) *Cave Bears and Modern Human Origins*. University Press of America, Inc., Lanham, Maryland.

Gladilin, V.N. and Demidenko, YU.E. (1989) Upper Palaeolithic Stone Tool Complexes from Korolevo. *Anthropologie* XXVII(2–3), 143–78.

Hayden, B. (1980) Confusion in the bipolar world: Bashed pebbles and splintered pieces. *Lithic Technology* 9(1), 2–7.

Jones, P. (1981) Experimental implement manufacture and use: A case study from Olduvai Gorge. *Philosophical Transactions of the Royal Society (London)* B292, 189–95.

Kadić, O. (1916) Ergebnisse der Erforschung der Szeletahöhle. *Mitteilungen aus dem Jahrbuch der Könglichischen Ungarischen Geologischen Reichanstalt* XXIII (4).

Karavanić, I. and Smith, F. H. (2000) More on the Neanderthal problem: The Vindija case. *Current Anthropology* 41, 838–40.

Klein, R.G. (1989) *The Human Career*. The University of Chicago Press, Chicago.

Kozlowski, J.K. (1988) Problems of continuity and discontinuity between the Middle and Upper Paleolithic of Central Europe. In: Dibble, H.L. and Montet-White, A. (eds) *Upper Pleistocene Prehistory of Western Eurasia*, University Museum, University of Pennsylvania, pp. 349–60.

Kretzoi, M. and Dobosi, V.T. (eds) (1990) *Vértesszölös: Man, Site and Culture*. Akadémia Kiadó, Budapest.

Kurtén, B. (1968) *Pleistocene Mammals of Europe*. Weidenfeld and Nicolson, London.

Kurtén, B. (1976) *The Cave Bear Story*. Columbia University Press, New York.

Kuhn, S.L. (1995) *Mousterian Lithic Technology: An Ecological Perspective*. Princeton University Press, Princeton.

Lischka, L. (1969) A possible noncultural bias in lithic debris. *American Antiquity* 34, 483–5.

Markó, A., Bíró, K. and Kasztovaszky, Zs. (2003) Szeletian felsitic porphyry: Non-destructive analysis of a classical Palaeolithic raw material. *Acta Archaeologia Hungarica* 54, 297–314.

McBrearty, S., Bishop, L., Plummer, T., Dewar, R. and Conard, N. (1998) Tools underfoot: Human trampling as an agent of lithic artifact edge modification. *American Antiquity* 63(1), 108–29.

Reher, C.A. and Frison, G.C. (1991) Rarity, clarity, symmetry: Quartz crystal utilization in hunter-gatherer stone tool assemblages. In: Montet-White, A. and Holen, S. (eds) *Raw Material Economies Among Prehistoric Hunter-Gatherers*, University of Kansas, Publications in Anthropology 19. Lawrence, Kansas, pp. 375–97.

Ringer, Á. (1990) Le Szélétien dans le Bükk en Hongrie: chronologie, origine et transition vers le Paléolithique supérieur. In: *Paléolithique moyen et recent et Paléolithique supérieur ancien en Europe*. Actes du Colloque international de Nemours 9-10-11 Mai 1988. Mémoires du Musée de Préhistoire d'Ile de France No. 3, pp. 107–9.

Schick, K.D. and Toth, N. (1993) *Making Silent Stones Speak*. Simon and Schuster, New York.

Simán, K. (1986, May 20–22) Felsitic Quartz Porphyry. In: Katalin, T.B. (ed.) *International Conference on Prehistoric Flint Mining and Lithic Raw Material Identification in the Carpathian Basin-Sümeg*, Magyar Nemzeti Muzeum, Budapest, pp. 271–5.

Simán, K. (1991) Patterns of raw material use in the Middle Paleolithic of Hungary. In: Montet-White, A. and Holen, S. (eds) *Raw Material Economies among Prehistoric Hunter-Gatherers*, University of Kansas, Publications in Anthropology 19. Lawrence, Kansas, pp. 49–57.

Simek, J.F. (1991) Stone tool assemblages from Krapina (Croatia, Yugoslavia). In: Montet-White, A. and Holen, S. (eds) *Raw Material Economies among Prehistoric Hunter-Gatherers*, University of Kansas, Publications in Anthropology 19. Lawrence, Kansas, pp. 59–71.

Smith, F.H., Trinkaus, E., Pettitt, P.B., Karavanić, I. and Paunović, M. (1999) Direct radiocarbon dates for Vindija G1 and Velika Pećina Late Pleistocene hominid remains. *Proceedings of the National Academy of Sciences U.S.A.* 97, 7663–5.

Stiles, D. (1979) Early Acheulean and Developed Oldowan. *Current Anthropology* 20, 126–9.

Straus, L.G. (1978) Of Neanderthal hillbillies, origin myths, and stone tools: Notes on Upper Paleolithic assemblage variability. *Lithic Technology* 7(2), 35–9.

Straus, L.G. (1980) The role of raw materials in lithic assemblage variability. *Lithic Technology* 9, 68–72.

Straus, L.G. (1993) Saint Césaire and the Debate on the Transition from the Middle to Upper Paleolithic. In: *Context of a Late Neandertal*, Monographs in World Archaeology No. 16. Prehistory Press, Madison, Wisconsin, pp. xi–xii.

Svoboda, J. and Simán, K. (1989) The Middle-Upper Paleolithic transition in southeastern Central Europe (Czechoslovakia and Hungary). *Journal of World Prehistory* 3(3), 283–322.

Svoboda, J., Ložek, V., and Vlček, E. (1996) *Hunters Between East and West*, Plenum Press, New York and London.

Takács-Biró, K. (1986a, May 20–22) The raw material stock for chipped stone artefacts in the Northern Mid-Mountains Tertiary in Hungary. In: Katalin, T.B. (ed.) *International Conference on Prehistoric Flint Mining and Lithic Raw Material Identification in the Carpathian Basin-Sümeg*, Magyar Nemzeti Muzeum, Budapest, pp. 183–95.

Takács-Biró, K. (1986b) Sources of raw materials used for the manufacture of chipped stone implements in Hungary. In: Sievking, G.De C. and Hart, M.B. (eds.) *The Scientific Study of Flint and Chert*, Cambridge University Press, Cambridge, pp. 121–32.

Tieu, L.T. (1991) *Palaeolithic Pebble Industries in Europe*. Akadémia Kiadó, Budapest.

Valoch, K. (1986) Stone industries of the Middle/Upper Palaeolithic transition. In: *The Pleistocene Perspective*, vol. I, The World Archaeological Congress, 1–7 September 1986. Allen and Unwin, Southhampton, pp. 263–8.

Vértes, L. (1955) Neure Ausgrabungen und Paläolithische Funde in der Höhle von Istállóskő. *Acta Archaeologica Academiae Scientiarum Hungaricae* V(3–4), 111–31.

Vértes, L. (1959) *Untersuchungen an Höhlensedimenten*. Régészeti Füzetek II (7). Magyar Nemzeti Múzeum-Történeti Muzeum, Budapest.

Vértes, L. (1964) *Tata: Eine Mittelpaläolithische Travertin-Siedlung inUngarn*. Akadémia Kiadó, Budapest.

Vértes, L. (1968) Szeleta-Symposium in Ungarn. *Quartär* 19, 381–90.

Vértes, L. and Toth, L. (1963) Der Gebrauch de Glasigen Quarzporphyrs in Paläolithikum des Bükk-Gebirges. *Acta Archaeologica Academiae Scientiarum Hungaricae* XV, 3–10.

Zilhão, J. and D'Errico, F. (2003) The Chronology of the Aurignacian and Transitional Technocomplexes. Where do We Stand? In: Zilhãom, J. and D'Errico, F. (eds) *The Chronology of the Aurignacian and of the Transitional Technocomplexes: Dating, Stratigraphies, Cultural Implications*, Arqueologia 33. Lisboa, Instituto Português de Arqueologia, Portugal, pp. 313–48.

Chapter 19

RAW MATERIAL DURABILITY, FUNCTION, AND RETOUCH IN THE UPPER PALEOLITHIC OF THE TRANSBAIKAL REGION, SIBERIA

Karisa Terry[1], William Andrefsky, Jr.[1], and Mikhail V. Konstantinov[2]

[1]Department of Anthropology, Washington State University
[2]Chita State Pedagogical University

ABSTRACT

Stone tool reduction intensity and raw material availability have often been correlated with human organizational strategies. Assemblages from Early, Middle, and Late Upper Paleolithic sites in the Transbaikal region of Siberia were compared and assessed for retouch amount using several techniques. These studies reveal that stone tool functional requirements play an important role in the amount and intensity of retouch in the region. It is also shown that artifact function correlates with raw material selection, where certain kinds of activities, such as scraping, are most effectively accomplished with specific raw material characteristics. Human organizational patterns in the Transbaikal region are only evident when these factors are considered in assessing retouch measures.

Lithic Materials and Paleolithic Societies, 1st edition. Edited by B. Adams and B.S. Blades, ©2009 Blackwell Publishing. ISBN 978-1-4051-6837-3.

Retouch intensity on stone tools has long been associated with artifact curation (Binford 1973; Bamforth 1986; Nelson 1991; Shott 1996). It has been shown that artifact curation has much to do with raw material availability (Bamforth 1986; Andrefsky 1994a; Kuhn 1995), and human land-use practices (Hiscock 1994; Daniel 2001; Blades 2003) as these relate to location and quality of toolstone, and relationships with other key resources. Several stone tool assemblages from Upper Paleolithic sites in the Transbaikal region of Siberia, however, do not directly correlate with raw material availability and human land-use practices. In this study, we applied various measures of retouch amount on excavated and experimentally replicated assemblages in an effort to isolate other factors influencing stone tool retouch.

In addition to human land-use practices, we found that raw material package size and artifact function play important roles in stone tool retouch amounts.

PROJECT BACKGROUND

Our exploration of retouch amount evaluates eight sites from the Siberian Transbaikal region, the area southeast of Lake Baikal to the Russian–Mongolian border. Figure 19.1 shows the project area and locations of sites in this study. Three periods of the Upper Paleolithic are included within this analysis; Early Upper Paleolithic (EUP), 35 000–27 000 years ago; Middle Upper Paleolithic (MUP), 27 000–18 000 years ago, and Late Upper Paleolithic (LUP), 18 000–12 000

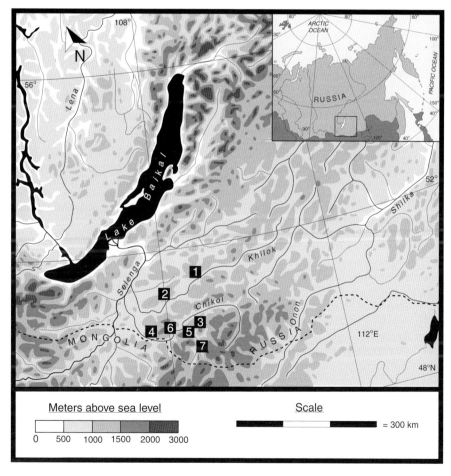

Figure 19.1   Map of the Transbaikal region, Russia showing locations of sites in this study. 1: Tolbaga, 2: Kunalei, 3: Priiskovoe, 4: Chitkan, 5: Melnichnoe 2, 6: Studenoe 1 and 2, 7: Ust' Menza 1 and 2.

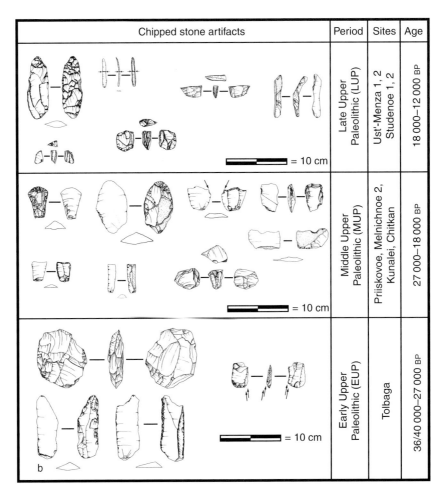

Figure 19.2 Diagram showing Upper Paleolithic cultural periods, ages, sites, and representative artifacts discussed in this study.

years ago (Figure 19.2). Study sites, along with frequencies of artifact types, are listed in Table 19.1. Artifacts were classified based on morphological attributes developed by Bordes (1961) and de Sonneville-Bordes and Perrot (1954–56) for the European Middle and Upper Paleolithic periods, as well as Russian terminology described by Medvedev *et al.* (1974). For this analysis we made comparisons of flake tools including raclettes, notches, scrapers, knives, and denticulates.

Previous research in Siberia (Goebel 1999, 2002, 2004) has characterized each time period in terms of Binford's (1980) Forager-Collector model. Goebel's interpretations are based on characteristics of site function, artifact assemblages, and faunal assemblages representative of each time period. For this study,

we chose to describe details about sites located in the Transbaikal region during the entire Upper Paleolithic and how they relate to more extensive patterns in Siberia, as this provides details about the variability of past settlement organization in the Transbaikal.

The adaptive strategy during the EUP has been depicted as residentially sedentary groups occupying locations for long periods of time. Goebel (1999) and Goebel *et al.* (2001) posit that groups were positioning relatively long-term camps in resource rich areas where diverse game animals and lithic raw materials were abundant. EUP sites are characterized by large, semi-subterranean dwellings with stone-lined hearths and storage pits. Tolbaga, the type site for the EUP in the Transbaikal, is a permanent residential location with

Table 19.1 Artifact type frequencies from Early (EUP), Middle (MUP), and Late (LUP) Upper Paleolithic sites in the study area.

Time period	Site	Point	Raclette	Notch	Scraper	Knife	Denticulate	Burin	Percoir	Wedge
EUP	Tolbaga (N = 173)	4(2%)	4(2%)	24 (14%)	121 (70%)	2 (1%)	6 (4%)	8 (5%)	2 (1%)	2 (1%)
MUP	Melnichnoe 2 (N = 26)	0	2 (8%)	7 (27%)	6 (23%)	1(4%)	0	6 (23%)	3 (11%)	1(4%)
	Priiskovoe (N = 94)	0	0	17 (18%)	72 (77%)	0	5 (5%)	0	0	0
	Chitkan (N = 30)	0	3 (10%)	13 (43%)	6 (20%)	0	0	7 (23%)	1 (4%)	0
	Kunalei (N = 89)	0	2 (2%)	11 (12%)	70 (79%)	0	6 (7%)	0	0	0
LUP	Ust-Menza 1 (N = 26)	1 (4%)	0	1 (4%)	20 (77%)	0	0	4 (15%)	0	0
	Ust-Menza 2 (N = 11)	0	0	2 (18%)	2 (18%)	0	0	3 (28%)	2(18%)	2 (18%)
	Studenoe 1 (N = 9)	0	0	0	8 (89%)	0	0	1 (11%)	0	0
	Studenoe 2 (N = 46)	1 (2%)	2 (3%)	1 (2%)	21 (47%)	0	1(2%)	14 (31%)	4 (9%)	2 (4%)

dwellings containing numerous stone-lined hearths and storage pits (Konstantinov 1994). Chipped stone artifacts during the EUP maintained a few elements of earlier Levallois technology (Konstantinov 1994; Konstantinov *et al.* 2003; Brantingham *et al.* 2004), but there was more emphasis on large end-struck blades (Goebel 2004). In addition to sub-prismatic cores, at Tolbaga large blades were produced from Levallois-like cores and classic Levallois cores. A wide variety of tools at Tolbaga and other EUP sites in the Transbaikal were manufactured on large-blade blanks (Konstantinov 1994; Derev'anko *et al.* 1998; Goebel 1999; Dolukhanov *et al.* 2002; Lbova 2002). Faunal remains from Tolbaga (Table 19.2) represent diverse species that inhabit open landscapes (Konstantinov 1994). The EUP corresponds to the last interglacial of the Pleistocene, locally referred to as the Karginsk; the few pollen data currently available indicate that birch and pine forests were present during the earliest human occupation of the site (Konstantinov 1994). The environment in general may have been less patchy, allowing for relatively long-term human use of resource rich areas.

The MUP adaptive strategy has been described as logistical foraging from semi-permanent residential locations (Goebel 1999, 2002). There is evidence that groups were hunting one or two fauna species from small campsites. In the Transbaikal both MUP semi-permanent residential and short-term extraction

sites exist. Priiskovoe, for example, is a semi-permanent residential site with a dwelling and hearth features (Konstantinov 1994). Kunalei, Chitkan, and Melnichnoe 2, on the other hand, were likely campsites used relatively briefly during resource extraction. Kunalei contained a small stone-lined hearth, but features at Chitkan and Melnichnoe 2 were limited to charcoal smears. Chitkan was reoccupied several times and may have been used seasonally (Konstantinov 1994). Faunal assemblages at these MUP sites, where data exists, are less diverse than the EUP (Table 19.2). Unlike EUP assemblages, there is a complete absence of Levallois technology at MUP sites. MUP toolkits are still blade based, but the size of the artifacts generally decreased and the diversity of forms increased compared to EUP industries (Table 19.1 and Figure 19.2) (Konstantinov 1994; Derev'anko *et al.* 1998; Goebel 1999, 2000; Dolukhanov *et al.* 2002). This reduction of tool size may reflect increased mobility.

Sites during the LUP were repeatedly occupied by mobile groups for short periods of time (Goebel 1999, 2002, 2004). LUP sites in the Transbaikal, such as Studenoe 1 and 2 and Ust' Menza 1 and 2, reveal dwellings clearly demarcated by river cobbles with central stone-lined hearths. These dwellings are thought to have been occupied briefly, perhaps on a seasonal basis as indicated by very thin living floors with few artifacts (Goebel 1999; Buvit 2000; Goebel *et al.* 2000).

Table 19.2 Fauna recovered from Early (EUP), Middle (MUP), and Late (LUP) Upper Paleolithic sites in the study area (Konstantinov 1994).

Time period	Site	Wooly rhinoceros	Woolly mammoth	Bison	Bear	Cave hyena	Horse	Saiga	Wild ass	Argali sheep	Antelope	Gazelle	Roe deer	Red deer
EUP	Tolbaga	X	X	X		X	X	X	X	X	X	X		X
MUP	Priiskovoe			X	X		X							X
	Kunalei	X		X			X	X						X
LUP	Ust-Menza 1												X	X
	Studenoe 1&2												X	X

Inhabitants focused on hunting reindeer and roe deer (Table 19.2) (Konstantinov 1994; Derev'anko *et al.* 1998; Goebel 1999; Dolukhanov *et al.* 2002). Groups were likely moving relatively frequently but reoccupying the same base camp during certain times of year to extract targeted resources (red deer and roe deer). Microblade technology appears in the Transbaikal during this time period, sometime between 21 000 and 18 000 years ago (Kuzmin and Orlova 1998; Goebel *et al.* 2000; Goebel 2002; Buvit *et al.* 2004; Kuzmin and Keates 2005). This technology is regarded as a means of reducing risk by conserving raw material and maximizing the amount of cutting edge on tools (Flenniken 1987; Rasic and Andrefsky 2001; Elston and Brantingham 2002). Other stone tools found in the Transbaikal during the LUP were made from blades and flakes (see Table 19.1). To gain a better understanding of how these groups were using the landscape during the Upper Paleolithic, our studies of lithic raw material use incorporated source location, raw material quality, and retouch amounts into the above models.

RAW MATERIAL AVAILABILITY AND USE

Previous studies have shown that more sedentary foragers should rely more heavily on local lithic raw materials, provided adequate material is abundant. Conversely, if groups are more mobile, they should transport raw materials from more distant sources to reduce the risk of not having sufficient toolstone at new locations (Parry and Kelly 1987; Bamforth 1990, 1991; Andrefsky 1994a,b; Kuhn 1995). Therefore, in the Transbaikal during the EUP, when groups occupied residences for relatively long time periods, frequencies of non-local raw materials should be low if abundant adequate raw materials exist near these sites. As foragers began using different strategies during the MUP, and especially the LUP in which they moved around on the landscape more frequently, use of non-local high quality materials should increase, while reliance on local stone should decrease.

All sites discussed here contain local very fine- to medium- grained argillite, as well as more durable raw materials, such as quartzite, diorite, gabbro, and basalt. These materials are fairly abundant near the sites. Non-local high chipping quality cherts are found in assemblages, but are not known at sources close to the sites (i.e., <5 km from each site). Additionally, the locally available raw materials can be obtained in relatively

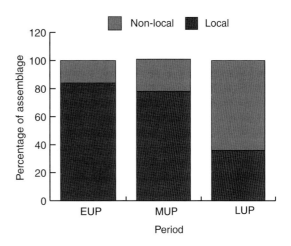

Figure 19.3 Frequencies of flake tool raw material types for the Early (EUP), Middle (MUP), and Late (LUP) Upper Paleolithic periods in the Transbaikal region, Russia.

larger specimen sizes than the non-local cherts. Figure 19.3 shows frequencies of different raw materials found in artifact form during each time period. The frequency of local and non-local raw materials differs significantly in the Late Upper Paleolithic when compared to the Early and Middle Upper Paleolithic ($X^2 = 56.218$, $df = 2$, $p < 0.0005$). During the EUP local raw materials dominate the assemblages. Again, local toolstone is more abundant during the MUP, but the occurrence of non-local cherts increases slightly. LUP assemblages, in contrast, are dominated by high quality non-local chert. Considering traditional land-use and raw material economy models, tools of non-local raw materials should exhibit more retouch than those of locally available stone (Bamforth 1986, 1990, 1991; Andrefsky 1994b), implying that non-local stone is more heavily curated than local stone. Conversely, tools of local stone are replaced more readily than tools on non-local material.

STONE TOOL RETOUCH AND RAW MATERIAL AVAILABILITY

The amount of retouch on tools has traditionally been tied to how often foragers move around on the landscape. The logic is that as mobility increases folks will more intensively retouch tools of non-local high quality raw material (Marks *et al.* 1991; Kuhn 1995; Shott 1995). However, recent studies reveal that other factors must also be considered when evaluating retouch amounts on stone tools including the behavior of game animals, whether they are dispersed or aggregated (Blades 2003), and if processing of resources is intense (Hayden *et al.* 1996; Tomka 2001).

It has been shown that using several retouch measures is preferred as different techniques reflect different dimensions of reduction (Eren *et al.* 2005; Andrefsky 2006). The amount of retouch for this study was calculated using two measures. One measure we used was Kuhn's (1990) flake tool index of retouch intensity, which incorporates edge angle, retouch invasiveness, and thickness of the specimen to evaluate how much resharpening the tool has undergone. As the tool edge is retouched, a portion of this edge will be lost and the relationship of the edge angle to the tool's thickness will change. This measure has been shown to be an especially effective technique for side scrapers (Hiscock and Clarkson 2005). The second index we used was a measure of retouch extent that calculates the amount of the tool margin exhibiting retouch (Barton 1988). This measure was calculated as the percentage of tool edge with retouch. Presumably, as a stone tool is used and reused, the amount of edge that is modified should also increase. This index has proven useful in many studies on stone tool variability and reduction models (Barton 1988; Kuhn 1995; Hiscock and Attenbrow 2003).

Retouch intensity and extent reveal unexpected patterns with regard to raw material availability (Figure 19.4). Locally available raw materials, such as argillite, quartzite, diorite, and gabbro, show similar retouch intensity and extent rates as non-local high quality cherts during each time period. Although land-use practices change through time there is no significant change in retouch intensity and extent between local and non-local raw materials (Table 19.3). These trends suggest that some factor other than availability of raw materials was influencing retouch intensity and extent. However, retouch extent and intensity are

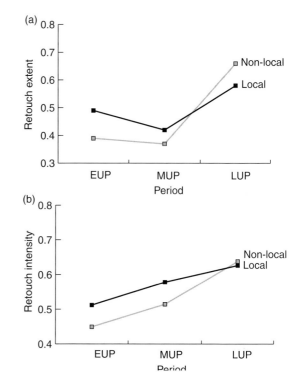

Figure 19.4 Retouch extent (a) and retouch intensity (b) of local (black) and non-local (gray) raw materials during the Early (EUP), Middle (MUP), and Late (LUP) Upper Paleolithic in the Transbaikal region, Russia.

Table 19.3 Significance values comparing retouch extent and intensity of local and non-local raw material by time period (Values from Figure 19.4).

Time period	Retouch index (local vs. non-local stone)	*T*	*Df*	*P*
EUP	Extent	1.302	151	0.195
	Intensity	1.165	151	0.246
MUP	Extent	0.997	202	0.320
	Intensity	1.370	202	0.172
LUP	Extent	−0.563	54	0.576
	Intensity	0.186	54	0.853

not exactly the same over time. Retouch extent drops from EUP to MUP and then increases again during the LUP. Retouch intensity gradually increases from the EUP through LUP. These trends suggest that retouch extent and intensity may be sensitive to different kinds of influences. In any case, it is also apparent that some factor other than lithic raw material availability was influencing both retouch extent and intensity.

TOOL FUNCTION AND RETOUCH INTENSITY

We suspected that stone tool function was important for understanding retouch extent and intensity, and that retouch was not simply a matter of raw material availability. Studies based on amounts of tool reduction and use, microwear analysis, and ethnographic studies have shown that tool function does not necessarily equate to tool form (Ahler 1971; Odell 1981; Dibble 1987; Barton 1991; Andrefsky 1997). For example, Odell (1981) compared artifact forms to microscopic use-wear on tools. His study revealed that some tool forms, such as burins and microlithic points, and their single inferred function correlated. Other forms, such as scrapers and knives, however, were used for several tasks. Andrefsky (1997) also notes that ethnographically toolmakers equate working edge, rather than tool form, with function. Even though tool form may not correlate with function in some cases, it is reasonable to expect that steep edge angled tools, such as scrapers, have a different functional effectiveness than more acute angled tools such as knives. Nor is it unreasonable to assume that "saw toothed" or serrated tools might be more effective for things like sawing than smooth edged tools such as raclettes. Similarly, it is possible that tool forms with the same edge angles and/or

edge designs, but of considerably different sizes, might be used to perform different kinds of tasks. The archeological literature is full of such assumptions relating an increase in tool-form diversity to increases in site function (Shott 1986; Chatters 1987; Andrefsky 2005).

To explore the relationship between tool function and retouch amount, we first compared tool forms (functional proxy) to amounts of retouch extent and intensity (Table 19.4). These data clearly show that retouch intensity and extent are greater on certain tool forms than others. Scrapers and denticulates, which typically exhibit high edge angles, have the highest amount of retouch in terms of extent and intensity. On the other hand, knives, which usually exhibit low edge angles, have the lowest amounts of retouch. If we use tool form as a possible proxy for tool function, our data show a significant relationship between tool function and retouch (extent $F = 14.229$, $df = 4$, $p < 0.0005$; intensity $F = 32.677$, $df = 4$, $p = 0.032$). However, embedded in this relationship is another characteristic that needs further exploration. Tools such as scrapers and denticulates are by definition recognized as scrapers and denticulates because of the type and extent of modification they undergo. These classes of stone tools are defined based upon more extensive formal variability derived from retouching them into particular shapes by the tool makers and users. It is little wonder that these tool forms have greater retouch values than knives, notches, and raclettes.

Since scrapers consistently had higher retouch values than all other tool forms we examined the population of scrapers more closely, thinking that perhaps tool function may be related to tool size and raw material differences. Table 19.5 compares large and small scrapers against locally and non-locally available lithic raw materials. In the archeological data, locally available raw materials were favored for large scrapers, whereas

Table 19.4 Retouch indices for tool types during the Early, Middle, and Late Upper Paleolithic in the Transbaikal.

Retouch measure	Tool type				
	Raclette mean (SD)	Knife mean (SD)	Notch mean (SD)	Scraper mean (SD)	Denticulate mean (SD)
Extent	0.38 (0.21)	0.29 (0.07)	0.22 (0.17)	0.51 (0.30)	0.56 (0.41)
Intensity	0.4279 (0.4177)	0.2585 (0.1365)	0.4874 (0.3163)	0.5822 (0.3356)	0.5465 (0.2636)

Table 19.5 Comparison of raw material availability and scraper size for the Early, Middle, and Late Upper Paleolithic periods in the Transbaikal.

Raw material	Scraper	
	Small (>10 g)	Large (<10 g)
Locally available	18 (32%)	224 (89%)
Non-locally available	47 (72%)	28 (11%)

$X^2 = 107.140$ df $= 1$ $p < 0.0005$.

small scrapers tended to be made of non-local cherts. Obviously, certain raw materials were selected for the manufacture of certain types of scrapers and possibly for certain tasks. We explored functional differences in tool size and raw material type by conducting a series of controlled experiments.

THE EXPERIMENT: TOOL FUNCTION AND LITHIC RAW MATERIAL

The assemblages from the Transbaikal show that different scraper sizes are significantly associated with variability in lithic raw material types. This suggests that either lithic raw material type may have been selected for its functional effectiveness, or that lithic raw material type may have been selected because of some aspect of abundance or size. To assess these possibilities, we conducted a series of experiments that tested the effectiveness of raw material type as scraping tools of varying sizes. Four small (<10 g) and four large (>10 g) scrapers were manufactured of chert and basalt, two of each raw material type, to mimic the small and large scrapers from the Transbaikal assemblages. Figure 19.5 shows some of the experimentally manufactured scrapers compared to samples of the excavated scrapers.

In the experiment, small scrapers were used to scrape bark from four 5 cm-diameter oceanspray branches. Bark on this type of wood is relatively thin and soft, requiring an acute edge angled tool. Large scrapers were used to scrape thick and hard bark from four 12 cm-diameter alder branches. Scraping on this type of material requires a high edge angle. Each branch was selected based on uniformity in diameter and

Figure 19.5 Comparison of scrapers from archeological assemblages from the Transbaikal region, Russia (a) and experimentally manufactured for functional studies (b).

then cut to equal lengths for each set of comparisons (all oceanspray = 160 cm; alder 2 pieces = 45 cm and 2 pieces = 73.5 cm).

After blanks were made and selected, each flake was shaped into either a large or small scraper. Each tool was used until it became ineffective and required resharpening. At this point, the length of bark scraped from the pieces of wood was marked and recorded for large and small scrapers of chert and basalt. Next, the tools were resharpened, and scraping continued on the wood using the same tool until it became ineffective again. Four large scrapers, two of each raw material type, and two small scrapers, one of each raw material type, were used to scrape bark twice. Only one small chert and one small basalt scraper underwent a third set of retouch and use. Each set of tool resharpening and use produced a scraping efficiency measure. Scraping efficiency of each tool was calculated by comparing the total area of bark scraped (linear cm of bark scraped multiplied by the diameter of each piece of wood) each time the tool was used until it needed resharpening.

When comparing the efficiency against raw material and scraper size, small scrapers show no significant difference regardless of raw material type ($t = 0.46$, $df = 8$, $p = 0.658$) (Figure 19.6). Small basalt scrapers approach the efficiency of small chert scrapers, especially if the edge is smooth and its angle is relatively low. However, large basalt scrapers are more

efficient than large chert scrapers ($t = -3.302$, $df = 6$, $p = 0.016$). Edges of large chert scrapers dulled very quickly, while the bulkier sizes of basalt scrapers allowed more pressure for productive scraping. Our experiments indicate that selection of locally abundant raw materials, especially durable stone types, for manufacture of large scrapers is based upon functional effectiveness. However, raw material package size may also be a factor. Experimentally produced large basalt scrapers (mean weight = 65 g) were more massive than experimental chert scrapers (mean weight = 35 g). At the archeological sites, locally available materials are found in larger sizes than non-local stone. Therefore, locally available stone types were possibly selected for use as large scrapers because of higher scraping efficiency and greater sizes.

DISCUSSION

Previous researchers have established that mobility range and frequency of movement increased throughout the Upper Paleolithic in Siberia (Goebel 1999, 2002, 2004). Such increases have often been associated with increases in amounts of non-local materials and retouch extent on tools. Our analysis shows that traditional interpretations about retouch amount and raw material availability must be tempered with aspects of tool function and effectiveness of raw material types for completion of certain tasks. During the EUP, when groups are thought to have more intensively exploited a variety of local resources during longer stays at the sites, local raw materials dominated assemblages. When land-use patterns shifted to more semi-permanent residences and small extraction campsites during the MUP, non-local cherts found in assemblages increase but local stone types still dominate. Use of non-local cherts dramatically increased in the LUP, when groups began shifting and reusing residences often and intensively exploiting key resources such as roe and red deer. Generally, retouch increased on tools made from non-local raw material types during the LUP when residential mobility increased, except in small scrapers (Figure 19.7).

Our investigations revealed that different raw material types were differentially effective for curtain task performances, and that such "functional" preferences influence raw material selection as much as or more than changes in mobility patterns in some tool forms. For instance, if we look more closely at the most

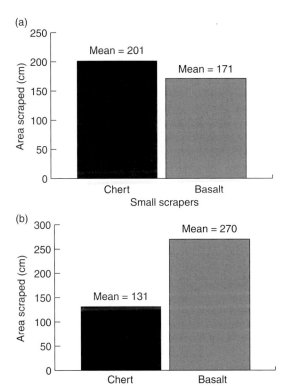

Figure 19.6   Comparison of scraping efficiency for small (a) and large (b) chert and basalt scrapers.

(a)

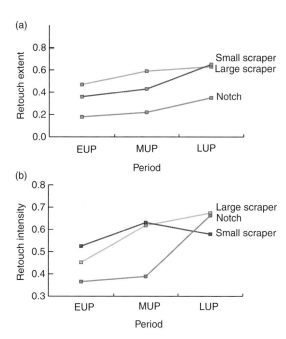

(b)

Figure 19.7 Retouch extent (a) and retouch intensity (b) of non-local raw materials during the Early (EUP), Middle (MUP), and Late (LUP) Upper Paleolithic in the Transbaikal region, Russia.

Table 19.6 Percentage of scraper assemblage using local and non-local lithic raw materials.

	Time period	Local (%)	Non-local (%)
Small scrapers	EUP	78	22
	MUP	35	65
	LUP	7	93
Large scrapers	EUP	86	14
	MUP	95	5
	LUP	73	27

heavily retouched tool category (scrapers) we see that large scrapers were primarily made from local raw materials throughout the entire Upper Paleolithic. Our experiments showed that this was probably related to the functional properties of more coarse-grained rocks such as quartzite and basalt for heavy scraping chores. It is also probable that the larger package size of these locally available quartzites and basalts influenced the selection of these rock types for large scraper production (Table 19.6). Similarly, retouch amounts on large scrapers are similar for local and non-local materials during all three time periods, a pattern that we would expect if local materials were favored based on function (extent EUP $t = 0.720$ $df = 110$ $p = .475$, MUP $t = -1.472$ $df = 117$ $p = 0.144$, LUP $t = -0.166$ $df = 20$ $p = 0.870$; intensity EUP $t = 1.459$ $df = 110$ $p = 0.142$, MUP $t = -0.364$ $df = 117$, $p = 0.717$, LUP $t = -0.22$ $df = 19$ $p = 0.827$).

The pattern of raw material selection for the production of small scrapers differed significantly from the larger scrapers. Local raw materials were used to make the majority of small scrapers in the EUP (78%), when the settlement pattern was relatively sedentary. During the MUP and into the LUP aboriginal populations increasingly became more residentially mobile. We see a drop in the use of local raw materials for the production of small scrapers to 35% and 7% respectively for the MUP and LUP (Table 19.6). We believe this trend deviates from the large scraper production trend because large scrapers are more closely linked to functional efficiency of raw material types and small scrapers are more closely linked to relative mobility patterns that afforded access to non-local lithic raw materials. Our experiments showed that there were no significant differences in small scraper efficiency based upon raw material type.

Since small scraper production is more likely to have been associated with residential mobility than large scrapers we would expect to see a greater amount of retouch on small scrapers from the EUP through the LUP. Since 93% of small scrapers were made from non-local lithic raw materials during the LUP we evaluated retouch extent using only local lithic raw materials. Small scrapers show an increase in retouch extent from 36% to 43% to 65% respectively for EUP, MUP, and LUP. Values of retouch intensity on non-local small scrapers do not increase during the LUP as we had expected. This is a pattern that we are currently unable to explain. However, intensity of non-local and local raw material use does change. Intensity values in the LUP are significantly greater for non-local materials than local materials ($t = -4.741$, $df = 28$, $p = 0.004$), a trend not seen in earlier time periods. Not only are local raw materials used less often for small scrapers, they are also resharpened less. This indicates that during the LUP non-local raw materials were relied on more heavily than local sources as mobility increased.

CONCLUSION

Our investigations have shown that retouch amount is specific to different tool forms. As such, it is not wise or reasonable to compare the different tool forms against one another using uniform retouch measures. Retouch should be compared among similar tool forms to gain a better understanding of tool curation and land use. In our study we found that large and small scrapers were sensitive to different external influences and, as such, it was not reasonable to expect these two different tool forms to express the same kinds of retouch patterns. Similarly, our data hint that different retouch indices may be more effective measures of retouch for specific kinds of tools. We also found that lithic raw material types (such as chert and more coarse-grained igneous rocks) may be significantly related to the effectiveness of performing various functions. The excavated assemblages, as well as experimental results, showed a significant association between functional variables and raw material types. We showed that raw material efficiency and effectiveness for performing certain functional tasks were related to the properties of the stone and also to raw material package size. Local, more durable, stone types are preferable for large scrapers based on scraping efficiency and large package size during the Upper Paleolithic. This had much to do with the function of large scrapers. Finally we have shown that raw material effectiveness for task completion is not the only factor involved with toolstone selection. Toolstone selection and stone tool use and retouch is related to human land-use practices for some stone tool types and not others. Our investigations suggest that human land-use practices are intimately related to artifact provisioning strategies but that artifact functional requirements are also influencing provisioning tactics, particularly where raw material properties such as size and consistency act upon artifact function.

REFERENCES

Ahler, S. (1971) *Projectile Point Form and Function at Rodgers Shelter, Missouri.* Missouri Archaeological Society Research Series no. 8, Missouri Archaeological Society, Columbia.

Andrefsky, W., Jr. (1994a) Raw material availability and the organization of technology. *American Antiquity* 59, 21–35.

Andrefsky, W., Jr. (1994b) The geological occurrence of lithic material and stone tool production strategies. *Geoarchaeology: An International Journal* 9, 345–62.

Andrefsky, W., Jr. (1997) Thoughts on stone tool shape and inferred function. *Journal of Middle Atlantic Archaeology* 13, 125–44.

Andrefsky, W., Jr. (2005) *Lithics: Macroscopic Approaches to Analysis*, 2nd edn. Cambridge University Press, Cambridge.

Andrefsky, W., Jr. (2006) Experimental and archaeological verification of an index of retouch for hafted bifaces. *American Antiquity* 71, 743–57.

Bamforth, D. (1986) Technological efficiency and tool curation. *American Antiquity* 51, 38–50.

Bamforth, D. (1990) Settlement, raw material, and lithic procurement in the Central Mojave Desert. *Journal of Anthropological Archaeology* 9, 70–104.

Bamforth, D. (1991) Technological organization and hunter-gatherer land use: A California example. *American Antiquity* 56, 216–34.

Barton, C.M. (1988) *Lithic Variability and Middle Paleolithic Behavior: New Evidence from the Iberian Peninsula.* BAR International Series 408, Oxford.

Barton, C.M. (1991) Retouched tools, fact or fiction?: Paradigms for interpreting Paleolithic chipped stone. In: Clark, G. (ed.) *Perspectives on the Past: Theoretical Biases in Mediterranean Hunter-Gatherer Research*, University of Pennsylvania Press, Philadelphia, pp. 143–63.

Binford, L.R. (1973) Interassemblage variability: The Mousterian and the "functional" argument. In: Renfrew, C (ed.) *The Explanation of Culture Change: Models in Prehistory*, Duckworth, London, pp. 227–54.

Binford, L.R. (1980) Willow smoke and dog's tails: Hunter-gatherer settlement systems and archaeological site formation. *American Antiquity* 45, 4–20.

Blades, B.S. (2003) End scraper reduction and hunter-gatherer mobility. *American Antiquity* 68, 141–56.

Bordes, F. (1961) *Typologie du Paleolithique Ancien et Moyen.* Publication de l'Institute de Prehistoire de l'Universite de Bordeaux, Bordeaux.

Brantingham, P.J., Kerry, K.W., Krivoshapkin, A.I. and Kuzmin, Y.V. (2004) Time-space dynamics in the early Upper Paleolithic of northeast Asia. In: Madsen, D.B. (ed.) *Entering America*, University of Utah Press, Salt Lake City, pp. 255–83.

Buvit, I. (2000) *The Geoarchaeology and Archaeology of Stud'onoye, a Late Upper Paleolithic Site in Siberia.* M.A. thesis, Department of Anthropology, Texas A&M University, College Station, Texas.

Buvit, I., Terry, K., Konstantinov, A.V., and Konstantinov, M.V. (2004) Studenoe 2: An update. *Current Research in the Pleistocene* 21, 1–3.

Chatters, J.C. (1987) Hunter-gatherer adaptations and assemblage structure. *Journal of Anthropological Research* 6, 336–75.

Daniel, I.R., Jr. (2001) Stone raw material availability and Early Archaic settlement in the southeastern United States. *American Antiquity* 66, 237–66.

Derev'anko, A.P., Shimkin, D.B., and Powers, W.R. (eds) (1998) *The Paleolithic of Siberia: New Discoveries and Interpretations*. University of Illinois Press, Urbana.

Dibble, H. (1987) Interpretation of Middle Paleolithic scraper morphology. *American Antiquity* 52, 109–17.

Dolukhanov, P.M., Shukurov, A.M., Tarasov, P.E., and Zaitseva, G.I. (2002) Colonization of northern Eurasia by modern humans: Radiocarbon chronology and environment. *Journal of Archaeological Sciences* 29, 593–606.

Elston, R.G. and Brantingham, P.J. (2002) Microlithic technology in northern Asia: A risk-minimizing strategy of the Late Paleolithic and Early Holocene. In: Elston, R.G. and Kuhn, S.L. (eds) *Thinking Small: Global Perspectives on Microlithization*, Archaeological Papers of the American Anthropological Association Number 12, Washington, DC, pp. 103–16.

Eren, M.I., Dominguez-Rodrigo, M., Kuhn, S.L., Adler, D.S., Le, I., and Bar-Yosef, O. (2005) Defining and measuring reduction in unifacial stone tools. *Journal of Archaeological Science* 32, 1190–206.

Flennikan, J. (1987) The Paleolithic Dyuktai pressure blade technique of Siberia. *Arctic Anthropology* 24, 117–32.

Goebel, T. (1999) Pleistocene colonization of Siberia and peopling the Americas. *Evolutionary Anthropology* 8, 208–27.

Goebel, T. (2000) Siberian middle Upper Paleolithic. In: Peregrine, P. and Ember, M. (eds) *Encyclopedia of Prehistory*. Kluwer Academic/Plenum Publishers, New York, pp. 192–6.

Goebel, T. (2002) The "microblade adaptation" and recolonization of Siberia during the late Upper Pleistocene. In: Elston, R.G. and Kuhn, S.L. (eds) *Thinking Small: Global Perspectives on Microlithization*, Archaeological Papers of the American Anthropological Association Number 12, Washington, DC, pp. 117–31.

Goebel, T. (2004) The early Upper Paleolithic of Siberia. In: Brantingham, P.J., Kuhn, S.L. and Kerry, K.W. (eds) *The Early Upper Paleolithic Beyond Western Europe*, University of California Press, Los Angeles, pp. 162–95.

Goebel, T., Waters, M. and Meshcherin, M. (2001) Masterov' Kliuch and the early Upper Paleolithic of the Transbaikal, Siberia. *Asian Perspectives* 74, 47–70.

Goebel, T., Waters, M.R., Buvit, I., Konstantinov, M.V. and Konstantinov, A.V. (2000) Studenoe-2 and the origins of microblade technologies in the Transbaikal, Siberia. *Antiquity* 74, 567–75.

Hayden, B., Franco, N. and Spafford, J. (1996) Evaluating lithic strategies and design criteria. In: Odell, G.H. (ed.) *Stone Tools: Theoretical Insights into Human Prehistory*, Plenum Press, New York, pp. 9–50.

Hiscock, P. (1994) Technological responses to risk in Early Holocene Australia. *Journal of World Prehistory* 8, 267–92.

Hiscock, P. and Attenbrow, V. (2003) Early Australian implement variation: A reduction model. *Journal of Archaeological Science* 30, 239–49.

Hiscock, P. and Clarkson, C. (2005) Experimental evaluation of Kuhn's geometric index of reduction and the flat flake problem. *Journal of Archaeological Science* 32, 1015–22.

Konstantinov, M.V. (1994) *The Stone Age of the Eastern part of Baikal Asia*. Chita State Pedagogical Institute, Chita, Russia (in Russian).

Konstantinov, M.V., Konstantinov, A.V., Vasiliev, S.G., Ekimova, L.V. and Razgil'deeva, I.I. (2003) *Under Protection of the Great Shaman*. Chita State Pedagogical University and Chita Institute of Natural Resources, Chita, Russia (in Russian).

Kuhn, S.L. (1990) A geometric index of reduction for unifacial stone tools. *Journal of Archaeological Science* 17, 585–93.

Kuhn, S.L. (1995) *Mousterian Lithic Technology*. Princeton University Press, Princeton, NJ.

Kuzmin, Y. and Keates, S. (2005) Dates are not just data: Paleolithic settlement patterns in Siberia derived from radiocarbon records. *American Antiquity* 70, 773–89.

Kuzmin, Y. and Orlova, L. (1998) Radiocarbon chronology of the Siberian Paleolithic. *Journal of World Prehistory* 12, 1–53.

Lbova, L. (2002) The transition from the Middle to Upper Paleolithic in the Western Trans-Baikal. *Archaeology, Ethnology and Anthropology of Eurasia* 1, 59–75.

Marks, A.E., Shokler, J. and Zilhao, J. (1991) Raw material usage in the Paleolithic: The effects of local availability on selection and economy. In: Montet-White, A. and Holen, S. (eds) *Raw Material Economies Among Prehistoric Hunter-Gatherers*, University of Kansas Publications in Anthropology 19, Lawrence, Kansas, pp. 127–40.

Medvedev, G.E., Mikhniuk, G.N. and Leshenko, I.L. (1974) On the nomenclatural designation and morphology of cores in the Preceramic Complexes of Preangara. In: Aksenov, M.P. (ed.) *Ancient History of Peoples of South Eastern Siberia*, Irkutsk State University, Irkutsk, Russia, pp. 60–90 (In Russian).

Nelson, M.C. (1991) The study of technological organization. In: Schiffer, M.B. (ed.) *Archaeological Method and Theory*, vol. 3. University of Arizona Press, Tucson, pp. 57–100.

Odell, G. (1981) The morphological express at function junction: Searching for meaning in lithic tool types. *Journal of Anthropological Research* 37, 319–42.

Parry, W. and Kelly, R. (1987) Expedient core technology and sedentism. In: Johnson, J. and Morrow, C. (eds) *The Organization of Core Technology*, Westview Press, Boulder, Colorado, pp. 285–304.

Rasic, J. and Andrefsky, Jr., W. (2001) Alaskan blade cores as specialized components of mobile toolkits: Assessing design parameters and toolkit organization through debitage analysis. In: Andrefsky, W., Jr. (ed.) *Lithic Debitage: Context,*

Form, and Meaning, University of Utah Press, Salt Lake City, pp. 61–79.

Shott, M. (1986) Technological organization and settlement mobility: An ethnographic example. *Journal of Anthropological Research* 42, 15–51.

Shott, M. (1995) How much is a scraper? Curation, use rates, and the formation of a scraper assemblage. *Lithic Technology* 20, 53–72.

Shott, M. (1996) An exegesis of the curation concept. *Journal of Anthropological Research* 52, 259–80.

de Sonneville-Bordes, D. and Perrot, J. (1954–56) Lexique Typologique du Peolithique Superieur Outillage Lithique. *Bulletin de la Societe Prehistorique Francaise* 51, 327; 52, 76; 53, 408; 53, 547.

Tomka, S. (2001) The effect of processing requirements on reduction strategies and tool form: A new perspective. In: Andrefsky, W., Jr. (ed.) *Lithic Debitage: Context, Form, and Meaning*, University of Utah Press, Salt Lake City, pp. 207–25.

Chapter 20

CLOVIS AND DALTON: UNBOUNDED AND BOUNDED SYSTEMS IN THE MIDCONTINENT OF NORTH AMERICA

Brad Koldehoff[1] *and Thomas J. Loebel*[2]

[1]Illinois Transportation Archeological Research Program, University of Illinois, Urbana-Champaign
[2]Department of Anthropology, University of Illinois, Chicago

ABSTRACT

In this chapter, information from recent lithic procurement studies is used to make inferences about patterns of land use and social interaction. Clovis and Dalton patterns are compared and contrasted highlighting important differences that support the conclusion that Dalton culture does not represent a continuation of Paleoindian lifeways, but rather, represents a new, more sedentary lifestyle that created bounded landscapes and laid the foundation for subsequent cultural developments.

INTRODUCTION

In the archeological record of North America, the sparse remnants of Late Pleistocene fluted-point cultures are succeeded by rich accumulations left by regional traditions focused on Holocene resources. The emergence of regional traditions has been used by researchers to help demarcate the end of the Paleoindian period and the beginning of the Archaic (or Mesoindian) period (Anderson and Sassaman 1996; Hofman 1996; Emerson *et al.* 2009). However, debate continues about how to define or separate Paleoindian and Archaic traditions. For example, among scholars there is a continent-wide trend towards downplaying change. Continuity rather than discontinuity is emphasized, and Paleoindian and Archaic sequences are merged

Lithic Materials and Paleolithic Societies, 1st edition. Edited by B. Adams and B.S. Blades, ©2009 Blackwell Publishing. ISBN 978-1-4051-6837-3.

into a single homogenized continuum (e.g., Meltzer and Smith 1986; Beck and Jones 1997; Blackmar and Hofman 2006). Such an approach masks variability and presupposes gradual change (Hofman 1996; Winters 1974; Koldehoff and Walthall 2009) and is often based more on theoretical assumptions than on detailed information about regional sequences (Emerson and McElrath 2009).

For instance, despite the absence of detailed information about Early Paleoindian (Clovis/Gainey) groups in the Midcontinent, Meltzer and Smith's (1986: 5) notion of long-term continuity has gained widespread acceptance: "in the eastern forests ... there is demonstrable continuity in adaptation from Paleoindian through the Archaic." Furthermore, Meltzer (1988) outlines two Paleoindian adaptations: a specialized caribou hunting pattern for groups operating in northern areas recently deglaciated and dominated by tundra and spruce parkland (particularly the northeast), and a generalized foraging pattern for groups operating south of this zone, especially in forested regions. The specialized groups, he believes, were derived from the generalized groups, which he also believes were the forebears of the Archaic tradition.

Based on recent research in the Midcontinent, we reject the notion of continuity (Walthall 1998, 1999; Walthall and Koldehoff 1998, 1999; Amick *et al.* 1999; Koldehoff and Walthall 2004, 2009; Loebel 2005). We see a disjuncture between Clovis and Dalton: Clovis land-use patterns reflect long-distance settlement relocations covering hundreds of kilometers that likely targeted caribou; conversely, Dalton land-use patterns reflect a more sedentary and likely river-oriented lifestyle focused on deer, fish, waterfowl, and other localized seasonal resources. Although Dalton culture is often classified as Late Paleoindian (e.g., Morse 1997; Ellis *et al.* 1998; Goodyear 1999), evidence of reduced mobility, increased population densities, increased woodworking, and budding social complexity support the argument that Dalton is an initial Archaic florescence, not a Late Paleoindian climax. The Archaic tradition is more than a cultural-historical unit demarcated by point types and radiocarbon dates. In the Midcontinent, it denotes new economies and social strategies geared towards not only new habitats and resources but also new and increasingly complex social landscapes (Anderson 2002; Gibson and Carr 2004; Sassaman 2005; Emerson *et al.* 2009).

In this chapter, we use information from recent lithic procurement studies to make inferences about patterns of land use and social interaction. We compare and contrast Clovis and Dalton patterns to highlight differences between these two cultures. The differences support our contention that Dalton culture does not represent a continuation of Paleoindian lifeways, but rather, represents a new, more sedentary lifestyle that created bounded landscapes and laid the foundation for subsequent cultural developments.

THE LITHIC LANDSCAPE

Detailed information about the "lithic landscape," the raw materials available across a region, is critical when delineating patterns of lithic procurement and, in turn, patterns of settlement mobility and material exchange. Paleoindian researchers have shown that lithic materials were frequently procured from distant sources (100–400 km), underscoring the need to document the range of variability and distribution of raw materials across vast areas that typically encompass several physiographic regions and crosscut modern political boundaries (e.g., Ellis and Lothrop 1989; Amick 1996).

In this and previous studies, we draw upon decades of lithic resource investigations to model the lithic landscape of Illinois, Wisconsin, and adjoining states (Koldehoff 1999, 2006; Loebel 2005). A key feature of this landscape is the Illinois Basin: a bedrock structure composed primarily of chertless Pennsylvanian formations, capped by glacial layers that contain chert cobbles poorly suited for the manufacture of large bifacial tools (>5 cm in length). Sources of moderate- to high-quality chert and orthoquartzite are scattered around the margins of the basin, often in massive exposures associated with prominent landforms (Figure 20.1). Our research has show that when early hunter-gatherers moved into or across the Illinois Basin they "geared-up" for such a journey at one or more of these bedrock sources. Consequently, movements into and across the basin can be reconstructed by linking raw materials in site assemblages to known bedrock sources.

However, social networks and ritual practices also shaped lithic procurement (Gould and Saggers 1985; Hampton 1999; Topping and Lynott 2005). Lithic sources were often part of ritual landscapes, and like other important landscape features (e.g., springs and overlooks) and economic resources, they were incorporated into oral traditions. Apprentice stoneworkers received instruction at lithic sources; thus, quarries

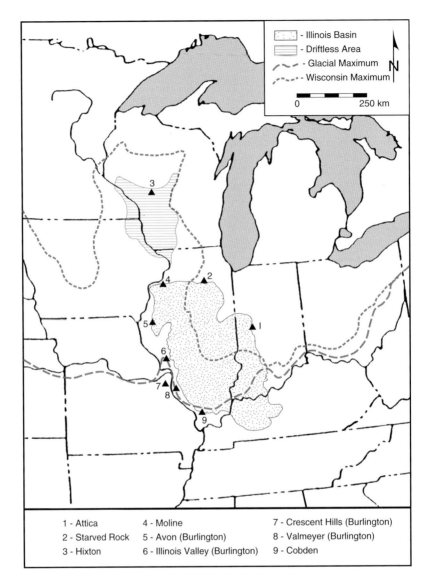

Figure 20.1   Major Midwestern lithic sources discussed in text.

were also classrooms. We believe lithic sources, from the inception of stone-tool technology, were key elements of a group's cultural geography: their mental map of a region and its resources (Schick and Toth 1993; Mithen 1996). Specific raw materials likely became associated with landmarks or regions. Hence, stone from a well-known source could have been used to intentionally signal regional or group affiliation. Such signaling likely became important as population densities increased, territories were defended, and material goods were exchanged to forge alliances (Ellis 1989; Topping and Lynott 2005; Whallon 2006). Conversely, for thinly populated, highly mobile foragers, major lithic sources were renowned landmarks for retooling and training and possibly for gathering places. With low population densities, especially in colonizing situations, dispersed populations need designated meeting places for scheduled and unscheduled (emergency) gatherings.

Prominent landforms on or near transportation corridors (rivers and trails) were ideal rendezvous points (Anderson 1995; Metlzer 2002).

In sum, major lithic sources were predictable resource concentrations: fixed points on the landscape that afforded access to great quantities of workable stone. As such, they were also landmarks that played a key role in the economic, social, and ritual cycles of stone-age societies. Clues to how these fixed resources were incorporated into systems of land use and social interaction should be evident in the raw materials in site assemblages. Meltzer (1989) outlines patterns indicative of direct procurement versus indirect procurement (exchange). Lithic raw materials, however, provide only a partial measure of group movements and interactions. Several factors limit their inferential power: (i) not all activities involved stone tools; (ii) the exact route a particular tool traveled or the number of hands it passed through cannot be easily reconstructed; and (iii) most site assemblages are probably the end result of several occupational episodes over the course of several years or decades; thus, they should be considered time-averaged measures of movements and material exchanges. In other words, lithic assemblages typically reflect the most common or redundant patterns of raw material procurement and discard. Despite these limitations, lithic assemblages (and their raw materials) provide unparalleled insights into the direction, distance, and frequency of movements and interactions. By gathering data from multiple sites, regional patterns can be delineated and then used to make inferences about land-use patterns and social networks.

Paleoindian researchers, especially those focused on Folsom, have honed this approach (e.g., Hofman 1994, 1999, 2003; Amick 1996, 2000). Patterns of lithic procurement indicate Folsom groups were highly mobile, and bison killsites indicate this mobility stemmed in part from hunting a mobile prey. Folsom groups are seen as having been unconstrained by neighboring groups. They operated within home ranges that were place-oriented (contra Kelly and Todd 1988), but their ranges shifted freely. Amick (2000: 139) outlines the circumstances that created and transformed this system of land use:

> For Paleoindian groups in the Southwest, group mobility rates were exceptionally high and population levels were exceptionally low. These initial low population densities encouraged high mobility and social cooperation ...

whereas population increases over the next several thousand years eventually resulted in territorial circumscription, social boundaries, and competition for resources.

UNBOUNDED AND BOUNDED SYSTEMS IN THE MIDCONTINENT

Clovis and Dalton land-use patterns unfolded against a backdrop of environmental change. As the climate warmed and ice sheets melted, Late Pleistocene plant and animal communities, which have few modern analogs, were reorganized into the basic biomes of today. These changes occurred over centuries and moved northward in a zonal mosaic. Large mammals, like mammoths and mastodons, became extinct, while deer flourished and caribou shifted their range northward. Once the flow of meltwater stopped, the Mississippi River began to down-cut and meander, creating resource-rich lakes and wetlands. As this meandering river regime migrated up the valley, so did temperate deciduous forest. These new habitats held abundant, seasonal resources (e.g., deer, fish, waterfowl, and nuts). In the Mississippi Valley and adjacent Ozarks, by 10 000 RCYBP, Dalton groups with new technologies, such as the adze for heavy-duty woodworking, were literally carving out a new way of life. Riverine resources and river travel, via dugout canoes, likely played a pivotal role in this new economy (Walthall and Koldehoff 1998, 1999; Koldehoff and Walthall 2004, 2009; McElrath *et al.* 2009).

Clovis: unbounded land use

The established archeological record of the Midwest begins with Clovis culture. Whether the region was inhabited before Clovis is a contested issue that is beyond the scope this chapter. We do know from decades of research that Clovis groups were small and dispersed. Thus, like much of North America, Midwestern Late Pleistocene landscapes were largely devoid of people. Even so, there is a growing database of documented Clovis discoveries. In the following pages, Clovis assemblages from six sites in Illinois and four sites in Wisconsin are summarized. Distances between sites and lithic sources are given as straight-line or minimal distances.

1. Mueller-Keck. The Mueller and Keck sites occupy adjacent upland ridges near the mouth of Prairie du

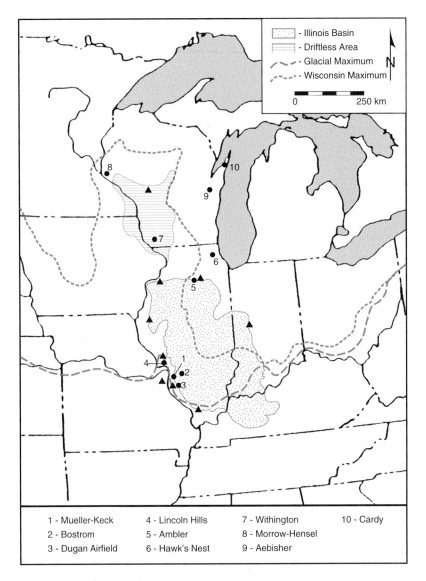

Figure 20.2   Clovis sites discussed in text.

Pont Creek, a natural entryway into the broad expanse of Mississippi River floodplain known as the American Bottom (Figure 20.2). Hundreds of Clovis artifacts have been recovered: 40 fluted points, 60 bifacial preforms, 30 cores and tabular pieces, 274 unifacial scrapers and flake tools, and 352 waste flakes (Koldehoff and Walthall 2004; Amick and Koldehoff 2005). With few exceptions, these artifacts are made from a single raw material, Attica chert from the Wabash Valley, 320 km

to the northeast. Regional lithic sources are poorly represented, but several points and tools are made from Holland chert from southern Indiana, 290 km to the east. The predominance of Attica chert is intriguing because of its distance from Mueller-Keck and because the Mueller-Keck assemblage contains numerous Clovis preforms (production failures and rejects), as well as broken and expended points and tools (Figure 20.3). When moving from the Wabash

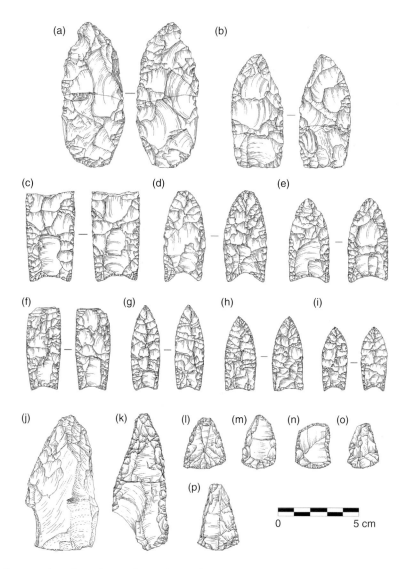

Figure 20.3 Clovis artifacts from the Mueller-Keck complex: (a) mid-stage Clovis preform, (b) late-stage Clovis preform, (c–i) Clovis points, (j–k) side scrapers, (l–p) end scrapers.

Valley to the Mississippi Valley, Clovis groups were well prepared for traversing the chert-poor interior of Illinois, and they arrived with fully stocked toolkits. This long-distance preparation was likely linked to time-sensitive subsistence activities, such as intercepting caribou or other game. Although chert sources are common in the region, the Mueller-Keck complex is not situated next to a chert source; it is located in an area

well suited for monitoring game moving in and out of the American Bottom.

2. Bostrom. Located in the chert-poor uplands, the Bostrom site occupies a ridge overlooking Silver Creek at a stream crossing along an early overland trail. The assemblage of 25 fluted points and preforms, and numerous scrapers and other tools, is not dominated by a single raw material (Tankersley 1995;

Morrow 1996; Koldehoff and Walthall 2004). Rather, an array of materials from distant sources to the north (Burlington and Payson), south (Cobden, Kaolin, and Mill Creek), and east (Attica and Holland) are represented. The greatest distance is 300 km to the Attica source. A single scraper made from Knife River Flint (KRF) is noteworthy because the KRF source is located in North Dakota, 1400 km to the northwest. This distance is exceptional, indicating the scraper may have moved through social networks, rather then having been directly procured (see below). The Bostrom site likely represents a campsite and meeting place used for monitoring and intercepting game crossing Silver Creek.

3. Dugan Airfield. The Dugan Airfield site covers the south slope of a prominent knob along high ground that was the route of an early overland trail. The assemblage contains 5 Clovis points, 3 fluted preforms, unfluted bifaces, and numerous unifacial tools (Koldehoff and Walthall 2004). Local cherts are common (Fern Glen and Salem), as are extraregional cherts: Cobden and Kaolin from southern Illinois; Attica from the Wabash Valley; Payson and Moline from the Mississippi Valley in northern Illinois; and Oneota and Platteville/Galena from the Illinois Valley in northern Illinois. The latter materials are from the Starved Rock locale, some 350 km distant. Like the Bostrom site, Dugan appears to have been a well-known hunting camp and meeting place along an ancient footpath.

4. Lincoln Hills. Situated on high ground overlooking the confluence of the Mississippi and Illinois Rivers, the Lincoln Hills (or Ready) site is a Clovis workshop and campsite associated with local exposures of Burlington chert. The large assemblage is dominated by Clovis bifaces, especially production failures: 26 fluted points, 202 fluted preforms, numerous unfluted preforms, and about 80 unifacial tools (Morrow 1995, 1996; Koldehoff and Walthall 2004). As expected, the vast majority are fashioned from Burlington chert. Several points and tools are made from extraregional cherts: Cobden, Kaolin, and Kinkaid from southern Illinois, and Attica from the Wabash Valley. These items likely represent remnants of earlier toolkits that were discarded after retooling with Burlington chert, and Burlington was likely used for instructing young knappers, given the many fluted and unfluted production failures.

5. Ambler. The Ambler site is located in the uplands near Starved Rock, a high bluff and well-known landmark on the Illinois River. The site has yielded two Clovis points, several preforms, and more than 100 tools and waste flakes all made from Hixton Silicified Sandstone (Loebel 2004, 2005). Hixton is a fine-grained orthoquartzite with a source area in central Wisconsin, 350 km to the north. A high incidence of core-reduction platforms and cortex within the assemblage indicates Hixton materials were directly procured from the source as part of a planned move to the Starved Rock locale. The transport of a well-stocked and minimally modified toolkit into a relatively lithic-rich area, like the Starved Rock locale, echoes the pattern displayed at the Mueller-Keck complex. Interestingly, a second site ("Site 2") located 1.2 km to the west has yielded several Clovis points and preforms made from locally available Platteville/Galena and Oneota cherts, in addition to unifacial tools made of Moline, Burlington, KRF, and a single flake of Hixton (Loebel 2004). These tools denote broader patterns of movement and interaction, similar to those represented by the Bostrom and Dugan assemblages.

6. Hawk's Nest. Located 120 km northeast of Starved Rock, in a chert-poor area southwest of Lake Michigan, the Hawk's Nest site occupies an upland rise next to glacial bogs and wetlands. The assemblage includes 48 Clovis preforms, mostly mid-to-late stage manufacturing failures, 150 unifacial tools, and hundreds of waste flakes primarily made from one raw material – Burlington chert, probably from the Avon source, 300 km to the southwest (Amick and Loebel 2002; Loebel 2005). Minor amounts of other raw materials are represented: Moline from the Mississippi Valley, Oneota from Starved Rock, and Attica and Plummer from Indiana. Like the Ambler and Mueller-Keck assemblages, the Hawk's Nest assemblage represents another example of fully stocked toolkits made from a single raw material being transported several hundred kilometers.

7. Withington. Situated in the driftless region of southwest Wisconsin, the Withington site occupies a high point overlooking the Platte River valley, a likely spot for observing game. The assemblage includes 6 Clovis points, 15 preforms, 91 unifacial tools, and 170 waste flakes made from a single raw material, Hixton, which outcrops 170 km to the north (Stoltman 1993; Mason 1997, 2003; Loebel 2005). Like Ambler, the Withington assemblage displays a high incidence of cortex and core-reduction platforms, indicating on-site reduction of early-stage blanks and cores. Many of the discarded tools are seemingly unused or minimally used, suggesting the site residents were unconcerned about maximizing the utility of their toolkits. Local Galena chert is represented by two points, several tools, and waste flakes. Minor amounts of other raw materials are present. Of particular interest is an end scraper made from Cobden chert. Because the Cobden source is located nearly 600 km to the south, this end scraper likely arrived at Withington not by direct procurement but by intergroup interaction (see below).

8. Morrow-Hensel. Located 125 km northwest of Hixton, the Morrow-Hensel site is situated in the uplands 10 km from the Mississippi Valley. Made primarily from Hixton, the assemblage includes 25 fluted points, 34 fluted preforms, 28 unfluted preforms, more than 100 scrapers and flake tools, in addition to over 10 000 waste flakes (Amick *et al.* 1999). Unlike the Withington assemblage, the Morrow-Hensel assemblage displays a high degree of fragmentation, reworking, and recycling – hallmarks of an exhausted assemblage. This intensive use and reduction may reflect a lengthy residential stay and/or the pursuit of subsistence tasks that limited opportunities for replenishment of toolkits. Although nearby raw materials were utilized, it appears that the Morrow-Hensel residents were sufficiently prepared for the tasks at hand with their transported Hixton toolkits. In fact, the Morrow-Hensel assemblage indicates that other assemblages, like those at Withington and Mueller-Keck, were substantially under-utilized.

9. Aebischer. The Aebischer site is situated along the Killsnake River basin, a marshy area west of Lake Michigan. Investigations have produced a Clovis assemblage made almost entirely from Moline chert, which outcrops 330 km to the southwest. The assemblage includes at least 17 fluted points, 11 preforms, 65 scrapers and flake tools, and more than 200 waste flakes (Stoltman 1993; Mason 1997, 2003; Loebel 2005). Three points are made from Burlington chert, which is available 50 km south of the Moline source. The Aebischer assemblage indicates that groups in the Moline area moved to the northeast and arrived well-stocked at a spot selected for subsistence reasons. The Aebischer site is well positioned for monitoring the Greater Horicon marsh system, a locale that was probably prime habitat for caribou calving during the late Pleistocene.

10. Cardy. The Cardy site is located 120 km northeast of Aebischer and is linked to Aebischer through it assemblage, which is predominately made from Moline chert (Mason 1997, 2003; Loebel 2005). At least

6 points, 40 tools, and hundreds of waste flakes have been recovered. An additional fluted point and several dozen unifacial tools are made from local Maqouketa chert. The Cardy site may postdate Aebischer on the basis of its more northerly location, if Clovis groups were tracking the northward shifting ranges of caribou herds. Because Maqouketa chert artifacts are present in the Cardy assemblage but are absent in the Aebischer assemblage, groups at Aebischer may have been unaware of this nearby lithic resource. Eventually, the groups frequenting the Cardy site incorporated this local raw material into their toolkits. Nevertheless, fully stocked toolkits were still transported to the Cardy site.

Dalton: bounded land use

After Clovis there is evidence of regionalism in point styles, which we believe reflects a partitioning of the landscape (see also Anderson 1995). An early example of this growing regionalism is the Dalton horizon (Goodyear 1999; Koldehoff and Walthall 2009), which encompasses of a series of point type clusters. Dalton points and toolkits from the central Mississippi Valley and the Ozarks represent a distinct regional expression (Figures 20.4 and 20.5). Early dates (10 500–10 000 RCYBP) indicate this region was likely the hearth of Dalton culture. Some believe Dalton is a Late Paleoindian development derived directly from Clovis and that it was coeval with Folsom (e.g., Wyckoff and Bartlett 1995; Morse 1997). While this scenario is open to debate, researchers do agree that Dalton groups spread across the landscape, populating some areas more densely than other, such as the central Mississippi Valley. In the valley an expanding Holocene resource base likely promoted reduced mobility, population growth, and bounded landscapes. Evidence of this more sedentary lifestyle is evident in lithic procurement patterns. In the following pages, Dalton lithic data from one site in Arkansas, six sites in Illinois, and one site in Wisconsin are summarized (Figure 20.6).

1. Sloan. Located in the Mississippi Valley in northeast Arkansas, the Sloan site is a Dalton cemetery. Excavations resulted in the discovery 439 artifacts, many of which were found in groupings representing grave offerings (Morse 1997). Like most Dalton assemblages from northeast Arkansas, the majority of points and tools are made from local gravels. For instance, about 80% of the Sloan site points are made from gravel

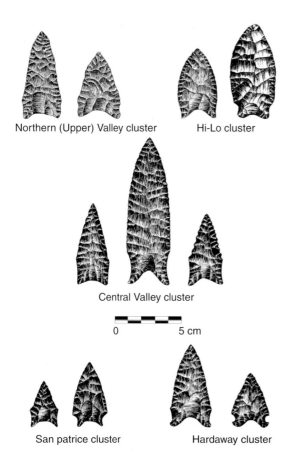

Northern (Upper) Valley cluster   Hi-Lo cluster

Central Valley cluster

0   5 cm

San patrice cluster   Hardaway cluster

Figure 20.4 Dalton horizon point-type clusters. (*Source*: Adapted from Koldehoff and Walthall 2009.)

cherts, 12% are made from Ozark cherts, available 50 km to the west, and 8% are made from Burlington chert, with sources in Illinois and Missouri (Figure 20.1). The Burlington points appear to be derived from the Crescent Quarries, 300 km to the north (Morse 1997; Walthall and Koldehoff 1998). These points, on average, are longer than the other Sloan site points, the longest (19 cm) being two to three times longer. Morse (1997) used this point, and other examples of extra-large Daltons, to establish the Sloan type. Subsequent research by Walthall and Koldehoff (1998) delineated the so-called Sloan Zone, a 400 km stretch of the Mississippi Valley that contains all known Sloan Dalton discoveries (Figure 20.7). Because Sloan Daltons are oversized, finely crafted, and typically lack use-wear and reworking, they likely represent non-utilitarian,

ritually charged blades that were exchanged among neighboring groups (Walthall and Koldehoff 1998). Smaller likely utilitarian Daltons made from various cherts were probably exchanged as well. Largely absent at Sloan and at nearby Dalton sites are tools, preforms, and waste flakes made from Burlington chert. Entire toolkits made from Burlington, or other extraregional raw materials, have not been reported.

2. Olive Branch. Situated 200 km upstream from Sloan, the Olive Branch site covers a terrace within Thebes Gap next to a bedrock river crossing. Excavations uncovered a series of rich Dalton deposits, including caches of Sloan points, a human cremation, and midden accumulations (Gramly 2002). Initial work resulted in the recovery of 223 Dalton points, 242 preforms, 38 adzes, and hundreds of tools and waste flakes (Gramly and Funk 1991). Many additional artifacts have been found, but final totals are unavailable (Gramly 2002: 116). Olive Branch is an intensively occupied Dalton settlement that appears to be a well-used Dalton workshop, aggregation site, and possible year-round habitation area. While the vast majority of Dalton points and tools are made from local southern Illinois cherts, particularly Bailey, cherts from Arkansas have been identified, in addition to Burlington chert. These distant raw materials occur primarily as finished and expended points.

3. George. Located east of the American Bottom at the base of a high knob in the chert-poor uplands, the George site appears to be an aggregation site (Koldehoff and Walthall 2004). The recovery of 36 Dalton points, 9 adzes, and numerous scrapers and flake tools, made from an assortment of chert types, supports this idea. For example, 69% of the points are made from American Bottom cherts (Burlington, Fern Glen, and Salem, 85–110 km distant), 17% are made from Blair/St. David and Kinkaid cherts from the nearby Marys River drainage and Mississippi River bluffs (20–50 km distant), and 14% are made from Bailey and Cobden cherts from the western Shawnee Hills (80–90 km distant). The adzes are made from some of the same cherts but include a Mounds Gravel specimen and a St. Louis chert specimen. Mounds Gravel has exposures across extreme southern Illinois (100–120 km distant), and St. Louis chert has exposures along the American Bottom. Locally available glacial chert cobbles were utilized but not for formal tools. Thus, all of the points and adzes were transported to the site from western and southern sources, with some sources being more than 100 km distant.

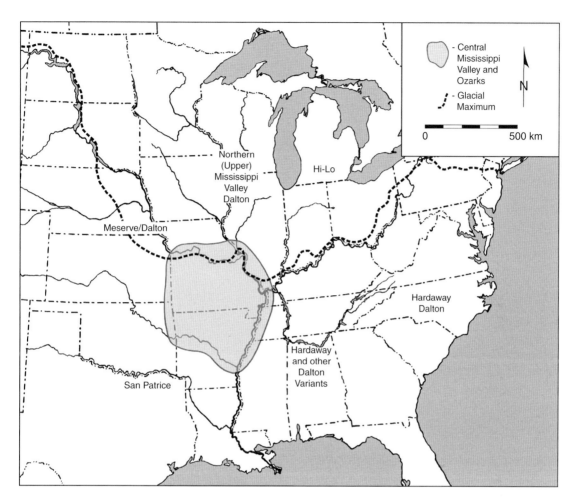

Figure 20.5   Location of Dalton horizon cultural complexes. (*Source*: Adapted from Koldehoff and Walthall 2009.)

4. Dugan Airfield. As previously discussed, the Dugan Airfield site is a high knob along an overland trail. Clovis materials are common along its south slope, while Dalton points and adzes have been found across the site. The Dalton occupation is represented by 11 points and 7 adzes, with all but one point being manufactured from Burlington chert and other nearby cherts (Salem, St. Louis, and Fern Glen). The single non-local point is made from Bailey chert, which outcrops 90 km to the south.

5. New Valmeyer. A series of blufftop sites, littered with local chert (Burlington, Fern Glen, and Salem), make up the New Valmeyer complex (Koldehoff and Walthall 2004). The bluffs at Valmeyer form a prominence known as Salt Lick Point, which provides a commanding view of the American Bottom. In all, 31 Dalton points and preforms and 256 adzes have been recovered. The points and adzes are evidence of regular visits and retooling episodes. Seventy per cent of the adzes are production failures and rejects, which indicates Salt Lick Point was not only a workshop but also a classroom for young Dalton knappers. Five broken and expended Dalton points and adzes are made from non-local cherts: one point and two adzes are made from Blair/St. David chert; one point is made from Mounds Gravel, and one adze is made from Cobden chert. These cherts have sources 60–150 km southeast of the Valmeyer. Because these few items are the only

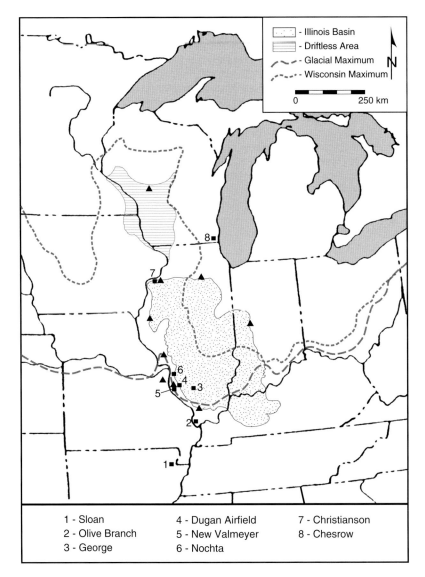

Figure 20.6   Dalton horizon sites discussed in text.

artifacts made from non-local cherts, it is likely that they were obtained through exchange, or perhaps, they were discarded by individuals who temporarily resided with the groups that routinely visited and retooled at Salt Lick Point. The George site, 85 km to the east, is a likely spot where such items, or individuals, moved between groups.

6. Nochta. The Nochta site is located in the American Bottom on a Pleistocene terrace. Excavations revealed

Early and Middle Holocene occupation areas (Higgins 1990; McElrath *et al.* 2009). The Dalton occupation represents a base camp with tools made primarily from Burlington chert, which has source areas 30–50 km away. Unlike the Valmeyer area, chert is scarce in the nearby bluffs. Thirty-one Dalton points are made from the following chert types: 22 Burlington, 2 Ste. Genevieve, 2 Salem, 1 Fern Glen, 1 Kinkaid, 1 Bailey, and 2 indeterminate (Koldehoff 2006). Kinkaid and

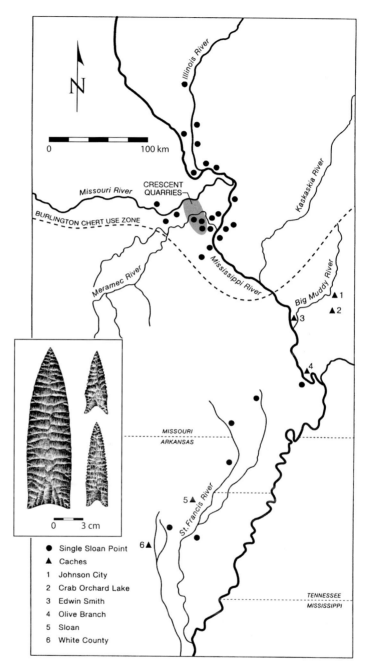

Figure 20.7 Distribution of Sloan Dalton discoveries in the central Mississippi Valley with insert showing a Sloan Dalton (left) and two standard Daltons. (*Source*: Adapted from Walthall and Koldehoff 1998.)

Bailey cherts have source areas 90 and 130 km to the south, respectively.

7. Christianson. Located about 300 km upstream from the American Bottom, near Moline, the Christianson site occupies a Pleistocene terrace near

the confluence of the Rock and Mississippi Rivers. Investigations have recovered 40 Dalton points, several adzes, numerous preforms, and large amounts of debitage (Evans and Womac 1998). The Dalton points are distinct from those found in the central Mississippi

Valley: they are shorter, more triangular, and their blade edges are less strongly beveled (Figure 20.4). They are similar to Hi-Lo points (Ellis 2004), but they are considered a separate type, "Northern Valley" or "Upper Valley" Daltons (Koldehoff and Walthall 2009). About 100 km north of the American Bottom, this Upper Valley Dalton type dominates, marking what we believe to be a cultural divide (Figure 20.5). Two lithic materials account for the bulk of the Christianson assemblage: Moline chert is available in the adjacent bluffs, and Burlington chert is available 50 km to the south.

8. Chesrow. The Chesrow site is located 250 km northeast of the Moline area. It and several surrounding sites make up the Chesrow complex (Overstreet 1993). While the age of this complex has been debated, we concur with Ellis (2004) that it is part of the Hi-Lo complex. Although some Chesrow points resemble Upper Valley Daltons, most resemble Hi-Lo points. Chesrow and related sites occupy old Lake Michigan beach lines a few kilometers west of the current beach front. In total, from these sites, 50 points, 84 preforms, and numerous flake tools, cores, and waste flakes have been recovered (Overstreet 1993). All of these items are made from locally occurring glacial cobbles. No Burlington, Moline, or other extraregional raw materials have been identified.

DISCUSSION

In the previous section, we used lithic data from a handful of the largest Clovis and Dalton sites in the Midwest to outline two strikingly different patterns or scales of raw material movement. The Clovis assemblages, with the exception of Lincoln Hills, which is quarry related, are dominated by extraregional materials transported 100–400 km. These materials are often minimally modified and represented by all artifact classes, including preforms and debitage. The assemblages match Meltzer's (1989) expectations for direct procurement and are likely the result of scheduled movements linked to subsistence activities. Seven assemblages document long-distance transport of complete toolkits made from a single raw material (Figure 20.8): Attica chert to Mueller-Keck (320 km), Burlington chert to Hawk's Nest (300 km), Moline chert to Aebischer (300 km) and Cardy (420 km), and Hixton orthoquartzite to Ambler (350 km), Withington (170 km), and Morrow-Hensel (125 km). The Dugan and Bostrom assemblages

contain multiple extraregional raw materials, indicating that these sites were visited by individuals or groups coming from different directions. The seven assemblages dominated by a single long-distance raw material are fascinating because they represent planned moves from specific lithic sources to specific spots on distant landscapes. These distant spots are often unrelated to or removed from lithic sources. Clovis groups arrived at these distant locations with fully stocked toolkits. Thus, these long-distance moves were apparently driven by time-sensitive subsistence tasks, like intercepting caribou or other game.

Subsistence remains, however, have not been recovered from these sites, and similar raw material patterns have been interpreted by others as evidence of one-way colonizing movements (e.g., Tankersley 1988, 1991; Dincauze 1993). But one-way colonizing movements and related raw material patterns, if present in the archeological record, may be difficult to detect because such colonizing sites and assemblages were likely overprinted by subsequent occupations with lithic assemblages characteristic of regional land-use patterns (Loebel 2005). The Clovis raw material patterns in our sample are fully consistent with seasonal or annual movements within large, flexible territories or home ranges. One-way raw material movements are common in Folsom assemblages, and Hofman (2003: 234) provides an explanation:

> This model suggests directional movement of lithics away from key source locations toward bison hunting areas. Return movements would have de-emphasized lithics in favor of other critical products and equipment. Hunting equipment may have seen limited service after a successful series of hunts. If Folsom groups repeatedly utilized one or more high-quality lithic sources in this manner, their long-term pattern of land use would have resulted in lithic distribution patterns suggesting one-way movement, even if people moved in complex patterns which had them eventually or recurrently returning to the original lithic source location.

We suspect Clovis groups were hunting caribou, in addition to other large game, like deer and mastodon (Graham *et al.* 1981), as well as taking small game and plant foods. Although we cannot be certain about Clovis subsistence, we are certain that Clovis groups periodically trekked hundreds of kilometers, probably to intercept caribou at specific locations (Jackson and Thacker 1997). Caribou hunting was likely part of a land-use pattern that encompassed large territories

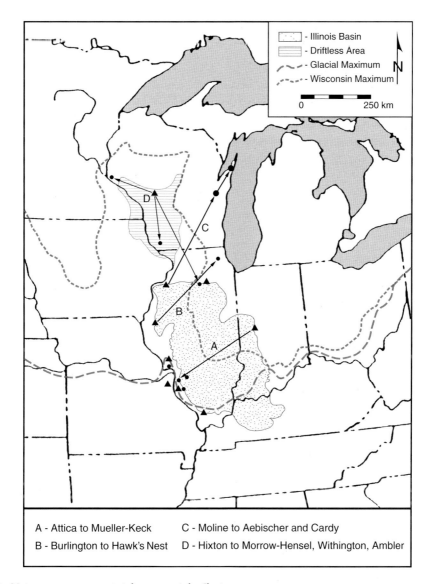

A - Attica to Mueller-Keck   C - Moline to Aebischer and Cardy

B - Burlington to Hawk's Nest   D - Hixton to Morrow-Hensel, Withington, Ambler

Figure 20.8   Major one-way raw material movements by Clovis groups.

and varied resources. Using additional Midwestern datasets, Loebel (2005) has identified possible circulation patterns between lithic sources that may outline band territories (or home ranges). For instance, a single group over time may have repeatedly moved between Burlington, Moline, and Hixton sources, generating sites like Hawk's Nest, Aebischer, Cardy, Withington, and Ambler. With retreating ice sheets and shifting vegetation zones, Hawk's Nest, Aebischer, and Cardy may track the northerly shift of a single band as it hunted caribou herds following retreating habitats. An additional (or earlier?) band range may have encompassed Holland, Attica, Burlington, and Cobden sources and overlapped with the northern band range in the Starved Rock locale. These interpretations warrant further investigation.

Nonetheless, raw material patterns indicate Clovis groups operated within large, flexible territories unimpeded by neighboring groups. But we are not implying that neighbors were avoided or unimportant. In fact, with low population densities, neighbors were a critical resource, and periodic gatherings were essential (Meltzer 1989, 2002; Hofman 1994; Anderson 1995). Gatherings would have afforded opportunities to exchange information, mates, and materials. The best evidence of such interactions is the small amounts of exotic raw materials found substantially outside of their normal distribution ranges, for example, the Cobden scraper in the Withington assemblage, the KRF scraper in the Bostrom assemblage, and the Moline, Oneota, and Platteville/Galena points and scrapers in the Dugan assemblage. Additional examples include Cobden points found in northern Illinois, Hixton points found in southern Illinois and Indiana, and KRF points found in Wisconsin, Missouri, and Illinois (Tankersley 1988; Loebel 2005). These and other items are the likely result of interband transfers of people and/or material goods. Although dispersed, Clovis groups appear to have been connected and coordinated as they traveled great distances targeting caribou herds and other resource concentrations. Periodic gatherings at specific landmarks likely facilitated their integration.

In contrast, the Dalton assemblages in our sample are dominated by local and regional raw materials, typically procured within 50 km. Extraregional raw materials occur primarily as finished points and adzes that likely moved through social networks. Unlike Clovis, entire toolkits are not made from distant raw materials. More importantly, lithic materials did not move outside of the central Mississippi Valley. While we found evidence of materials moving within the valley, and into and out of the Ozarks, we found no evidence of materials moving out of or into the valley from northern Illinois, southern Wisconsin, or other areas, like the Wabash Valley. This apparent lack of interaction corresponds with changes in point styles, which may mark social boundaries. The absence of Sloan Daltons outside the central Mississippi Valley may also mark social boundaries.

Sloan Daltons, as argued by Walthall and Koldehoff (1998), moved through a network of alliances. This network linked Dalton populations and helped mediate intergroup conflicts stemming from reduced settlement mobility and increased population density. For example, in most areas researchers find five to ten times more Dalton sites and points than they do earlier sites

and points, and these points and toolkits are typically made from local raw materials, except for Sloan Daltons (Morse 1997; Koldehoff and Walthall 2004, 2009). All known Sloan points are made from Burlington chert, which indicates Burlington possessed special technical properties or cultural significance. There are other cherts in the region with similar technical properties. Thus, Burlington, a distinctive whitish colored chert, likely held special meaning, and the production and distribution of Sloan Daltons may have been controlled by certain groups or individuals.

Sloan Daltons are items of ritualized exchange that have no known parallels in Midwestern Clovis assemblages, but they do have parallels in subsequent Archaic and Woodland cultures (Walthall and Koldehoff 1998). Sloan Daltons, coupled with evidence of bounded landscapes, may indicate the emergence of tribal formations and expressions of individual or group identity (Anderson 2002; Sassaman 2005). Following Kelly (1991), Sloan Daltons might be markers of incipient leaders: with reduced mobility and less contact between individuals in neighboring groups, certain individuals rose to positions of authority and group representation; these leaders were likely bound together through alliances and ritualized exchanges. Conversely, in more mobile groups, contacts between individuals in neighboring groups were not superseded by emerging leaders; individuals negotiated their own relationships and alliances with neighbors. We suspect such face-to-face relationships were common among most Clovis groups, and standard (utilitarian) points or tools may have been exchanged to cement relations. Western caches of extra-large Clovis bifaces (Frison 1991) may denote the presence of more complex social strategies and relationships. This possibility reinforces our point about the importance of emphasizing diversity, not only across time but also across space.

SUMMARY

Striking differences between Clovis and Dalton mobility patterns, indicated by the movement of lithic raw materials, demonstrate discontinuity rather continuity during the Pleistocene–Holocene transition. Midwestern Clovis groups were not localized bands of generalized foragers that seamlessly evolved into Archaic groups. To the contrary, they were mobile hunters and foragers that periodically trekked hundreds of kilometers to take advantage of resource concentrations, such as migrating caribou. Mobility was likely a key strategy not only for dealing with shifting and often unpredictable biotic resources but also for dealing with a largely vacant social landscape. In contrast, for Dalton groups, an expanding and more localized Holocene resource base promoted reduced mobility, population growth, and bounded landscapes. Emerging social complexity in the central Mississippian Valley during the Dalton era foreshadowed subsequent Archaic and Woodland developments.

ACKNOWLEDGMENTS

This chapter draws from ongoing research with Dan Amick and John Walthall, and we are indebted to them for freely sharing ideas and data. We are also indebted to a host of other researchers and institutions for providing information and access to collections. Research support was provided to Thomas Loebel by the Department of Anthropology at the University of Illinois, Chicago, and similar support was provided to Brad Koldehoff from the Illinois Transportation Archaeological Research Program (ITARP) at the University of Illinois, Urbana-Champaign. The figures were prepared with the assistance of Mera Hertel and are used here courtesy ITARP. However, the individual artifact drawings in Figure 20.3 were completed by Sarah Moore with funding provided by the Mulcahy Scholarship program at Loyola University, and they are used here courtesy of Dan Amick. We thank Brian Adams and Brooke Blades for inviting us to be part of this volume and for their comments on an earlier version of this chapter. Comments were also provided by Dan Amick, John Wathall, Tom Emerson, and an anonymous reviewer. All errors and weakness in this chapter are solely our responsibility.

REFERENCES

Amick, D. (1996) Regional patterns of Folsom mobility and land use in the American Southwest. *World Archaeology* 27, 411–26.

Amick, D. (2000) Regional approaches with unbounded systems: The record of Folsom land use in New Mexico and West Texas. In: Hegmon, M. (ed.) *The Archaeology of Regional Interaction: Religion, Warfare, and Exchange Across the American Southwest and Beyond*, Univserity Press of Colorado, Boulder, pp. 119–47.

Amick, D. and Koldehoff, B. (2005) Systematic field investigations at the Mueller-Keck Clovis site complex in southwestern Illinois. *Current Research in the Pleistocene* 22, 39–41.

Amick, D. and Loebel, T. (2002) Final field report from the Hawk's Nest Clovis (Gainey) site in northeastern Illinois. *Current Research in the Pleistocene* 19, 1–4.

Amick, D., Boszhardt, R., Hensel, K., Hill, M., Loebel, T., and Wilder, D. (1999) *Pure Paleo in Western Wisconsin*. Report of Investigations No. 350. Mississippi Valley Archaeology Center at the University of Wisconsin, La Crosse.

Anderson, D. (1995) Paleoindian interaction networks in the eastern woodlands. In: Nassaney, M. and Sassaman, K. (ed.) *Native American Interactions: Multiscalar Analyses and Interpretations in the Eastern Woodlands*, University of Tennessee Press, Knoxville, pp. 3–26.

Anderson, D. (2002) The evolution of tribal social organization in the southeastern United States. In: Parkinson, W. (ed.) *The Archaeology of Tribal Societies*, Archaeological Series 15, International Monographs in Prehistory, Ann Arbor, Michigan, pp. 246–77.

Anderson, D. and Sassaman, K. (ed.) (1996) *The Paleoindian and Early Archaic Southeast*. University of Alabama Press, Tuscaloosa.

Beck, C. and Jones, G. (1997) The terminal Pleistocene/early holocene archaeology of the Great Basin. *Journal of World Archaeology* 11, 161–236.

Blackmar, J. and Hofman, J. (2006) The Paleoarchaic of Kansas. In: Hoard, R. and Banks, W. (ed.) *Kansas Archaeology*, University of Kansas Press, Lawrence, pp. 46–75.

Dincauze, D. (1993) Pioneering in the Pleistocene: Large Paleoindian sites in the Northeast. In: Stoltman, J. (ed.) *Archaeology of Eastern North America: Papers in Honor of Stephen Williams*, Archaeological Report No. 25, Mississippi Department of Archives and History, Jackson, pp. 43–60.

Emerson, T. and McElrath, D. (2009) The eastern woodlands archaic and the tyranny of theory. In: Emerson, T., McElrath, D., and Fortier, A. (ed.) *Archaic Societies: Diversity and Complexity Across the Midcontinent*. State University of New York, Albany.

Emerson, T., McElrath, D., and Fortier, A. (ed.) (2009) *Archaic Societies: Diversity and Complexity Across the Midcontinent*. State University of New York, Albany.

Ellis, C. (1989) The explanation of Northeastern Paleoindian lithic procurement patterns. In: Ellis, C. and Lothrop, J. (ed.) *Eastern Paleoindian Lithic Resource Use*. Westview Press, Boulder, pp. 139–64.

Ellis, C. (2004) Hi-Lo: An early lithic complex in the Great Lakes Region. In: Jackson, L. and Hinshelwood, A. (ed.) *The Late Palaeo-Indian Great Lakes: Geological and Archaeological Investigations of Late Pleistocene and Early Holocene Environments*, Mercury Series, Archaeological Paper 165, Canadian Museum of Civilization, Quebec, pp. 57–83.

Ellis, C., Goodyear, A., Morse, D., and Tankersley, K. (1998) Archaeology of the Pleistocene–Holocene Transition in Eastern North America. *Quaternary International* 49/50, 151–66.

Ellis, C. and Lothrop, J. (ed.) (1989) *Eastern Paleoindian Lithic Resource Use.* Westview Press, Boulder.

Evans, J. and Womac, K. (1998) The Christianson Site (11-RI-42): A Paleoindian Occupation in the Lower Rock River Valley, Illinois. *Illinois Archaeology* 10, 331–55.

Frison, G. (1991) The clovis cultural complex: New data from caches of flaked stone and worked bone Artifacts. In: Montet-White, A. and Holen, S. (ed.) *Raw Material Economies Among Prehistoric Hunter-Gatherers,* University of Kansas Press, Lawrence, pp. 321–33.

Gibson, J. and Carr, P. (ed.) (2004) *Signs of Power: The Rise of Cultural Complexity in the Southeast.* University of Alabama Press, Tuscaloosa.

Goodyear, A. (1999) The early holocene occupation of the southeastern United States: A geoarchaeological summary. In: Bonnichsen, R. and Turnmire, K. (ed.) *Ice Age People of North America: Environments, Origins, and Adaptions,* Oregon State University Press, Corvallis, pp. 432–81.

Gould, R. and Saggers, S. (1985) Lithic procurement in central Australia: A closer look at Binford's idea of embeddedness in archaeology. *American Antiquity* 50, 117–36.

Graham, R.W., Haynes, C.V. Johnson, D.L., and Kay, M. (1981) Kimmswick: A Clovis-Mastodon association in eastern Missouri. *Science* 213, 1115–7.

Gramly, R. (2002) Olive branch: A very early archaic site on the Mississippi River. *Amateur Archaeologist* 8, 5–232.

Gramly, R. and Funk, R. (1991) Olive branch: A large Dalton and Pre-Dalton Encampment at Thebes Gap, Alexander County, Illinois. In: McNutt, C. (ed.) *The Archaic Period in the Mid-South,* Archaeological Report No. 24, Mississippi Department of Archives and History, Jackson, pp. 23–33.

Hampton, O. (1999) *Culture of Stone: Sacred and Profane Uses of Stone among the Dani.* Texas A&M University Press, College Station, Texas.

Higgins, M. (1990) *The Nochta Site: The Early, Middle, and Late Archaic Occupations.* American Bottom Archaeology FAI-270 Site Reports 21, University of Illinois Press, Urbana.

Hofman, J. (1994) Paleoindian Aggregations on the Great Plains. *Journal of Anthropological Archaeology* 13, 341–70.

Hofman, J. (1996) Early Hunter-Gatherers of the central great plains: Paleoindian and mesoindian (Archaic) cultures. In: Hofman, J. (ed.) *Archeology and Paleoecology of the Central Great Plains,* Research Series No. 48, Arkansas Archeological Survey, Fayetteville, pp. 41–100.

Hofman, J. (1999) Unbounded hunters: Folsom Bison hunting on the southern plains circa 10,500 BP, the lithic evidence. In: Brugal, J., David, F., Enloe, J. and Jaubert, J. (ed.) *Le Bison: Gibier et moyen de subsistance des hommes du Paléolithique aux Paléoindiens des Grandes Plaines,* Actes du colloque international, Toulouse. Éditions APDCA, Antibes, pp. 383–415.

Hofman, J. (2003) Tethered to Stone or Freedom to Move: Folsom Biface Technology in Regional Perspective. In: Soressi, M. and Dibble, H. (ed.) *Multiple Approaches to the Study of Bifacial Technologies,* University of Pennsylvania Museum of Archaeology and Anthropology Press, Philadelphia, pp. 229–50.

Jackson, L. and Thacker, P. (ed.) (1997) *Caribou and Reindeer Hunters of the Northern Hemisphere.* Avebury, Aldershot, Great Britain.

Kelly, R. (1991) Sedentism, Sociopolitical Inequality, and Resource Fluctuations. In: Gregg, S. (ed.) *Between Bands and States,* Occasional Paper No. 9, Center for Archaeological Investigation, Southern Illinois University, Carbondale, pp. 135–58.

Kelly, R. and Todd, L. (1988) Coming into the country: Early paleoindian hunting and mobility. *American Antiquity* 52, 231–44.

Koldehoff, B. (1999) Attica chert and clovis land use in Illinois: The Anderson and Perkins sites. *Illinois Archaeology* 11, 1–26.

Koldehoff, B. (2006) *Paleoindian and Archaic Settlement and Lithic Procurement in the Illinois Uplands.* Illinois Transportation Archaeological Research Reports No. 108, Illinois Transportation Archaeological Research Program, University of Illinois, Urbana-Champaign.

Koldehoff, B. and Walthall, J. (2004) Settling in: Hunter-Gatherer mobility during the pleistocene-holocene transition in the central Mississippi valley. In: Cantwell, A., Conrad, L., and Reyman, J. (ed.) *Aboriginal Ritual and Economy in the Eastern Woodlands: Essays in Memory of Howard Dalton Winters,* Scientific Papers Vol. 30, Illinois State Museum, Springfield, pp. 49–72.

Koldehoff, B. and Walthall, J. (2009) Dalton and the early holocene midcontinent: Setting the stage. In: Emerson, T., McElrath, D., and Fortier, A. (ed.) Archaic societies: Diversity and complexity across the midcontinent, State University of New York, Albany.

Loebel, T. (2004) 11Ls981: A new fluted point site in northern Illinois. *Current Research in the Pleistocene* 21, 62–4.

Loebel, T. (2005) The *organization of early paleoindian economies in the western Great Lakes.* Unpublished PhD Dissertation, Department of Anthropology, University of Illinois, Chicago.

Mason, R. (1997) The Paleo-Indian Tradition. *The Wisconsin Archaeologist* 78, 78–110.

Mason, R. (2003) The cardy site: A fluted point camp in door county, Wisconsin. *Paper Presented at the 49th Midwest Archaeological Conference,* Milwaukee, Wisconsin.

McElrath, D., Fortier, A., Koldehoff, B. and Emerson, T. (2009) The American bottom: An archaic cultural crossroads. In: Emerson, T., McElrath, D., and Fortier, A. (ed.) *Archaic Societies: Diversity and Complexity Across the Midcontinent.* State University of New York, Albany.

Meltzer, D. (1988) Late pleistocene human adaptations in Eastern North America. *Journal of World Prehistory* 2, 1–52.

Meltzer, D. (1989) Was stone exchanged among Eastern North American Paleoindians? In: Ellis, C. and Lothrop, J. (eds) *Eastern Paleoindian Lithic Resource Use*, Westview Press, Boulder, pp. 11–39.

Meltzer, D. (2002) What do you do when no one's been there before? Thoughts on the exploration and colonization of new lands. In: Jablonski, N. (ed.) *The First Americans: The Pleistocene Colonization of the New World*, Memoirs of the California Academy of Sciences No. 27, San Francisco, pp. 27–58.

Meltzer, D. and Smith, B. (1986) Paleoindian and early archaic subsistence strategies in Eastern North America. In: Neusius, S. (ed.) *Foraging, Collecting, and Harvesting: Archaic Period Subsistence and Settlement in the Eastern Woodlands*, Occasional Paper No. 6, Center for Archaeological Investigations, Southern Illinois University, Carbondale, pp. 3–31.

Mithen, S. (1996) *The Prehistory of the Mind: The Cognitive origins of Art, Religion and Science*. Thames and Hudson, London.

Morrow, J. (1995) Clovis projectile point manufacture: A perspective from the Ready/Lincoln Hills site, 11JY46, Jersey County, Illinois. *Midcontinental Journal of Archaeology* 20, 167–91.

Morrow, J. (1996) The organization of early paleoindian lithic technology in the confluence region of the Mississippi, Illinois, and Missouri Rivers. Unpublished PhD dissertation, Department of Anthropology, Washington University, St. Louis, Missouri.

Morse, D. (1997) *Sloan: A Paleoindian Dalton Cemetery in Arkansas*. Smithsonian Institution Press, Washington, DC.

Overstreet, D. (1993) *Chesrow: A Paleoindian Complex in the Southern Lake Michigan Basin*. Great Lakes Archaeological Press, Milwaukee.

Sassaman, K. (2005) Structure and practice in the Archaic Southeast. In: Pauketat, T. and Loren, D. (eds) *North American Archaeology*, Blackwell Publishers, Oxford, pp. 79–107.

Shick, K. and Toth, N. (1993) *Making Silent Stones Speak: Human Evolution and the Dawn of Technology*, Simon and Schuster, New York.

Stoltman, J. (1993) A Reconsideration of fluted point diversity in Wisconsin. In: Stoltman, J. (ed.) *Archaeology of Eastern North America, Papers in Honor of Stephen Williams*, Archaeological Report No. 25, Mississippi Department of Archives and History, Jackson, pp. 61–72.

Tankersley, K. (1988) The exploitation frontier of Hixton quartzite. *Current Research in the Pleistocene* 5, 34–35.

Tankersley, K. (1991) A geoarchaeological investigation of distribution and exchange in the raw material economies of clovis groups in Eastern North America. In: Montet-White, A. and Holen, S. (eds) *Raw Material Economies among Prehistoric Hunter-Gatherers*, University of Kansas Press, Lawrence, pp. 285–304.

Tankersley, K. (1995) Paleoindian contexts and artifact distribution patterns at the Bostrom Site, St. Clair County, Illinois. *Midcontinental Journal of Archaeology* 20, 40–61.

Topping, P. and Lynott, M. (ed.) (2005) *The Cultural Landscape of Prehistoric Mines*. Oxbow Books, Oxford.

Walthall, J. (1998) Rockshelters and Hunter-Gatherer Adaptation to the Pleistocene/Holocene Transition. *American Antiquity* 63, 223–8.

Walthall, J. (1999) Mortuary behavior and early holocene land use in the North American Midcontinent. *North American Archaeologist* 20, 1–30.

Walthall, J. and Koldehoff, B. (1998) Hunter-Gatherer interaction and alliance formation: Dalton and the cult of the long blade. *Plains Anthropologist* 43, 257–73.

Walthall, J. and Koldehoff, B. (1999) Across the divide: Dalton land use in the Southern Till Plains. *Illinois Archaeology* 11, 27–49.

Whallon, R. (2006) Social networks and information: Non-"Utilitarian" mobility among Hunter-Gatherers. *Journal of Anthropological Archaeology* 25, 259–70.

Winters, H. (1974) Introduction to the new edition. In *Indian Knoll*, by W. Webb, University of Tennessee Press, Knoxville.

Wyckoff, D. and Bartlett, R. (1995) Living on the edge: Late Pleistocene–Early Holocene cultural interaction along Southeastern woodlands–plains border. In: Nassaney, M. and Sassaman, K. (eds) *Native American Interactions: Multiscalar Analyses and Interpretations in the Eastern Woodlands*, University of Tennessee, Knoxville, pp. 27–72.

INDEX